Lecture Notes in Mathematics

Edited by A. Dold, B. Eckmann and F. Takens

1443

Karl Heinz Dovermann
Reinhard Schultz

Equivariant Surgery Theories and Their Periodicity Properties

Springer-Verlag
Berlin Heidelberg New York London
Paris Tokyo Hong Kong Barcelona

Authors

Karl Heinz Dovermann
Department of Mathematics, University of Hawaii
Honolulu, Hawaii 96822, USA

Reinhard Schultz
Department of Mathematics, Purdue University
West Lafayette, Indiana 47907, USA

Mathematics Subject Classification (1980): Primary: 57R67, 57S17
Secondary: 18F25

ISBN 3-540-53042-8 Springer-Verlag Berlin Heidelberg New York
ISBN 0-387-53042-8 Springer-Verlag New York Berlin Heidelberg

Printed in Germany

Printing and binding: Druckhaus Beltz, Hemsbach/Bergstr.
2146/3140-543210 – Printed on acid-free paper

PREFACE

This book began as a pair of papers covering the authors' work on periodicity in equivariant surgery from late 1982 to mid 1984. Since our results apply to many different versions of equivariant surgery theory, it seemed desirable to provide a unified approach to these results based upon formal properties that hold in all the standard theories. Although workers in the area have been aware of these common properties for some time, relatively little has been written on the subject. Furthermore, it is not always obvious how the settings for different approaches to equivariant surgery are related to each other, and this can make it difficult to extract the basic properties and interrelationships of equivariant surgery theories from the literature. For these and other reasons we eventually decided to include a survey of equivariant surgery theories that would present their basic formal properties. Shortly after we began revising our papers to include such a survey, Wolfgang Lück and Ib Madsen began work on a more abstract—and in many respects more general—approach to equivariant surgery. It became increasingly clear that we should include their methods and results in our book, not only for the sake of completeness but also because their work leads to improvements in the exposition, simplifications of some proofs, and significant extensions of our main results and applications. The inclusion of Lück and Madsen's work and its effects on our own work are two excuses for the four year delay in revising our original two papers.

Our initial work on equivariant periodicity appears in Chapters III through V. The first two chapters summarize the main versions of equivariant surgery and the basic formal properties of these theories. A reader who is already familiar with equivariant surgery should be able to start with Chapter III and use the first two Chapters as reference material. On the other hand, a reader who wants to know what equivariant surgery is about should be able to use the first two chapters as an introduction to the subject. We shall assume the reader is somewhat familiar with the main results of nonsimply connected surgery and the basic concepts of transformation groups. Some standard references for these topics are discussed in the second paragraph of the introductory chapter entitled *Summary: Background Material and Main Results.*

Karl Heinz Dovermann
Honolulu, Hawaii

Reinhard Schultz
West Lafayette, Indiana
Evanston, Illinois and
Berkeley, California

March 1990

ACKNOWLEDGMENTS

Questions about periodicity in equivariant surgery first arose in the nineteen seventies. In particular, Bill Browder presented some crucial observations in lectures at the 1976 A.M.S. Summer Symposium on Algebraic and Geometric Topology at Stanford University but never put them into writing. We appreciate for the openness and good will with which he has discussed this unpublished work on several separate occasions. We are also grateful to Mel Rothenberg and Shmuel Weinberger for describing some of their unpublished work in enough detail so that we could outline their methods and conclusions in Section I.6 and for their generous attitude towards our inclusion of this material. Similarly, we wish to thank Min Yan for discussions regarding his thesis currently in preparation and the relation of his work to ours.

Andrew Ranicki has been very helpful in several connections, commenting on an old draft of Chapter IV, making suggestions on numerous points related to his research, and furnishing copies of some recent papers. We appreciate his repeated willingness to provide assistance. Preprints of work by other topologists have led to many enhancements and even some major improvements in this book; in particular we would like to mention Frank Connolly, Wolfgang Lück, Ib Madsen, Jan-Alve Svensson, Shmuel Weinberger and Bruce Williams and express our thanks to them.

Several typists from the Purdue University Mathematics Department deserve credit and thanks for converting the original hanndwritten text into typewritten drafts and preliminary TEX files at various times over the past six years.

It is also a pleasure to acknowledge partial research support and hospitality from several sources during various phases of our work on this book. In particular, we wish to express thanks for partial support by National Science Foundation Grants MCS 81-00751 and 85-14551 (to the first named author) and MCS 81-04852, 83-00669, 86-02543, and 89-02622 (to the second named author). Both authors also wish to thank Sonderforschungsbereich 170 „Geometrie und Analysis“ at the Mathematical Institute in Göttingen for its hospitality during separate visits, and the second named author is also grateful to the Mathematics Department at Northwestern University for its hospitality during portions of the work of this book and to the Mathematical Sciences Research Institute in Berkeley for partial support during the final stages of our work.

Finally, we are grateful to Albrecht Dold and the editorial personnel at Springer for their patience with the four year delay between their initial decision to publish this book in July of 1985 and our submission of a revised manuscript in December of 1989.

Table of Contents

Preface III
Acknowledgments IV

SUMMARY: BACKGROUND MATERIAL AND MAIN RESULTS 1

I. INTRODUCTION TO EQUIVARIANT SURGERY 9
1. Ordinary surgery and free actions 10
2. Strata and indexing data 13
3. Vector bundle systems 18
4. Stepwise surgery and the Gap Hypothesis 20
5. Equivariant surgery groups 22
6. Some examples 29
7. *Appendix.* Borderline cases of the Gap Hypothesis 31
References for Chapter I 33

II. RELATIONS BETWEEN EQUIVARIANT SURGERY THEORIES 37
1. Browder-Quinn theories 37
2. Theories with Gap Hypotheses 45
3. Passage from one theory to another 51
3A. Change of categories 52
3B. Rothenberg sequences in equivariant surgery 55
3C. Change of coefficients 56
3D. Change to pseudoequivalence 57
3E. Passage to Lück-Madsen groups 58
3F. Passage from Browder-Quinn groups 61
3G. Restrictions to subgroups 67
4. *Appendix.* Stratifications of smooth G-manifolds 67
References for Chapter II 75

III. PERIODICITY THEOREMS IN EQUIVARIANT SURGERY 80
1. Products in equivariant surgery 83
2. Statements of periodicity theorems 86
3. Permutation actions on product manifolds 92
4. Orbit sequences and product operations 97
5. Periodic stabilization 101
5A. Stable stepwise surgery obstructions 101
5B. Unstable surgery obstruction groups 104
5C. Splitting theorems 106
References for Chapter III 111

IV. TWISTED PRODUCT FORMULAS FOR SURGERY WITH COEFFICIENTS ... 115
1. Basic definitions and results ... 116
2. Algebraic description of geometric L-groups ... 121
3. Technical remarks ... 125
4. Projective and subprojective Wall groups ... 128
5. Cappell-Shaneson Γ-groups ... 131
6. Applications to periodicity theorems ... 133
References for Chapter IV ... 138

V. PRODUCTS AND PERIODICITY FOR SURGERY UP TO PSEUDOEQUIVALENCE ... 141
1. The setting ... 142
2. The main results ... 149
3. Stepwise obstructions and addition ... 151
4. Restriction morphisms ... 160
5. Projective class group obstructions ... 163
6. An exact sequence ... 171
7. Products and stepwise obstructions ... 175
8. Proofs of main results ... 180
9. *Appendix.* A result on Wall groups ... 187
References for Chapter V ... 189

Index to Numbered Items ... 192
Index to Notation ... 194
Subject Index ... 204

SUMMARY: BACKGROUND MATERIAL AND BASIC RESULTS

This book is about some surgery-theoretic methods in the theory of transformation groups. The main goals are to provide a unified description of equivariant surgery theories, to present some equivariant analogs of the fourfold periodicity theorems in ordinary surgery, and to study some implications of these periodicity theorems.

Since equivariant surgery theory involves both transformation groups and the theory of surgery on manifolds, we need to assume that the reader has some some familiarity with both subjects. Most of the necessary background on transformation groups appears in Chapters I–II and Sections VI.1–2 of Bredon's book [**Bre**] or in Chapter I and Sections II.1–2 of tom Dieck's book [**tD**]. The standard reference for surgery theory is Wall's book [**Wa**], particularly Chapters 1–6 and 9; most of the necessary background can also be found in survey articles by W. Browder [**Bro**], F. Latour [**Lat**], and J. A. Lees [**Lees**]. Additional background references for [**Wa**] are listed in a supplementary bibliography at the end of this Summary. Most of the algebraic topology that we use can be found in the books by Spanier [**Sp**] and Milnor and Stasheff [**MS**].

We have attempted to explain the specialized notation and terminology that we use; a subject index and an index for symbols can be found at the end of the book. For the sake of completeness we shall mention some standard conventions that are generally not stated explicitly. The unit interval $\{t \in \mathbf{R} | 0 \leq t \leq 1\}$ will be denoted by I or $[0,1]$, and the order of a finite group G will be denoted by $|G|$. Theorem X.88.99 will denote Theorem 99 in Chapter X, Section 88, and similary for Propositions, Corollaries, or Lemmas. References to the first and second author correspond to alphabetical order (so the first author is Dovermann and the second is Schultz).

Guide to the contents of this book

The first two chapters are a survey of equivariant surgery theory, and the last three chapters contain the periodicity theorems and some applications to analyzing the role of a basic technical condition in equivariant surgery called the Gap Hypothesis (this is defined in Section I.5). However, most of Chapter IV deals exclusively with A. Ranicki's theory of algebraic surgery (see [**Ra1**] and [**Ra2**]), and this portion of the book is independent of the remaining chapters.

As indicated by its title, Chapter I summarizes the main ideas in equivariant surgery theory. We do not assume any previous familiarity with equivariant surgery. In Section 1 we attempt to motivate the passage from ordinary to equivariant surgery by summarizing some applications of ordinary surgery to free actions of compact Lie groups; most of this work dates back to the nineteen sixties and seventies. Sections 2 through 5 are basically a survey of an equivariant surgery theory developed by the first author and M. Rothenberg [**DR**]. We look at this theory first because it illustrates all the main ideas but avoids various technical complications that appear in other equivariant surgery theories. In Section 5 we also include some extensions of [**DR**] that will be needed in subsequent chapters. Most of these results deal with relative or *adjusted* versions of the equivariant surgery obstruction groups in [**DR**]; a similar construction appears in work of W. Lück and I. Madsen [[**LM**, Part II, Section 1] where such groups are called *restricted.* Sections 5 and 7 also discuss some exceptional phenomena in limiting cases of the Gap Hypothesis that were discovered by M. Morimoto [**Mor**]. Finally, in Section 6 we describe some unpublished results of M. Rothenberg and S. Weinberger showing that a major result in Section 5—the equivariant $\pi - \pi$ theorem—fails to hold if one does not assume the Gap Hypothesis.

In Chapter II we summarize the various approaches to equivariant surgery in the literature and describe their relationships with each other. Section 1 introduces the *transverse-linear-isovariant* surgery theory constructed by W. Browder and F. Quinn in the early nineteen seventies [**BQ**]. This theory deals exclusively with equivariant maps satisfying very strong restrictions, but it does not require the Gap Hypothesis. The setting of [**BQ**] relies on the existence of a smooth stratification (in the sense of Thom and Mather) for the orbit space of a smooth manifold. We shall indicate how the necessary properties can be extracted from work of M. Davis [**Dav**], and in Section 4 we shall fill a gap in the literature by outlining a direct proof of the stratification theorem along the lines of W. Lellmann's 1975 *Diplomarbeit* [**Ll**]. In Section 2 we shall discuss several variants of the theory considered in Chapter I. The objective of the Chapter I theory is to convert an equivariant map of manifolds into an equivariant homotopy equivalence, where the manifolds in question have smooth group actions, the Gap Hypothesis holds, and certain simple connectivity conditions also hold. For certain problems in transformation groups it is preferable to have theories in which the objective or the conditions on the manifolds are modified. In particular, the objective might be to obtain an equivariant homotopy equivalence whose generalized equivariant Whitehead torsion (in the sense of [**DR2**] or [**I**]) is trivial, or the objective might be to obtain a map that is an equivariant homotopy equivalence after a suitable equivariant localization. Similarly, one might want to weaken the condition on the group action to piecewise linear or topological local linearity or to remove the simple connectivity assumption. However, in all these cases it is necessary to retain the Gap Hypothesis. Our discussion includes the relatively recent approach to equivariant surgery developed by Lück and Madsen [**LM**].

Most of the preceding material is expository. The new results of this book begin in Section II.3, where we study the relations between various theories and show that certain pairs of theories yield the same equivariant surgery obstruction groups. Much of this has been known to workers in the area for some time, but little has been written down. The central idea of equivariant surgery theory is to view an equivariant surgery problem as a sequence of ordinary surgery problems. As in the obstruction theory for extending continuous functions, one assumes that all problems up to a certain point can be solved and considers the next problem in the sequence. This problem determines a *stepwise surgery obstruction* that usually takes values in an ordinary surgery obstruction group; if this obstruction vanishes, one can solve the given problem in the sequence and proceed to the next one. The value groups for stepwise obstructions can also be viewed as adjusted obstruction groups for the equivariant surgery theory in question, and this suggests that a relation between equivariant surgery theories should involve a family of homomorphisms from the adjusted obstruction groups of one theory to the adjusted obstruction groups of another. Further analysis suggests that these homomorphisms should have compatibility properties like those of a natural transformation between homology or cohomology theories (see II.2.0 and II.3.1.A-B). In this setting it will follow that

> a transformation from one equivariant surgery theory to another is an isomorphism (modulo problems with borderline cases) if and only if the transformation induces isomorphisms on stepwise obstruction groups.

A precise statement of this principle appears in Theorem II.3.4. We apply this result to compare Browder-Quinn theories, the theory of Chapter I and some of its variants, and the Lück-Madsen theories. Specifically, we construct transformations between these theories and conclude that analogous theories of the different types determine the same groups (see Subsections II.3.A, E, and F). We shall also construct transformations between the different versions of Browder-Quinn and Lück-Madsen theories as well as the theory of Chapter I and its variants, and in Subsections II.B and C we shall use Theorem II.3.4 to obtain some qualitative information about the kernels and cokernels of those mappings.

In Chapter III we come to the main results of this book. Browder and Quinn showed in [**BQ**] that their transverse-linear-isovariant surgery theories have fourfold periodicity properties like those of ordinary surgery; in particular, the periodicity isomorphism is given geometrically by crossing with $\mathbf{CP}^2$. The link between algebraic and geometric periodicity arises from an integer valued invariant of an oriented manifold called the signature; for the complex projective plane this invariant equals $+1$. Similar results hold to a limited extent for Lück-Madsen theories and variants of the theories in Chapter I, and these periodicity isomorphisms are compatible with the transformations defined in Section II.3. However, there is one difficulty; namely, if one is not working with Browder-Quinn groups it is necessary to use G-manifolds X such that both X and $X \times \mathbf{CP}^2$ satisfy the Gap Hypothesis. Since it is always possible to find some $k(X) > 0$ such that $X \times (\mathbf{CP}^2)^k$ does not satisfy the Gap Hypothesis if $k \geq k(X)$, the $\mathbf{CP}^2$-periodicity for Lück-Madsen and Chapter I type theories is finite; in contrast, the $\mathbf{CP}^2$-periodicity properties of ordinary surgery obstruction groups and Browder-Quinn groups are infinite. On the other hand, Browder suggested an alternative in lectures at the 1976 A.M.S. Summer Symposium on Algebraic and Geometric Topology at Stanford. Specifically, if we let $\mathbf{CP}^2\!\uparrow\! G$ denote a product of $|G|$ copies of $\mathbf{CP}^2$ and let G act on $\mathbf{CP}^2$ by permuting the coordinates, then $\mathbf{CP}^2\!\uparrow\! G$ is a smooth G-manifold and for all $k > 0$ the product $X \times (\mathbf{CP}^2\!\uparrow\! G)^k$ will satisfy the Gap Hypothesis if X does. Furthermore, if G has odd order then $\mathbb{CP}^2\uparrow G$ satisfies an analog of the signature result for $\mathbb{CP}^2$; specifically, an equivariant refinement of the ordinary signature known as the *G-signature* of $\mathbb{CP}^2\uparrow G$ is the unit element of the real representation ring $RO(G)$. Thus if G has odd order it is reasonable to ask whether crossing with $\mathbb{CP}^2\uparrow G$ induces isomorphisms of equivariant surgery obstruction groups. The central results of Chapter III show this is true for many equivariant surgery theories. Statements of the periodicity theorems appear in Section III.2; the most important special cases are Theorems III.2.7–9. To prove such results we shall

(1) show that crossing with certain smooth G-manifolds can be viewed algebraically as a transformation of equivariant surgery theories,
(2) use the results of Chapter II to reduce the proofs to questions about the effects of products on stepwise obstructions,
(3) interpret the equivariant products as the twisted products in ordinary surgery that were studied by T. Yoshida [**Yo**],
(4) use the results of [**Yo**] and a few elementary computations to show that products induce isomorphisms of stepwise obstruction groups.

In his Stanford lectures Browder noticed one further property of $\mathbb{CP}^2\uparrow G$ when G has odd order. Namely, for each smooth G-manifold X one can find some positive integer $n(X)$ such that $X \times (\mathbb{CP}^2\uparrow G)^n$ satisfies the Gap Hypothesis if $n \geq n(X)$. Browder also noted that this yields equivariant surgery invariants for G-surgery problems outside the Gap Hypothesis range; it suffices to look at the product of such a G-surgery problem with $(\mathbb{CP}^2\uparrow G)^n$ for suitable values of n. Our periodicity theorems imply that these obstructions are essentially the same. In Section 5 we shall develop this systematically and obtain a few general results. For example, if G has odd order we show that the Browder-Quinn groups are direct summands of the Lück-Madsen groups. We shall view the process of crossing with $(\mathbb{CP}^2\uparrow G)^n$ for sufficiently large n as a *periodic stabilization* of an equivariant surgery problem. This appears to be a useful first step in analyzing G-equivariant surgery without the Gap Hypothesis if G has odd order; some further results in this direction are discussed in [**Sc**].

Recent work of M. Yan [**Ya**] yields periodicity theorems analogous to III.2.9 for the isovariant stratified surgery groups defined by S. Weinberger [**Wb**]. In fact, the results of [**Ya**] establish periodicity properties for the isovariant structure sequences of [**Wb**] that include the analog of III.2.9 and reduce to the usual fourfold periodicity of the topological surgery sequence from [**KiS**] (also see [**Ni**]) if G is the trivial group.

In the final two chapters we formulate and prove periodicity theorems for other equivariant surgery theories. Chapter IV considers equivariant homology surgery with coefficients, both in the elementary sense of [**An**] and the more sophisticated sense of [**CS**]. The methods of Chapter III show that such periodicity theorems hold if there are analogs of Yoshida's work on twisted products for surgery with coefficients. Precise statements of the periodicity theorems for equivariant homology surgery appear in Section IV.6; the remaining sections of Chapter IV establish the necessary generalizations of Yoshida's results. Since these extensions of [**Yo**] are potentially of interest in other contexts, these sections are written so that they can be read independently of the rest of the book.

Finally, in Chapter V we consider the theories of [**DP1**] and [**DP2**] for equivariant surgery up to pseudoequivalences (*i.e.*, a homotopy equivalence that is G-equivariant but not necessarily a G-homotopy equivalence). These theories are significantly more difficult to handle than the others, and special considerations are needed at many steps. For example, more complicated notions of reference data are required, and extra restrictions must be placed on the surgery problems to be considered; these include the Euler characteristic and connectivity hypotheses of [**DP1**] and [**DP2**]. Section 1 summarizes these points. In order to extend the periodicity results and their proofs it is necessary to examine some new algebraic invariants that do not arise in the other theories. These invariants take values in subquotients of the projective class group $\widetilde{K}_0(\mathbb{Z}[G])$, and the behavior of such invariants with respect to products must be analyzed. Our results on this question appear in Sections 4 and 5, the conclusions are weaker than we would like, but they suffice to prove the periodicity theorems in some important cases (for example, if G is abelian).

At various points throughout the book we mention questions that are related to the results of this book and seem to deserve further study. A few especially noteworthy examples appear in the next to last paragraph of the introduction to Chapter III and the final paragraphs of Sections III.4 and IV.4.

References

[An] G. A. Anderson, "Surgery with Coefficients," Lecture Notes in Mathematics Vol. 591, Springer, Berlin-Heidelberg-New York, 1977.

[Bre] G. Bredon, "Introduction to Compact Transformation Groups," Pure and Applied Mathematics Vol. 46, Academic Press, New York, 1972.

[Bro] W. Browder, *Manifolds and homotopy theory*, in "Manifolds—Amsterdam 1970 (Proc. NUFFIC Summer School, Amsterdam, Neth., 1970)," Lecture Notes in Mathematics Vol. 197, Springer, Berlin-Heidelberg-New York, 1971, pp. 17–35.

[BQ] W. Browder and F. Quinn, *A surgery theory for G-manifolds and stratified sets*, in "Manifolds–Tokyo, 1973," (Conf. Proc. Univ. of Tokyo, 1973), University of Tokyo Press, Tokyo, 1975, pp. 27–36.

[CS] S. Cappell and J. Shaneson, *The codimension two placement problem and homology equivalent manifolds*, Ann. of Math. **99** (1974), 277–348.

[tD] T. tom Dieck, "Transformation Groups," de Gruyter Studies in Mathematics Vol. 8, W. de Gruyter, Berlin and New York, 1987.

[DP1] K. H. Dovermann and T. Petrie, *G-Surgery II*, Memoirs Amer. Math. Soc. **37** (1982), No. 260.

[DP2] ______________, *An induction theorem for equivariant surgery (G-Surgery III)*, Amer. J. Math. **105** (1983), 1369–1403.

[DR] K. H. Dovermann and M. Rothenberg, *Equivariant Surgery and Classification of Finite Group Actions on Manifolds*, Memoirs Amer. Math. Soc. **71** (1988), No. 379.

[DR2] ______________, *An algebraic approach to the generalized Whitehead group*, in "Transformation Groups (Proceedings, Poznań, 1985)," Lecture Notes in Mathematics Vol. 1217, Springer, Berlin-Heidelberg-New York, 1986, pp. 92–114.

[I] S. Illman, *Equivariant Whitehead torsion and actions of compact Lie groups*, in "*Group Actions on Manifolds (Conference Proceedings, University of Colorado, 1983)*," Contemp. Math. Vol. 36, American Mathematical Society, 1985, pp. 91-106.

[KiS] R. C. Kirby and L. C. Siebenmann, "Foundational Essays on Topological Manifolds, Smoothings, and Triangulations," Annals of Mathematics Studies Vol. 88, Princeton University Press, Princeton, 1977.

[Lat] C. Latour, *Chirurgie non simplement connexe (d'après C. T. C. Wall)*, in "Séminaire Bourbaki Vol. 1970–1971," Lecture Notes in Mathematics Vol. 244, Exposé 397, Springer, Berlin-Heidelberg-New York, 1971, pp. 289–322.

[Lees] J. A. Lees, *The surgery obstruction groups of C. T. C. Wall*, Advances in Math. **11** (1973), 113–156.

[LL] W. Lellmann, *Orbiträume von G-Mannigfaltigkeiten und stratifizierte Mengen*, Diplomarbeit, Universität Bonn, 1975.

[**LM**] W. Lück and I. Madsen, *Equivariant L-theory I*, Aarhus Univ. Preprint Series (1987/1988), No. 8; [*same title*] *II*, Aarhus Univ. Preprint Series (1987/1988), No. 16.

[**MS**] J. Milnor and J. Stasheff, "Characteristic Classes," Annals of Mathematics Studies Vol. 76, Princeton University Press, Princeton, 1974.

[**Ni**] A. Nicas, *Induction theorems for groups of homotopy manifold structures*, Memoirs Amer. Math. Soc. **39** (1982). No. 267.

[**Ra1**] A. A. Ranicki, *The algebraic theory of surgery I: Foundations*, Proc. London Math. Soc. **3:40** (1980), 87–192.

[**Ra2**] ————, *The algebraic theory of surgery II: Applications to topology*, Proc. London Math. Soc. **3:40** (1980), 193–283.

[**Sc**] R. Schultz, *An infinite exact sequence in equivariant surgery*, Mathematisches Forschungsinstitut Oberwolfach Tagungsbericht 14/1985 (Surgery and L-theory), 4–5.

[**Sp**] E. H. Spanier, "Algebraic Topology," McGraw-Hill, New York, 1967.

[**Wa**] C.T.C. Wall, "Surgery on Compact Manifolds," London Math. Soc. Monographs Vol. 1, Academic Press, London and New York, 1970.

[**Ya**] M. Yan, *Periodicity in equivariant surgery and applications*, Ph. D. Thesis, University of Chicago, in preparation.

[**Yo**] T. Yoshida, *Surgery obstructions of twisted products*, J. Math. Okayama Univ. **24** (1982), 73–97.

Addendum: Additional references related to [**Wa**]

Wall's book [**Wa**] assumes the reader is familiar with a considerable amount of marial from differential and PL topology. The following references cover most of the aterial upon which [**Wa**] is based and might provide some helpful background information.

[**BJ**] Th. Bröcker and K. Jänich, "Introduction to Differential Topology," (Transl. by C. B. and M. J. Thomas), Cambridge University Press, Cambridge, U. K., and New York, 1982.

[**Bro2**] W. Browder, "Surgery on Simply Connected Manifolds," Ergeb. der Math. (2) 65, Springer, New York, 1972.

[**Coh**] M. Cohen, "A Course in Simple Homotopy Theory," Graduate Texts in Mathematics Vol. 10, Springer, Berlin-Heidelberg-New York, 1973.

[**HP**] A. Haefliger and V. Poenaru, *La classification des immersions combinatoires*, I. H. E. S. Publ. Math. **23** (1964), 75–91.

[**Hi1**] M. W. Hirsch, *Immersions of differentiable manifolds*, Trans. Amer. Math. Soc. **93** (1959), 242–276.

[**Hi2**] ———, "Differential Topology," Graduate Texts in Mathematics Vol. 33, Springer, Berlin-Heidelberg-New York, 1976.

[**Hud**] J. F. P. Hudson, "Piecewise Linear Topology," W. A. Benjamin, New York, 1969.

[**Krv**] M. Kervaire, *Le théorème de Barden-Mazur-Stallings*, Comment. Math. Helv. **40** (1965), 31–42.

[**Miln**] J. Milnor, "Lectures on the h-cobordism Theorem," Princeton Mathematical Notes No. 1, Princeton University Press, Princeton, N. J., 1965.

[**Mun**] J. R. Munkres, "Elementary Differential Topology (Revised Edition)," Annals of Mathematics Studies Vol. 54, Princeton University Press, Princeton, N. J., 1966.

[**Ph**] A. V. Phillips, *Submersions of open manifolds*, Topology **6** (1967), 171–206.

[**RS**] C. P. Rourke and B. J. Sanderson, "Introduction to Piecewise Linear Topology," Ergebnisse der Math. Bd. 69, Springer, Berlin-Heidelberg-New York, 1972.

CHAPTER I

INTRODUCTION TO EQUIVARIANT SURGERY

One of the main themes in topology is the development of techniques for passing back and forth between algebraic and geometric information. In particular, the phrase "surgery theory" generally refers to a collection of techniques for studying the structure of manifolds by means of homotopy theory and algebraic structures involving quadratic forms. For the most part, these techniques were developed during the past thirty years. At a very early point, topologists realized that surgery theory could be applied effectively to study a fairly wide range of problems involving group actions on manifolds; many ways of doing this have been developed. Several approaches can be described as *equivariant surgery theories* that are formally parallel to ordinary surgery theory as in Wall's book [**Wl**]. The underlying idea is simple: In ordinary surgery theory one considers maps between manifolds, and in equivariant surgery one attempts to consider equivariant maps between manifolds with group actions by similar methods. Some aspects of ordinary surgery theory extend easily to the category of manifolds with group actions. On the other hand, new types of difficulties appear when surgery theory is extended to manifolds with group actions, and effective means for dealing with such problems presently exist only if the manifolds or mappings satisfy some additional hypotheses. There are several possible choices of conditions that are useful in different contexts, and each choice has an associated version of equivariant surgery theory.

Although there are often major differences between the various approaches to equivariant surgery, such theories all satisfy some basic formal properties. In this chapter we shall describe a version of equivariant surgery that has relatively few technical complications but still illustrates the formal properties that such theories have in common. The specific theory to be considered is called $I^{ht,DIFF}$ in [**DR**], where it is studied at length; the same theory also appears in several other references including [**PR**]. A survey of other approaches to equivariant surgery appears in Chapter II of this book.

Applications of equivariant surgery

One of the principal reasons for introducing equivariant surgery theories is their usefulness in studying some basic problems in transformation groups. Unfortunately, an account of the applications would require a considerable amount of extra mathematical material that is not closely related to the main topics of this book, and consequently we have not attempted to treat this important aspect of equivariant surgery theory. Many of the various types of applications are presented in the books by M. Davis [**Dav2**] and T. Petrie and J. Randall [**PR**], a survey article by T. Petrie and the authors [**DPS**], and articles by the second author [**Sc3**], V. Vijums [**Vj**], the first author and L. Washington

[**DW**], and M. Rothenberg and G. Triantafillou [**RT**]. Of course, there are also other applications of surgery to transformation groups beyond the sorts described in these articles; the latter were chosen as a representative selection with references to other work in the area.

1. Ordinary surgery and free actions

Equivariant surgery theories evolved from earlier applications of ordinary surgery theory to transformation groups (see [**Brdr**], [**RS**]). In fact, some of these conclusions are essentially special cases of the equivariant surgery theories that were developed subsequently. In this section we shall describe some results of this type in order to provide motivation and background information.

An action of a group G on a space X is said to be *free* if for each $x \in X$ the isotropy subgroup G_x is trivial. If X is a reasonable space and G is a compact Lie group that acts freely and continuously on X, then the orbit space projection $X \to X/G$ is a principal G-bundle projection (compare [**Bre2**]). Furthermore, if X is a smooth manifold and G acts smoothly and freely on X, then X/G has a canonical smooth structure such that the orbit space projection is a smooth principal G–bundle (see [**GL**]).

These considerations "reduce" the study of free G–actions on spaces to the study of their orbit spaces and (the homotopy classes of) their classifying maps into the universal base space BG. The following special case of this reduction principle reflects the relevance of surgery theory to questions involving smooth G–manifolds:

PROPOSITION 1.1. *Let M^n be a closed smooth manifold, and let $G \to E \to X$ be a principal G–bundle over a finite complex X such that E is homotopy equivalent to M (perhaps $E = M$). Then passage to the orbit space defines a $1-1$ correspondence*

$$\left[\begin{array}{c}\textit{free smooth}\\ G-\textit{manifolds}\\ \textit{equivariantly}\\ \textit{homotopy equivalent}\\ \textit{to } E\end{array}\right] \cong \left[\begin{array}{c}\textit{smooth}\\ \textit{manifolds}\\ \textit{homotopy}\\ \textit{equivalent to}\\ X = E/G\end{array}\right]$$

■

COMPLEMENT 1.2. *A similar result holds in the topological category if we restrict attention to free actions that are locally linear (i.e., the orbit space is a topological manifold).*

REMARK: If G is finite, then every free action of a finite group on a manifold is locally linear. On the other hand, if G is a positive–dimensional compact Lie group, then the results of [**Lgr**] yield large, systematic families of free G–manifolds that are not locally linear (compare [**KS**]).

Proposition 1.1 and Complement 1.2 provide the basis for applying surgery theory to questions about manifolds with free actions of compact Lie groups. The special cases where $M = S^n$ have been studied extensively during the past three decades. Further information can be found in [**Hs**], [**Bru**, Thm. I.10], and [**Wng**] for the case $G = S^1$, [**LdM**] for the case $G = \mathbb{Z}_2$, and [**DM**] for G an arbitrary finite group.

REMARK: The preceding discussion illustrates the usefulness of viewing a space with a free G–action as an ordinary space (no group action) by passing to the orbit space. However, for some purposes it is very useful to study free G-spaces in their own right rather than by passage to the orbit space. In fact, it is is often illuminating to view an arbitrary space as the quotient space of a proper free action of some (possibly noncompact) Lie group G. Perhaps the most basic example of this involves the analysis of a nonsimply connected manifold Y in terms of its universal covering space $\widetilde{Y}$ and the free action of $\pi_1(Y)$ on $\widetilde{Y}$ by covering transformations.

Related applications to nonfree actions

During the nineteen sixties topologists used surgery theory to discover many new examples of free G-manifolds whose orbit spaces were homotopy equivalent but not homeomorphic/diffeomorphic to some fixed model M_0/G. The strength of these conclusions suggests that similar methods can be applied effectively to nonfree actions, and results along these lines were first obtained in the mid nineteen sixties.

A simple but very useful method of this type may be described as *equivariant cutting and pasting away form the group action's singular set.* In order to describe this construction, we shall assume that the compact Lie group G acts smoothly or locally linearly on the closed connected manifold M and that some point of M has a trivial isotropy subgroup; the Principal Orbit Theorem then implies that the set of points with trivial isotropy subgroup is open and dense (compare [**Bre2**], Chapter II). The *singular set* of M, denoted by Sing M, will be the set of all $x \in M$ with $G_x \neq \{1\}$, and the *nonsingular set* Nonsing M is the complement $M -$ Sing M; observe that Sing M and Nonsing M are G–invariant and G acts freely on Nonsing M. Furthermore Nonsing M is a manifold and has a canonical smooth structure if G acts smoothly.

If G acts smoothly on M we shall work in the category $CAT = DIFF$ of smooth manifolds, and if G acts locally linearly (but not specifically smoothly) on M we shall work in the category $CAT = TOP$ of topological manifolds.

Let X be a compact codimension zero CAT-submanifold of Nonsing M/G (usually with $\partial X \neq \emptyset$), and let

$$\mathcal{S}^{CAT,s}(X \text{ REL } \partial X)$$

be the simple relative CAT–manifold structure set of $(X, \partial X)$ as defined in [**WL**], Chapter 10. Given a class $\alpha \in \mathcal{S}^{CAT,s}(X \text{ REL } \partial X)$ represented by a map $h : (V, \partial V) \to (X, \partial X)$ we define a new CAT G–manifold $M\langle\alpha\rangle$, called the equivariant cutting and pasting of M via α, to be the identification space

$$M\langle\alpha\rangle := (M - \text{Int } \overline{X}) \cup_{\partial\overline{X} \sim \partial\overline{V}} \overline{V}$$

where $\overline{X} \to X$ is the restriction of the orbit bundle Nonsing $M \to$ Nonsing M/G and $\overline{V} \to V$ is the pullback of $\overline{X} \to X$ under h. If $\overline{h} : \overline{V} \to \overline{X}$ is the G–equivariant homotopy equivalence determined by the pullback or the orbit space projection p, then the identity map on $M - \text{Int}\,\overline{X}$ and $\overline{h}$ on $\overline{V}$ combine to yield an equivariant homotopy equivalence $M\langle\alpha\rangle \to M$ that is a CAT–isomorphism on a neighborhood of the singular set.

Important special cases

Over the past quarter century this construction has proved to be a simple but powerful method for constructing new group actions with the same singular set as a fixed model M. The models that have received the most attention are spheres with G-actions that are orthogonal and *semifree* (*i.e.*, $\{1\}$ and G are the only isotropy subgroups) with nonempty fixed point sets.

More precisely, let M be the unit sphere $S^{k\oplus W}$ in the representation $\mathbb{R}^{k+1} \oplus W$, where G acts freely and orthogonally on $W - \{0\}$ and trivially on $\mathbb{R}^{k+1}$. Following standard practice we shall denote the unit disk and sphere in W by $D(W)$ and $S(W)$ respectively. If we take the standard equivariant decomposition

$$M \approx S^k \times D(W) \cup_\partial D^{k+1} \times S(W)$$

such that

$$S^k \times D(W) \approx \{\, (x,w) \in \mathbb{R}^{k+1} \oplus W \mid \quad |x|^2 + |w|^2 = 1, |x| \geq |w| \,\}$$

$$D^{k+1} \times S(W) \approx \{\, (x,w) \in \mathbb{R}^{k+1} \oplus W \mid \quad |x|^2 + |w|^2 = 1, |x| \leq |w| \,\}$$

then the fixed point set is contained in the first summand and G acts freely on the second summand. The construction described above then assigns to each element α in the relative structure set

$$\mathcal{S}^{DIFF,s}(D^{k+1} \times S(W)/G \text{ REL } S^k \times S(W)/G)$$

a smooth G-manifold $S^{k\oplus W}\langle\alpha\rangle$ with the following properties:

(*i*) $S^{k\oplus W}\langle\alpha\rangle$ is homotopy equivalent to S^n (*i.e.*, $S^{k\oplus W}\langle\alpha\rangle$ is a homotopy sphere).
(*ii*) G acts smoothly and semifreely on $S^{k\oplus W}\langle\alpha\rangle$ with fixed point set S^k, and the equivariant normal bundle of the latter is $S^k \times W$.

These conditions imply that $S^{k\oplus W}\langle\alpha\rangle$ is in fact equivariantly homeomorphic to the linear model M if $k \geq 5$ and W is at least 3-dimensional (compare [**CMY**], [**I**], [**Ro**]). However, it is often possible to find examples for which $S^{k\oplus W}\langle\alpha\rangle$ is not equivariantly diffeomorphic to the linear model M. This can be seen as follows: Since $k > 2$ the relative structure set has a natural abelian group structure. In analogy with the work of Kervaire and Milnor in the nonequivariant case, one can define an abelian group $\Theta^{G,s}_{k\oplus W}$ of equivariantly oriented G-diffeomorphism classes of semifree G-manifolds that are homotopy n-spheres and also satisfy (a) the fixed point sets are homotopy k-spheres, (b) the representation of G on the tangent space of a fixed point is $\mathbb{R}^k \oplus W$. The group operation is given by equivariant connected sum along the fixed point set as in [**Sc2**] or [**RS**], using a suitable notion of equivariant orientation also defined in these papers.

This definition implies that a connected semifree G-manifold with connected fixed point set has at most four distinct equivalence classes of equivariant orientations. One advantage of defining these equivariant Kervaire-Milnor groups is that the construction $\alpha \to S^{k\oplus W}\langle\alpha\rangle$ defines a homomorphism

$$\gamma : S^{DIFF,s}(D^{k+1} \times S(W)/G \text{ REL } S^k \times S(W)/G) \to \Theta^{G,s}_{k\oplus W}$$

whose kernel consists of all α such that $S^{k\oplus W}\langle\alpha\rangle$ is G-diffeomorphic to M. Since the G-manifolds under consideration have only finitely many distinct equivalence classes of equivariant orientations, it follows that the construction $\alpha \to S^{k\oplus W}\langle\alpha\rangle$ will define an infinite family of differentiably distinct G-manifolds if the image of γ is infinite. The size of this image is effectively computable because γ embeds into a long exact sequence; in this sequence the objects adjacent to the domain and codomain of γ are

$$\Theta_{k+\varepsilon} \oplus \pi_{k+\varepsilon}(F_G(W)/C_G(W)) \qquad (\varepsilon = 0, 1)$$

where Θ_* denotes an ordinary nonequivariant Kervaire-Milnor group from [**KM**], the space $F_G(W)$ is the space of G-equivariant self-maps of $S(W)$, and $C_G(W)$ is the G-centralizer of the representation W in the orthogonal group of W (see [**Sc1**] and [**Sc2**] for details). It turns out that the image of γ is infinite in many cases, including the following:

(*i*) $G = S^1$, k is odd, and $\dim W \geq 6$.
(*ii*) $G = \mathbf{Z}_2$, k is odd, and $\dim W \geq 4$ is even.
(*iii*) $G = \mathbf{Z}_p$ (where $p > 2$ is prime), k is odd, and $\dim W < k$.
(*iv*) $G = \mathbf{Z}_p$ (where $p > 3$ is prime), $k \equiv 3 \bmod 4$, and $\dim W \geq 4$ is even.

Further information related to the first two computations can be found in [**Brdr**] and [**Ms2**] respectively, and further information related to the last two computations can be found in [**MSc**].

2. Strata and indexing data

The cutting and pasting construction of Section 1 illustrates how group actions on manifolds can be studied by applying surgery–theoretic methods over the nonsingular portions of G–manifolds. One of the goals of equivariant surgery theories is to develop methods that also apply to the singular portions of G–manifolds. In order to do this it is necessary to decompose the singular set into a G–invariant configuration of submanifolds that fit together decently. The purpose of this section is to develop formalism for working with such decompositions.

The most basic decomposition of the singular set is given by the following result (see [**Bre2**], Chapter II):

PROPOSITION 2.1. *Suppose that the compact Lie group G acts locally linearly on the compact manifold M, and let H be a closed subgroup of G. Then*

$$M_{(H)} = \{x \in M \mid G_x \text{ is conjugate to } H\}$$

is a finite disjoint union of connected manifolds. Furthermore, $M_{(H)}$ is the intersection of the closed set $G \cdot \mathrm{Fix}(H, M)$ and the open set

$$M - \bigcup_{K \supsetneq H} G \cdot \mathrm{Fix}(K, M).\blacksquare$$

It is well-known that different components of $M_{(H)}$ often have different dimensions (*e.g.*, consider the fixed sets of linear involutions on real projective spaces). Therefore the decomposition of M into the pairwise disjoint subspaces $\{M_{(H)}\}$ must be refined further in order to obtain a decomposition into manifolds. One way of doing this is to split each subset $M_{(H)}$ into components. Another, less drastic, way of splitting $M_{(H)}$ is by means of *normal slice types* (compare [**Dav1**]). Every point in $M_{(H)}$ has an invariant neighborhood of the form $G \times_H V_y$, where V_y is some orthogonal representation of H that is unique up to equivalence. If y and z lie in the same component of $M_{(H)}$ it is elementary to verify that $V_y \approx V_z$ (*i.e.*, linearly equivalent), and thus

$$M_{(H,V)} = \{y \in M_{(H)} | V_y \approx V\}$$

is a (finite) union of components. Furthermore, it follows that the dimensions of the components are constant (the dimension of the component of $M_{(H)}$ containing y is completely determined by the representation V_y). The decomposition of M into the pairwise disjoint family of locally closed, locally flat submanifolds $\{M_{(H,V)}\}$ is called the *canonical stratification* of the G-manifold M, and each submanifold $M_{(H,V)}$ is said to be a *stratum* of the G-manifold.

REMARKS: Our choice of terminology immediately suggests some relationship with the notions of smooth stratification due to Whitney, Thom [**T**], and Mather [**Mat**] and topological CS stratification due to Siebenmann [**Si**]. In fact, the orbit spaces $M_{(H,V)}/G$ are manifolds (smooth if the action is smooth), and the family of subsets $\{M_{(H,V)}/G\}$ defines a smooth stratification of M/G if G acts smoothly on M, or a CS stratification of M/G if G is only known to act continuously and locally linearly on M. Neither of these facts is needed in the present chapter. The proof of the assertion for smooth actions is discussed in Section II.4; verification of the assertion for topological actions is a straightforward exercise.

For our purposes it will be best to consider a decomposition of M closely related to the splitting into components of $M_{(H)}$, where H runs through the isotropy subgroups of G. In order to simplify the discussion we shall

assume G is FINITE for the rest of this chapter.

DEFINITION: If G acts locally linearly on the compact manifold M, the *closed substrata* of the action are the closures of the components of the subsets $M_{(H)}$, where H runs through the isotropy subgroups of G.

Local linearity considerations immediately yield the following:

PROPOSITION 2.2. *For each closed substratum C there is a subgroup H such that*
(i) H acts trivially on C,
(ii) there is an open, dense, connected subset $C_0 \subset C$ such that H is the isotropy subgroup at every point of C_0.■

NOTATION: The subgroup in Proposition 2.2 is called the *generic isotropy subgroup*. This subgroup can be characterized as follows: If the closed substratum in question is the closure $\overline{B}$ of a component B of $M_{(H)}$, then for all points of B the isotropy subgroup is some conjugate H' of H; the subgroup H' is the generic isotropy subgroup for $\overline{B}$.

It is often convenient to consider the closed substrata of a locally linear G-manifold M from a somewhat different perspective. In order to do this we shall need the following relatively weak condition:.

CODIMENSION ≥ 2 GAP HYPOTHESIS. *If C is a closed substratum of M, then the relatively singular set*

$$\text{RelSing}\, C = \{x \in C \mid G_x \text{ is not generic}\}$$

satisfies $\dim \text{RelSing}\, C \leq \dim C - 2$.

DEFAULT ASSUMPTION: Unless stated otherwise we assume the Codimension ≥ 2 Gap Hypothesis throughout the rest of this chapter.

REMARK: The Codimension ≥ 2 Gap Hypothesis always holds if the underlying group G has odd order.

DEFINITION: Let X be a compact G–space that is equivariantly dominated by a finite $G-CW$ complex (*e.g.*, X is a locally linear G–manifold or a finite $G-CW$ complex as defined in [**tD**]). The **geometric poset** $\pi(X)$ is given by

$$\pi(X) := \{D \subset X \mid D \text{ is a component of Fix}\,(H, X) \text{ for some subgroup } H \subset G\}$$

By compactness and equivariant finite domination it is immediate that $\pi(X)$ is finite. The set $\pi(X)$ is partially ordered by inclusion (this explains the "poset" label). Furthermore there is an order preserving G–action on $\pi(X)$ taking a component D of Fix (H, X) to the component gD of Fix (gHg^{-1}, X). This definition of a geometric poset coincides with the one in [**DP1**, 1.16].

The following result contains the desired alternate characterization of the closed substrata:

PROPOSITION 2.3. *Suppose that the compact locally linear G–manifold M satisfies the Codimension ≥ 2 Gap Hypothesis. Then the set of closed substrata is isomorphic to $\pi(M)$.*

PROOF: Suppose that C is a component of $M_{(H)}$ with isotropy subgroup H', and let $\overline{C}$ be the corresponding closed substratum. Since C is connected the same is true for

$\overline{C}$, and therefore $\overline{C}$ is contained in some component D of $\mathrm{Fix}(H', M)$; notice that D is a locally flat submanifold of M. If we remove the set of points in D with isotropy subgroup strictly greater than H', then the Codimension ≥ 2 Gap Hypothesis and duality imply that the remaining subset D^* is connected. Since every point in D^* has H' as its isotropy subgroup, it follows that $D^* \subset C$ and D^* is connected in $M_{(H)}$. Therefore D^* must be equal to C. This implies that the map $C \mapsto D$ must be one-to-one. To see that this map is onto, suppose that D is a component of $\mathrm{Fix}(K, M)$ for some subgroup K. By [**DP1**, Lemma 1.5, p. 33] we can write $D = \mathrm{Fix}(K', M)$ where K' is the isotropy subgroup of some point $y \in D$; if C is the component of $M_{(K')}$ containing y, it is immediate that $D = \overline{C}$.■

As in [**DP1**] and [**DR**], we shall frequently denote an element of $\pi(X)$ by X_α and say $\alpha \in \pi(X)$ in that case.

Indexing data

The definitions of equivariant surgery obstruction groups generally depend upon several types of data associated to the closed substrata of a manifold. The complete *indexing data* λ_X for X will consist of four items

$$\lambda_X := (\pi(X), d_X, s_X, w_X),$$

where $\pi(X)$ is the geometric poset and the remaining items are defined as follows (compare [**DP1**], Section 1, and [**DR**, Section 1]:

(2.5) *The dimension indices* d_X.

Every closed substratum X_α of X is a smoothly embedded submanifold, and $d_X(\alpha) = \dim X_\alpha$.

(2.6) *The orientation data* w_X.

Let G^α be the generic isotropy subgroup, and let $N(X_\alpha)$ be the set of all g in the normalizer of G^α in G such that $g(X_\alpha) = X_\alpha$. Then we obtain an induced action of the finite group

$$W(\alpha) = N(X_\alpha)/G^\alpha$$

on X_α. Assume that *each* X_α *is orientable.* Then one has a homomorphism

$$w_Y(\alpha) : W(\alpha) \to \{\pm 1\}$$

defined by a group element's effect on orientation.

NOTE ON ODD ORDER GROUPS. Of course, if G has odd order then $w_Y(\alpha)$ is trivial and the only significant point is the orientability of each X_α; in this case one can always find orientations for every X_α if X is orientable in the usual sense. Therefore we shall generally ignore (2.6) when discussing odd order groups and assume X is orientable (often each X_α will be 1-connected and hence orientability will be automatic).■

(2.7) *The oriented normal slice representation data* s_X.

Let G^α be the generic isotropy subgroup of X_α, and let $x \in X_\alpha$ satisfy $G_x = G^\alpha$. Then the local representation of G^α at x splits as the sum of a trivial representation and a representation $|s_X(\alpha)|$ with 0-dimensional fixed point set. This representation admits an equivariant G^α-orientation in the sense of **[RS]** or **[Sc2]**, and $s_X(\alpha)$ consists of $|s_X(\alpha)|$ together with a preferred G^α-orientation (compare **[DR]**, Remark 3.15, p. 27).

NOTE ON ORIENTATIONS. An orientation on X and the equivariant orientations from (2.7) determine unique orientations for all of the closed substrata.

Maps of geometric posets

In certain contexts it is necessary to know how geometric posets behave with respect to equivariant maps of G–manifolds. If $f : M \to X$ is a map of compact smooth G-manifolds, there is an **induced map of closed substrata** $\check{f} : \pi(M) \to \pi(X)$ (see **[DP1**, Definition 1.12]): Let M_α be a closed substratum of M with isotropy subgroup $G^\alpha = H$. Then $f(M_\alpha)$ is contained in a unique component Q of X^H. *Since X is a G-manifold* it follows that Q is a closed substratum X_β (use **[DP1**, 1.5, again]; we define $\check{f}(\alpha) = \beta$.

If G^β is the generic isotropy subgroup of Q, then it is immediate that $G^\beta \supseteq G^\alpha$. We shall say that f is **isogeneric** if $G^{\check{f}(\alpha))} = G^\alpha$ for all M_α. If $\check{f}$ induces an isomorphism of closed substrata, then f is isogeneric; this property follows from the results of **[DP1]**, Section 1.

Notational conventions involving indexing data

1. If λ is indexing data as above, then following **[LM]** we let $\mathbf{S}^{+k}\lambda$ denote the same data *except* that all dimensions are raised by $k \in \mathbb{Z}$; if $k < 0$ we shall often replace $\mathbf{S}^{+k}\lambda$ by $\mathbf{S}^{-|k|}\lambda$. The superscript will often be suppressed when $k = 1$. A similar concept appears in **[DR1]**, where it is denoted by λ^{+k}.

2. If λ and λ' are indexing data for X and X', then $\lambda \times \lambda'$ will denote the indexing data for $X \times X'$.

3. Frequently we denote the indexing data for a smooth G-manifold Y by data(Y).

4. If $f : X \to Y$ is an isogeneric map we shall frequently use $\check{f}$ to identify $\pi(X)$ and $\pi(Y)$. In particular, we shall often write $Y_{\check{f}(\alpha)}$ simply as Y_α.

3. Vector bundle systems

The definitions and results of Section 2 provide a framework for dealing with individual pieces of a group action's singular set. It is also necessary to have some means for understanding how the nonsingular set and the various pieces of the singular set fit together. If G acts smoothly and semifreely (*i.e.*, G and $\{1\}$ are the only isotropy subgroups), the equivariant tubular neighborhood theorem [**Bre2**, Section VI.2] provides a simple but effective way of reconstructing the G–manifold from its closed substrata. In this section we shall indicate how the equivariant tubular neighborhood theorem and the machinery of Section 2 can be combined to deal with more general actions.

If the finite group G acts smoothly on the compact manifold M, then each closed substratum in $\pi(M)$ is a smooth submanifold and has a suitably invariant tubular neighborhood; specifically, if $\alpha \in \pi(M)$ has generic isotropy subgroup G^α and W_α denotes the subgroup of all elements in $N(G^\alpha)/G^\alpha$ that map X_α into itself, then one has a W_α–equivariant tubular neighborhood. Furthermore, the tubular neighborhoods for the various closed substrata satisfy some compatibility conditions corresponding to the standard relationship

$$\text{normal bundle } (A \subset C) \cong$$

$$\text{normal bundle } (A \subset B) \oplus \text{pullback of normal bundle } (B \subset C),$$

which holds in ordinary and equivariant differential topology.

The conditions may be formalized as follows:

DEFINITION: Let G be a finite group acting smoothly on a closed connected manifold M, and let $\Pi \subset \pi(M)$ be a G–invariant subset. A *vector bundle system* over Π (in the notation of [**DP1**] or [**DR**], a Π–*bundle*) is a triple

$$(\Xi, \Phi, \Gamma)$$

where $\Xi = \{\xi_\alpha\}$ is a collection of vector bundles over the closed substrata $\alpha \in \Pi, \Phi = \{\varphi_\alpha^\beta\}$ is a collection of vector bundle surjections

$$\xi_\alpha \to \xi_\beta | \alpha$$

for each pair $\alpha, \beta \in \Pi$ with $\alpha < \beta$, and $\Gamma = \{\gamma_\alpha(g)\}$ is a collection of vector bundle isomorphisms $\xi_\alpha \to \xi_{g\alpha}$ covering the maps $g : \alpha \to g\alpha$ (where $\alpha \in \Pi, g \in G$) such that the following conditions hold:

(i) $\gamma_\alpha(gh) = \gamma_{h\alpha}(g)\gamma_\alpha(h)$ for all $g, h \in G$, and $\gamma_\alpha(1) = 1_\alpha$.
(ii) φ_α^β commutes with $\gamma(g)$ for all g such that $g\alpha = \alpha$ and $g\beta = \beta$.
(iii) The kernel of φ_α^β is the fixed point set of G^β in ξ_α (note that G^α acts linearly and fiber-preservingly on ξ_α and $G^\alpha \supset G^\beta$).

The standard example is given by the equivariant normal bundles of the proper closed substrata of M. Specifically, take $\Pi = \pi(M)$ and let ξ_α be the normal bundle of M_α in M (for convenience, one can choose an invariant riemannian metric to describe this precisely). By construction the total spaces of the bundles ξ_α are embedded in the equivariant tangent bundle $T(M)$, and we use the action of G of $T(M)$ to define the maps $\gamma_\alpha(g)$. Finally, the projections φ^β_α can be defined using an invariant metric and the splitting

$$\xi_\alpha = \xi_\beta \,|\, \alpha \oplus \nu_{\alpha\beta},$$

where $\nu_{\alpha\beta}$ denotes the normal bundle of M_α in M_β. This vector bundle system will be denoted by $\Pi(TM)$ or $\nu_\Pi M$; as noted in [**DR**, Example 3.2], a similar construction is valid for an arbitrary G–vector bundle ξ, yielding a vector bundle system $\Pi(\xi)$.

DEFINITION: If (Ξ, Φ, Γ) and (Ξ', Φ', Γ') are vector bundle systems over Π, a *homomorphism* of vector bundle systems is a collection of vector bundle endomorphisms

$$f_\alpha : \xi_\alpha \to \xi'_\alpha \qquad (\alpha \in \Pi)$$

(f_α maps fibers linearly to themselves)

such that

(i) $f_{g\alpha}\gamma_\alpha(g) = \gamma'_\alpha(g) f_\alpha$ for all $\alpha \in \Pi$ and $g \in G$,

(ii) there are invariant riemannian metrics on $\{\xi_\alpha\}$ and $\{\xi'_\alpha\}$ such that for each $\alpha \subset \beta$ we have $f_\alpha((\text{Ker } \varphi^\beta_\alpha)^\perp) \subset (\text{Ker } \varphi'^\beta_\alpha)^\perp$ and $f_\alpha|(\text{Ker } \varphi^\beta_\alpha)^\perp$ corresponds to f_β under the identifications $(\text{Ker } \{\varphi \text{ or } \varphi'\}^\beta_\alpha)^\perp \cong \{\xi \text{ or } \xi'\}_\beta$ determined by φ^β_α and φ'^β_α.

As in [**DR**], the second condition may be paraphrased as "f_α restricted to $\xi_\beta|\alpha$ is equal to f_β restricted to α."

Constructions on vector bundle systems

Given a vector bundle system Ξ over $\pi(M)$ and an isogeneric map $h : M' \to M$, a *pullback* vector bundle system $h^*\Xi$ over $\pi(M')$ may be defined using the pullbacks of the bundles ξ_α, projections φ^β_α, and bundle isomorphisms $\gamma_\alpha(g)$; verification of the relevant properties is a routine exercise (compare [**DR**, p. 18].

If Ξ and Ξ' are two vector bundle systems over $\pi(M)$, then one can define their direct sum $\Xi \oplus \Xi'$ via $(\{\xi_\alpha \oplus \xi'_\alpha\}, \{\varphi^\beta_\alpha \oplus \varphi'^\beta_\alpha\}, \{\gamma_\alpha(g) \oplus \gamma'_\alpha(g)\})$; once again it is elementary to verify that this defines a vector bundle system. It is also elementary to show that direct sum commutes with the construction Π: (*bundles*) $\to$ (*bundle systems*) given above; in other words, $\Pi(\alpha \oplus \beta) \cong \Pi(\alpha) \oplus \Pi(\beta)$.

Finally, we shall need the notion of *stable equivalence* for vector bundle systems. Specifically, Ξ and Ξ' are stably equivalent if and only if there are G–modules V and V' such that

$$\Xi \oplus \Pi(M \times V) \cong \Xi' \oplus \Pi(M \times V').$$

The role of these concepts in equivariant surgery theory will be explained in the next section.

4. Stepwise surgery and the Gap Hypothesis

The basic strategy of equivariant surgery is to view an equivariant surgery problem as an inductive sequence of ordinary surgery problems. At each step one encounters a problem of the following type.

(4.1) STEPWISE SURGERY PROBLEM. *Let $f : N \to M$ be an isogeneric map of closed smooth G–manifolds such that $\check{f} : \pi(N) \to \pi(M)$ is bijective, assume that each induced map $f_\alpha : N_\alpha \to M_\alpha$ has degree ± 1, and let $b : T(N) \to \xi$ be a stable G–vector bundle map covering f (as usual, $T(N)$ denotes the tangent bundle). Let $\Sigma \subset \pi(N)$ be closed ($\alpha \in \Sigma$ and $\beta \leq \alpha \Rightarrow \beta \in \Sigma$) and G–invariant, and assume that if $\alpha \in \Sigma$ then $f_\alpha : N_\alpha \to M_\alpha$ is a homotopy equivalence (the inductive hypothesis). Let β be a minimal element of $\pi(N) - \Sigma$. Is (f, b) cobordant to a pair (f^*, b^*) that agrees with (f, b) over the subsets in Σ and such that $f^*_\beta : N_\beta \to M_\beta$ is a homotopy equivalence?*

If the answer to (4.1) is yes, then by equivariance the map f^* satisfies the inductive hypothesis for $\Sigma \cup G\{\beta\} = \Sigma'$ instead of Σ, and thus one can ask (4.1) for a minimal element of $\pi(M) - \Sigma'$. Of course, if the answer to (4.1) is yes at every step, then we obtain a G–map $h : N \to M$ that is a homotopy equivalence on every closed substratum, and therefore h is an equivariant homotopy equivalence by an elementary argument involving the equivariant Whitehead Theorem (see [**Bre1**, Section II.5] or [**tD**, Prop. II.2.7, p. 107]) and equivariant pushouts (*i.e.*, the equivariant analog of [**Brwn**, Thm. 7.4.1, pp. 240–242]).

Motivated in part by the last half of Section 1, we shall begin by considering the following closely related question:

(4.2) RESTRICTED STEPWISE SURGERY PROBLEM. *Let $f : N \to M$ be an isovariant map satisfying the conditions of (4.1). Is it possible to do surgery equivariantly on N_β away from its relatively singular set $N_\beta \cap (\cup_{\alpha \in \Sigma} N_\alpha)$ and obtain an equivariant map that is a homotopy equivalence?*

As indicated in Section 3, the group that acts on N_β is the group W_β consisting of all elements in $W(G^\beta) := \text{Normalizer}_G(G^\beta)/G^\beta$ that map N_β into itself (recall that the normalizer quotient acts on the fixed set of G^β).

The statement of (4.2) suggests that we would like to perform surgery without touching the singular part of N_β; more precisely, if $q : S^k \to N_\beta$ represents a homotopy class that we wish to kill by surgery, we would like to approximate q by an embedding q' such that $q'(S^k)$ does not meet the singular part of N_β. Geometrically speaking, the easiest way of avoiding a subset is to impose some general position hypothesis. Such conditions were first introduced into transformation groups during the early nineteen seventies (compare Straus [**St**]) and became a fixture in the subject through the work of Petrie [**P1**], [**P2**]. Here is the version we shall need.

STANDARD GAP HYPOTHESIS: A locally linear G–manifold M is said to satisfy the *Standard Gap Hypothesis* if for each pair of closed substrata $\alpha, \beta \in \pi(M)$ with $\alpha < \beta$ we have

$$(\ddagger) \qquad \dim M_\alpha < \tfrac{1}{2}(\dim M_\beta).$$

If we use the term "Gap Hypothesis" without any modifiers (*e.g.*, "Standard" or "Codimension ≥ 2" or "Codimension ≥ 3") we mean the Standard Gap Hypothesis.

If we assume the Standard Gap Hypothesis we can take a step towards answering (4.2):

PROPOSITION 4.3. *Assume the setting of 4.2 and also assume that $M \times D^2$ satisfies the Standard Gap Hypothesis (in fact, it suffices to assume (‡) for $N_\beta \times D^2$ and $\alpha \in \Sigma$ (where Σ is viewed as a subset of $\pi(M \times D^2) \approx \pi(M)$). Furthermore, assume $\dim N_\beta \geq 5$. Then*

(*i*) *one can do surgery on β equivariantly and away from the singular set to obtain a map $f_0 : N'_\beta \to M_\beta$ that is connected up to the middle dimension,*
(*ii*) *if f_0 is as in (i), there is an obstruction $\sigma(f_0) \in L^h_{\dim \beta}(\mathbb{Z}[E_\beta], w)$ such that the answer to (4.2) is yes for f_0 if the class $\sigma(f_0)$ is trivial; in fact, if $\sigma(f_0) = 0$ one can obtain an equivalence $f_1 : N''_\beta \to M_\beta$ by doing surgery on N'_β equivariantly away from the singular set.*

EXPLANATION: L^h denotes the homotopy Wall group, E_β denotes the fundamental group of $(M_\beta - \cup_{\alpha \in \Sigma} M_\alpha)/W_\beta$, and w is given by the first Stiefel–Whitney class of this manifold.

The proof of 4.3 is standard, the idea being that surgery is done by means of embedded k–spheres where $k \leq \frac{1}{2} \dim N_\beta$ and these spheres can be deformed to miss the singular set by the Gap Hypothesis and general position. Further details see Sections 5 and 9 of [DR].■

Proposition 4.3 goes a long way towards answering (4.1) if the Gap Hypothesis holds. However, this result only yields an equivalence over the closed substrata in Σ', and there is no claim that this Σ'–equivalence extends to an equivariant degree 1 map into M. In order to ensure that an extension exists, it is often necessary to introduce additional data. Vector bundle systems provide one method for doing this.

DEFINITION: Let $f : N \to M$ be an equivariant degree 1 map of closed smooth G–manifolds, and assume that $\check{f}$ is bijective and f is isogeneric. A *system of equivariant bundle data for f* is a triple (b_0, c, φ) where

(i) ξ is a stable G–vector bundle over M,
(ii) $b_0 : T(N) \to f^*\xi$ is a stable G–vector bundle isomorphism,
(iii) η is a vector bundle system over $\pi(M)$,
(iv) $c : \Pi(T(N)) \to f^*\eta$ is an isomorphism of vector bundle systems
(v) $\varphi : \eta \oplus \Pi(M \times V) \to \Pi(\xi)$ is a stable isomorphism of vector bundle systems such that

$$\begin{array}{ccc} \Pi(T(N)) \oplus \Pi(N \times V) & \xrightarrow{c \oplus V} & f^*\eta \oplus \Pi(N \times V) \\ \Big\downarrow = & & \Big\downarrow f^*\varphi \\ \Pi(T(N) \oplus V) & \xrightarrow{\Pi(b_0)} & \Pi(f^*\xi) \end{array}$$

commutes.

NOTE: In **[DP1]** and **[DR]** the isomorphism φ is not listed explicitly as part of the structure. One reason for including φ in the definition is that bordism classes of data

(f, b_0, c, φ) can be described homotopy–theoretically using results of F. Connolly and V. Vijums [**CV**]. Another reason is that explicit identifications of the normal fibers with fixed representations are needed to define the group operation described in [**DR**].

DEFINITIONS: The data (f, b_0, c, φ) given by an equivariant degree 1 map of manifolds with boundary and bundle data (as above) will be called a *G–equivariant normal map* (or a *G-equivariant surgery problem) of type* $(ht, DIFF)$. As noted before, there are several different approaches to equivariant surgery theory, each having its own concept of normal map, and we have included the type specification here for the sake of completeness. Since this chapter deals mainly with one version of equivariant surgery theory, references to type $(ht, DIFF)$ will be omitted if no chance of confusion seems likely.

For each type of normal maps there is a standard procedure for defining a *normal cobordism* between two such maps; it suffices to take a map of manifold triads $(W; \partial_0 W, \partial_1 W) \rightarrow M \times (I; \{0\}, \{1\})$ equipped with the appropriate extra data, and perhaps satisfying some appropriate restrictions. The precise definition for the type $(ht, DIFF)$ appears in [**DR**, p. 24].

With these definitions we can state the following result on the Stepwise Surgery Problem formulated in (4.1):

THEOREM 4.4. *Suppose we are given the setting of* (4.1) *and Proposition* 4.3, *and let* (b_0, c, φ) *be bundle data for* f *with* b *equal to the composition*

$$T(N) \xrightarrow{b_0} f^*\xi \xrightarrow{\text{pullback}} \xi.$$

Then the following hold:

(i) (f, b_0, c, φ) *is cobordant to data* (f', b'_0, c', φ') *such that the cobordism is a product of the original data with* $[0, 1]$ *over* Σ *and* $f'_\beta : N'_\beta \rightarrow M_\beta$ *is connected up to the middle dimension.*

(ii) There is an obstruction $\sigma(f'_\beta) \in L^h_{\dim \beta}(\mathbb{Z}[E_\beta], w)$ *such that* (f', b'_0, c', φ') *is nicely cobordant to data* $(f'', b''_0, c'', \varphi'')$ *with* f''_β *a homotopy equivalence if* $\sigma(f'_\beta) = 0$.

PROOF: Proposition 4.3 yields an equivariant normal cobordism from $f_\beta : N_\beta \rightarrow M_\beta$ to a highly connected map $f'_\beta : N'_\beta \rightarrow M_\beta$, and if $\sigma(f'_\beta) = 0$ the same result yields an equivariant normal cobordism from f'_β to an equivariant homotopy equivalence $f''_\beta : N''_\beta \rightarrow M_\beta$. In each case the normal cobordism is constructed from a finite sequence of surgeries over M_β. Therefore it suffices to see that each of these surgeries over M_β can be extended equivariantly to an equivariant surgery over M. The additional bundle data carried by c and φ is used to construct the extension. Details of this procedure are given in [**DP1**], Section 5.■

5. Equivariant surgery groups

One objective of equivariant surgery theory is to define global equivariant surgery obstructions corresponding to the stepwise obstructions of Section 4. The reasons for doing this go beyond the search for an attractive formal setting. Global considerations sometimes imply that many of the stepwise obstructions can be disregarded (compare [**DH**]), and in some cases it is possible to analyze the global groups effectively by algebraic methods (for example, the induction theorem of [**DP2**]). The ordinary Wall groups associated to stepwise obstructions are related to the global obstruction groups by spectral sequences that begin with the Wall groups and converge to the global objects (compare [**BQ**]). In this section we shall describe the general ideas underlying the construction of the global obstruction groups and spectral sequences as they apply to the theory $I^{ht,DIFF}$ of [**DR**].

STANDING HYPOTHESES: G is a finite group and M is a compact smooth G–manifold. All closed substrata of M and ∂M will be assumed to be simply connected with dimension ≥ 5, and we shall also assume a *Codimension* ≥ 3 *Gap Hypothesis:*

$$\text{If } \alpha < \beta \text{ are closed substrata of } M, \text{ then } \dim M_\beta - \dim M_\alpha \geq 3.$$

In [**WL**, Chapter 9] Wall presented a bordism–theoretic definition of the surgery obstruction groups $L_*^s(\mathbf{Z}[\pi], w)$; a small correction to this definition is described in [**FH**], pp. 102–105. Similar definitions exist for other surgery groups such as the homotopy groups $L_*^h(\mathbf{Z}[\pi], w)$ and the Cappell–Shaneson Γ–groups [**CS**]. In fact, the methods of [**WL**, Chapter 9] apply in any reasonable setting for which one has an analog of Wall's $\pi - \pi$ theorem [**WL**, Chapter 4]. Most approaches to equivariant surgery theory are based upon this principle.

We begin by formulating the appropriate version of the $\pi - \pi$ theorem.

THEOREM 5.1. *Let M be a compact smooth G–manifold as above, and assume in addition that*
(i) $\check{\imath} : \pi(\partial M) \to \pi(M)$ is bijective,
(ii) each closed substratum $\alpha \in \pi(M)$ has dimension ≥ 6,
(iii) ∂M satisfies the Standard Gap Hypothesis.

If $f : (N, \partial N) \to (M, \partial M)$ is a degree 1 map of pairs and (b_0, c, φ) is a system of bundle data for f, then (f, b_0, c, φ) is equivariantly cobordant to an equivariant homotopy equivalence (of pairs).

This can be established using a step by step argument as in Section 4. Assume that one can find an equivariantly cobordant object that is an equivariant homotopy equivalence over the closed substrata in Σ, and let β be minimal in $\pi(M) - \Sigma$. Then it is possible to find an equivariantly cobordant object that is an equivariant homotopy equivalence over $\Sigma' = \Sigma \cup G\{\beta\}$ if a stepwise surgery obstruction vanishes. But in this case the Gap Hypothesis and the proof of the $\pi - \pi$ theorem for L^h imply that the obstruction is trivial. Proceeding inductively over the closed substrata, we ultimately conclude that the original object is equivariantly cobordant to an equivariant homotopy equivalence.■

Let $\lambda = \lambda_M$ where M satisfies the conditions of Theorem 5.1. Following Wall [**WL**, Chapter 9], we define $I^{ht,DIFF}(G, \lambda)$ as bordism classes of objects (f, b_0, c, φ, k) where $f : (Y, \partial Y) \to (X, \partial X)$ is an equivariant degree 1 map such that each induced map of a closed substratum has degree ± 1, the boundary map $\partial Y \to \partial X$ is a G–homotopy equivalence, (b_0, c, φ) is a system of bundle data for f, and $k : \lambda_X \approx \lambda$ identifies λ_X with $\lambda = \lambda_M$. The formalism of [**WL**] and Theorem 5.1 imply that $I^{ht,DIFF}(G, \lambda)$ can also be described as equivalence classes of objects satisfying suitable restrictions modulo bordisms of an appropriate type (see the example below). Furthermore, if $\mathbf{S}\lambda$ satisfies the Gap Hypothesis then $I^{ht,DIFF}(G, \lambda)$ has an abelian group structure with the usual sorts of geometric interpretations (compare [**DR**, Sections 7-8]).

EXAMPLE: Here is one particularly important method for describing the equivariant obstruction groups $I^{ht,DIFF}(G, \lambda)$ in terms of restricted objects and bordisms. Suppose that V is a closed smooth G-manifold such that $\lambda_V \approx \mathbf{S}^{-1}\lambda_M$ (*i.e.*, all dimensions reduced by 1; see the end of Section 1). Then every class in $I^{ht,DIFF}(G, \lambda)$ is represented by an object $(f; b_0, c, \varphi; k)$ where

$$f : (W, \partial W) \to (V \times I, V \times \partial I)$$

is an equivariant degree one map such that

$$\partial_i W = f^{-1}(V \times \{i\}) \quad i = 0, 1$$

with $f|\partial_0 W$ a G-diffeomorphism and $f|\partial_1 W$ a G-homotopy equivalence. Furthermore, two such objects $(f_0; \textit{other data}), (f_1; \textit{other data})$ are equivalent if and only if they are related by a cobordism (F; etc.) where

$$F : X \to V \times I \times I$$

such that $F = f_i$ over $V \times I \times \{i\}$, F is a G-diffeomorphism over $V \times \{0\} \times I$, and F is a G-homotopy equivalence over $V \times \{1\} \times I$ (compare [**DR**, Lemma 8.2, p. 66] and [**WL**, Chapters 9 and 10]).

Relating global and stepwise obstructions

The first step in this direction is to give yet another set of definitions:

DEFINITION: Let M satisfy the standing hypotheses, assume further that M satisfies the Gap Hypothesis, and let $\Sigma \subset \pi(M)$ be closed and G-invariant. An isogeneric equivariant map $f : X \to M$ is said to be Σ-*adjusted* (or Σ-*good*, as in [**DP1–2**] or [**DR**]) if for each closed substratum α with $\alpha \in \Sigma$ the induced map $M_\alpha \to N_\alpha$ is a homotopy equivalence.

REMARKS: 1. This condition played a central role in Section 4. In fact, the stepwise surgery question can be rephrased to ask whether a Σ-adjusted map can be modified to a Σ'-adjusted map, where $\Sigma' = \Sigma \cup G\{\beta\}$ and β is minimal in $\pi(M) - \Sigma$.

2. If $\Sigma = \pi(M)$, then a Σ-adjusted map in the category of $G - CW$ complexes is an equivariant homotopy equivalence (as noted in Section 4, this follows from considerations involving pushouts and the equivariant Whitehead Theorem).

3. It is possible to take $\Sigma = \varnothing$, and for this choice all maps are Σ-adjusted.

Our immediate goal is to construct analogs of the group $I^{ht,DIFF}(G, \lambda)$ for Σ-adjusted maps. In order to do this, we need the following:

COMPLEMENT TO 5.1. *If the map f in 5.1 is Σ–adjusted, then one can find a Σ–adjusted bordism to a G–homotopy equivalence (of pairs).*

This follows from the same step by step procedure that was described for Theorem 5.1.■

The equivariant surgery obstruction groups of Σ–adjusted maps

$$I^{ht,DIFF}(G;\lambda;\Sigma)$$

can now be defined as bordism classes of objects (f, etc.) such that f is Σ–adjusted. All cobordisms that enter into the definition of bordism equivalence are also assumed to be Σ–adjusted. As noted in Remark 3 above, the groups $I^{ht,DIFF}(G;\lambda;\emptyset)$ are precisely the groups $I^{ht,DIFF}(G;\lambda)$ considered previously. In fact, *all the basic properties of these groups extend directly to their Σ–adjusted analogs.* For example, the complement to 5.1 implies that elements of $I^{ht,DIFF}(G;\lambda;\Sigma)$ have nice Σ–adjusted representatives, and the geometric construction of the abelian group structure also extends. Furthermore, if $\Sigma \subset \Sigma'$ there is an obvious forgetful map

$$j_{\Sigma,\Sigma'} : I^{ht,DIFF}(G;\lambda;\Sigma') \to I^{ht,DIFF}(G;\lambda;\Sigma)$$

and this map is a homomorphism.

We would like to extend this formalism even further to describe relative sets

$$I^{ht,DIFF}(G;\lambda;\Sigma \subset \Sigma')$$

and construct an exact sequence involving such groups and the forgetful homomorphism $j_{\Sigma,\Sigma'}$. In order to define these groups we must assume

($\star$) $\lambda = \mathbf{S}\lambda_0$ *for some appropriate* λ_0.

The elements of the relative sets will be bordism classes of objects (f, b_0, c, φ, k) where

$$f : (Y; Y_0, Y_1) \to (X; X_0, X_1)$$

is a degree 1 map of manifold triads such that

$$f_0 : (Y_0, \partial Y_0) \to (X_0, \partial X_0)$$

is a G–homotopy equivalence of pairs,

$$f_1 : (Y_1, \partial Y_1) \to (X_1, \partial X_1)$$

is Σ'–adjusted (and a G–homotopy equivalence on the boundary—recall $\partial\partial_0 = \partial\partial_1$), the inclusion induces a bijection $\pi(X_1) \cong \pi(X)$, and f itself is Σ–adjusted.

If λ desuspends to λ_0 as above, then there is a canonical interpretation of the absolute groups $I^{ht,DIFF}(G;\lambda;\Sigma)$ as the relative groups $I^{ht,DIFF}(G;\lambda;\Sigma \subset \pi(X))$. By the definitions there is a forgetful homomorphism $J_\Sigma : I^{ht,DIFF}(G;\lambda;\Sigma \subset \pi(X)) \to I^{ht,DIFF}(G;\lambda;\Sigma)$. Furthermore, the map J_Σ is $1-1$ and onto by the previously mentioned results on defining $I^{ht,DIFF}(G;\lambda;\Sigma)$ with objects satisfying suitable restrictions.

The following result is a direct consequence of the definitions and the preceding discussion:

PROPOSITION 5.2. *If all the closed substrata of M are at least 6–dimensional and there is a manifold V with $\lambda_V = \mathbf{S}^{-1}\lambda$, then the following sequence is exact:*

$$I^{ht,DIFF}(G;\lambda;\Sigma') \xrightarrow{j} I^{ht,DIFF}(G;\lambda;\Sigma) \xrightarrow{\ell} I^{ht,DIFF}(G;\lambda;\Sigma\subset\Sigma') \xrightarrow{\partial_1}$$
$$[\xrightarrow{\partial_1}]\ I^{ht,DIFF}(G;\mathbf{S}^{-1}\lambda;\Sigma') \xrightarrow{j} I^{ht,DIFF}(G;\mathbf{S}^{-1}\lambda;\Sigma).$$

REMARKS: 1. The map j is the forgetful map described above, the map ∂_1 is given by taking the top half of the boundary (*i.e.*, apply ∂_1 to the triad), and the map ℓ is a forgetful map (use the identification of $I^{ht,DIFF}(G;\lambda;\Sigma)$ and $I^{ht,DIFF}(G;\lambda;\Sigma\subset\pi(X))$ described above; then the pair $\Sigma\subset\pi(X)$ contains the pair $\Sigma\subset\Sigma'$ and thus relatively adjusted maps of the first type are automatically relatively adjusted maps of the second type).

2. The usual arguments imply that the sets $I^{ht,DIFF}(G;\lambda;\Sigma\subset\Sigma')$ have abelian group structures, and it is a straightforward exercise to verify that the maps ℓ and ∂_1 are homomorphisms.

The proof of Proposition 5.2 is essentially the usual sort of argument for verifying that a relative bordism sequence is exact (compare [**CF**] or [**LM**, Part II, Section 1].■

Suppose now that $\Sigma' = \Sigma\cup G\{\beta\}$ where β is minimal in $\pi(M)-\Sigma$. If $(f;-)$ represents a class in $I^{ht,DIFF}(G;\lambda;\Sigma\subset\Sigma')$, then Proposition 4.3 yields a stepwise surgery obstruction $\sigma(f'_\beta)\in L^h_{\dim\beta}(\mathbb{Z}[W_\beta],w)$.

PROPOSITION 5.3. *In the setting above, assume that $\mathbf{S}^2\lambda$ also satisfies the Gap Hypothesis. Then the stepwise surgery obstruction defines an isomorphism*

$$\sigma_\beta : I^{ht,DIFF}(G;\lambda;\Sigma\subset\Sigma') \to L^h_{\dim\beta}(\mathbb{Z}[W_\beta],w).$$

PROOF: *(Sketch)* The first order of business is to show that σ_β is well–defined. Suppose we have two Σ–adjusted representatives f, f' of the same class such that f_β and f'_β are both highly connected. Let F be the map associated to the cobordism between f and f'. Since $\mathbf{S}^2\lambda$ satisfies the Gap Hypothesis we can do surgery to make F_β highly connected and ensure that all embedded spheres in the middle dimension can be deformed to miss the singular set of the cobordism. Reasoning as in Wall's book, we can use the surgery kernels of F_β and $(F_\beta,\partial F_\beta)$ to show that the algebraic elements $\sigma(f_\beta)$ and $\sigma(f'_\beta)$ are equal.

The additivity of σ_β can be established by formal geometric considerations as in [**DR**] (see in particular the second paragraph of the proof of 9.2 on page 81). Theorem 4.4 and additivity imply that σ_β is one to one.

Finally, the proof that σ_β is onto proceeds as follows: Take a G–manifold V such that $\lambda_V \cong \mathbf{S}^{-1}\lambda$, and let $\alpha\in L^h_{\dim\beta}(\mathbb{Z}[W_\beta],w)$. As in [**WL**, Sections 5–6] or [**DR**] we can add handles to $(V_\beta\times I)/W_\beta$ *away from the singular set* to obtain an equivariant normal map $F_\beta : U_\beta\to V_\beta\times I$ that is a W_β-equivalence on the boundary and near the singular set, and such that the relative stepwise surgery obstruction for the top closed substratum is α (here relative surgery means surgery away from the boundary).

By construction U_β is formed from $V_\beta\times[0,\varepsilon]$ by a finite sequence of equivariant surgeries. As in the proof of 4.4, one can use the $\pi(M)$–bundle system from the bundle data to extend these surgeries to equivariant surgeries over $V \times [0,\varepsilon]$, and this can be done so that the map $1_{V\times[0,\varepsilon]} \cup F_\beta$ and the bundle data also extend to the trace W of the equivariant surgeries on $V \times [0,\varepsilon]$. If $F : W \to V \times I$ is the map obtained in this way, it is immediate from the construction that $\sigma_\beta(F;\cdots) = \sigma(F_\beta;\cdots) = \alpha$.■

Propositions 5.2 and 5.3 yield the **Orbit Sequence:**

THEOREM 5.4. *If $\mathbf{S}^2\lambda$ satisfies the Gap Hypothesis, the set $\Sigma \subset \pi(X)$ is closed, and $\alpha \notin \Sigma$ is such that $\Sigma' = \Sigma \cup G\{\alpha\}$ is closed, then we have the following exact sequence:*

$$I^{ht,DIFF}(G;\mathbf{S}\lambda;\Sigma') \to I^{ht,DIFF}(G;\mathbf{S}\lambda;\Sigma) \xrightarrow{res(\alpha)} L^h_{\dim\alpha+1}(\mathbf{Z}[W_\alpha],w_\alpha) \xrightarrow{\partial}$$

$$[\xrightarrow{\partial}]\, I^{ht,DIFF}(G;\lambda;\Sigma') \xrightarrow{\beta} I^{ht,DIFF}(G;\lambda;\Sigma) \xrightarrow{res(\alpha)} L^h_{\dim\alpha}(\mathbf{Z}[W_\alpha],w_\alpha)\cdots$$

In this sequence β is a forgetful map and $res(\alpha)$ is given by restricting to the induced W_α-surgery problem over the closed substratum corresponding to α. ■

Associated surgery sequences. The central point of [**DR**] is the construction of a "rather long" equivariant surgery exact sequence generalizing the ordinary Sullivan-Wall sequences for homotopy structures [**WL**, Chapters 10 and 17] such that the equivariant groups $I^{ht,DIFF}$ correspond to the ordinary groups L^h. In this book we are mainly interested in certain properties of the equivariant surgery groups themselves, and therefore we have not included a detailed discussion of the equivariant surgery sequence; however, a working knowledge of this sequence is probably worthwhile for further reading about the uses of [**DR**] in transformation groups. One significant difference between the equivariant and ordinary sequences is that the ordinary sequence extends infinitely to the left but the equivariant sequence construction only yields a finite sequence. In particular, since $\mathbf{S}^{k+1}\lambda_X$ does not satisfy the Gap Hypothesis if $k > \dim X$, the sequence as constructed in [**DR**] will never have more than $3\dim X$ terms. One motivation for the periodicity theorems of Chapter II is the question of finding further extensions of this sequence that are algebraically tractable and to some extent geometrically meaningful. Some preliminary results along these lines are described in [**Sc4**].

Exceptional cases

In Chapter II of this book we shall need an analog of Proposition 5.3 when $\mathbf{S}\lambda$ satisfies the Gap Hypothesis but $\mathbf{S}^2\lambda$ does not.

PROPOSITION 5.5. *In the setting above, assume only that $\mathbf{S}\lambda$ satisfies the Gap Hypothesis. Then the stepwise surgery obstruction defines an isomorphism*

$$\sigma_\beta : I^{ht,DIFF}(G;\lambda;\Sigma \subset \Sigma') \to L^h_{\dim\beta}(\mathbf{Z}[W_\beta],w)/B$$

where B is a certain subgroup of $L^h_{\dim\beta}(\mathbf{Z}[W_\beta],w)$.

PROOF: *(Sketch)* If f is a Σ-adjusted map satisfying the conditions of (4.1), then by Theorem 4.4 one can find a highly connected map f' representing the same class as f (with the appropriate extra data), and in this manner we obtain the stepwise surgery obstruction $\sigma(f'_\beta)$. However, the argument of Proposition 5.3 does not show that the stepwise obstruction is well defined if $\mathbf{S}\lambda$ satisfies the Gap Hypothesis but $\mathbf{S}^2\lambda$ does not. In this case it is necessary to use methods from the proof of [**DR**, Theorem 9.2]. Specifically, the techniques employed in [**DR**, pp. 78–82] establish the existence of a subgroup $B \subset L^h_{\dim}(\mathbb{Z}[W_\beta], w)$ such that

(*i*) the image of $\sigma(f'_\beta)$ mod B is well defined,
(*ii*) the associated map σ_β is additive,
(*iii*) the class of f (plus extra data) has a Σ'-adjusted representative if and only if $\sigma_\beta(f)$ is zero mod B.

It follows that σ_β is a monomorphism.

The proof that σ_β is onto closely resembles the corresponding argument for Proposition 5.3. In this case one obtains a homomorphism

$$\gamma_\beta : L^h_{\dim\,\beta}(\mathbb{Z}[W_\beta], w_\beta) \to I^{ht,DIFF}(G; \lambda; \Sigma \subset \Sigma')$$

such that $\sigma_\beta\gamma_\beta$ is reduction mod B. Since the latter map is obviously onto, it follows that σ_β is also onto, and by the preceding paragraph the map σ_β is in fact an isomorphism.■

Results of M. Morimoto [**Mor**] yield a more precise description of the quotient group in Proposition 5.5 when $\mathbf{S}\lambda$ satisfies the Gap Hypothesis but $\mathbf{S}^2\lambda$ does not. Namely, the group in question is the Bak group

$$\mathbf{W}_{\dim\,\beta}(\mathbb{Z}[W_\beta], \Gamma W_\beta(X_\beta); w_\beta)$$

introduced in [**Ba**], where $\Gamma W_\beta(X_\beta)$ is a set defined as on page 467 of [**Mor**], and w_β is a first Stiefel-Whitney class. Frequently this group is isomorphic to the corresponding Wall group. In particular, this is true if G has odd order or $\dim(\alpha) \equiv 0$ or 1 mod 4 and the orientation homomorphism is trivial. However, the example in [**Mor**, Corollary C, page 468] shows that the projection from the Wall group to the Bak group has a nontrivial kernel in some cases.

REMARK: In the paragraph of [**Mor**] following Corollary C (and also the second paragraph on page 466) Morimoto discusses the validity of some statements in the literature. We shall consider some of the questions he raises in Section 7.

An exact couple

In analogy with the final paragraph of [**BQ**] and [**LM**, Part II, Section 1], the exact sequences for the forgetful maps

$$I^{ht,DIFF}(\lambda; G; \Sigma') \to I^{ht,DIFF}(\lambda; G; \Sigma)$$

define an exact couple, but there is one important drawback. Namely, the exact sequence can only be extended k degrees to the left when the indexing data $\mathbf{S}^{+k}\lambda$ satisfies the Gap Hypothesis because our analysis of $I^{ht,DIFF}(G; \lambda; \Sigma)$ requires this condition; of course, the sequence can only be extended to right when $\mathbf{S}^{-k}\lambda$ is geometrically meaningful, but this is less bothersome because exact couples of this type arise naturally in the theory of half exact functors [**Dld**]. Nevertheless, one does obtain a partial spectral sequence of sorts from the partial exact couple defined by the maps $I^{ht,DIFF}(\cdots\Sigma') \to I^{ht,DIFF}(\cdots\Sigma)$; by Theorem 5.4 and Proposition 5.5 the initial terms are either ordinary Wall groups or Bak groups (in exceptional cases only), and the usual considerations imply that the partial spectral sequence converges in some sense to the equivariant surgery groups $I^{ht,DIFF}(G;\mathbf{S}^{\pm *}\lambda)$.

FINAL NOTE: In the previous paragraph we mentioned that our construction for the groups $I^{ht,DIFF}(G;\mathbf{S}^{+k}\lambda;\Sigma)$ only works for some values of k; specifically, this is only true if $\mathbf{S}^{+k}\lambda$ satisfies the Gap Hypothesis, and there are only finitely many such values of k. Although the results of Section 6 show that the approach outlined above does not work if the Gap Hypothesis fails, it is still natural to ask if other methods can be used to define the groups $I^{ht,DIFF}(G;\mathbf{S}^{+k}\lambda;\Sigma)$ for all $k \geq 0$. In Chapter II we shall present two ways of doing this, one of which is due to Browder and Quinn [**BQ**] and the other to Lück and Madsen [**LM**]. By Propositions II.3.12 and II.3.15 each of these theories reduces to $I^{ht,DIFF}$ if one assumes the Gap Hypothesis and the appropriate restrictions on dimensions and fundamental groups (however, [**BQ**] and [**LM**] yield nonisomorphic groups if the Gap Hypothesis fails). For groups of odd order we shall describe yet another way of extending the definition of $I^{ht,DIFF}(G;\mathbf{S}^{k}\lambda)$ to all $k \geq 0$ in Section III.5; this process can be thought of as a *periodic stabilization*, and we shall use it to establish some basic relationships between the equivariant surgery obstruction groups obtained in [**BQ**] and [**LM**].

6. Some examples

If the Gap Hypothesis does not hold for a closed smooth G–manifold M, it is still possible to define a set of bordism classes $I^{ht,DIFF}(G;\lambda_M)$ as in Section 5. In principle the Gap Hypothesis is needed to ensure that these sets have good formal properties; for example, the existence of special types of representatives for each bordism class and a canonical abelian group structure. A construction for a group structure is often possible without the Gap Hypothesis (compare [**LM**]). However, one needs the Gap Hypothesis to prove the equivariant $\pi-\pi$ theorem (Theorem 5.1 and its complement), and one needs this version of the $\pi-\pi$ theorem to establish the good properties of $I^{ht,DIFF}(G;\lambda_M)$ by a suitable modification of Wall's approach in the nonequivariant case [**WL**, Chapter 9 again]. Thus one of the most basic questions regarding equivariant surgery outside the Gap Hypothesis range is the

UNSTABLE $\pi-\pi$ QUESTION. *To what extent do Theorem 5.1 and its complement generalize to G–manifolds not satisfying the Gap Hypothesis?*

Results of the first author [**Dov1**] show that the basic techniques of equivariant surgery can be modified to work effectively just outside the Gap Hypothesis range, so there is evidence that 5.1 might extend to such cases. However, M. Rothenberg and S. Weinberger (unpublished) have shown that the conclusion of 5.1 is false for a large class of examples. Their approach can be summarized as follows:

(6.1). *Let F be a 1-connected manifold of even dimension ≥ 6 such that ∂F is nonempty and simply connected, let $\mathbf{Z}_k$ (where $k > 1$) act semifreely on $F \times D^{2q}$ by the trivial action on the first coordinate and scalar multiplication on the second, and assume that $q \geq 2$ and $\dim F + 2q \equiv 0 \mod 4$. If $(f; b_0, c, \varphi)$ is a $\mathbf{Z}_k$–equivariant surgery problem with degree one map*

$$f : (N, \partial N) \to (F \times D^{2q}, \partial(F \times D^{2q}))$$

such that f is an homotopy equivalence (of pairs) on the fixed set and $(f; -)$ is $\mathbf{Z}_k$–cobordant to an equivariant homotopy equivalence, then the (ordinary) signature of a tubular neighborhood of the fixed set of N is trivial.■

(6.2). *Let $F := \mathbf{C}P^{n+1} - \mathrm{Int}\, D^{2n+2}$ in the setting of (6.1), and assume further that $n \geq q + 1$ and $n \geq 3$. Then there exist $\mathbf{Z}_k$–equivariant surgery problems with target $F \times D^{2q}$ such that the signature of a tubular neighborhood of the fixed point set of N (= the domain) is nontrivial.*■

In fact, Rothenberg and Weinberger observed that examples of this type exist for many other codomains besides the specific manifolds $F \times D^{2q}$ described in (6.2).

COROLLARY 6.3. *The examples of (6.2) satisfy all the hypotheses of Theorem 5.1 except the Gap Hypothesis, but they are not equivariantly normally cobordant to equivariant homotopy equivalences of pairs.*■

REMARKS: 1. If the Gap Hypothesis holds, examples with the properties in (6.2) cannot exist.

2. Specific examples for (6.2) may be constructed as follows: Various homotopy-theoretic considerations imply the existence of infinitely many distinct $2q$–dimensional vector bundles $\xi \downarrow \mathbf{C}P^n$ such that

(i) ξ has a complex structure for which the $\mathbf{Z}_k$–equivariant sphere bundle $(S(\xi)$, scalar mult.) is stably fiber homotopically trivial,
(ii) the Euler class of ξ is nontrivial.

For each choice of a stable equivariant fiber homotopy trivialization $S(\xi \oplus W) \to \mathbf{C}P^n \times S(V \oplus W)$, the standard theorems on equivariant transversality **[P1]** yield a $\mathbf{Z}_k$–surgery problem $(f; b_0, c, \varphi)$ such that

(i) $f : (N, \partial N) \to (F \times D^{2q}, \partial(F \times D^{2q}))$ has degree 1,
(ii) $b_0 : T(N) \to T(F) \oplus (\xi \otimes_{\mathbf{C}} \mathbf{V}_1)$ is a stable $\mathbf{Z}_k$–vector bundle map covering f (here $\mathbf{V}_1$ generates the representation ring $R(\mathbf{Z}_k)$),
(iii) $c : \nu_{(\mathrm{Fix}\, \mathbf{C}N)} \to \xi$ is the unstable $\mathbf{Z}_k$–vector bundle map ($\Leftrightarrow$ map of vector bundle systems in this case) consistent with b_0.

Using the nonequivariant $\pi - \pi$ theorem and the bundle data we can construct a representative of this type for which the map of fixed sets is a homotopy equivalence of pairs.

By construction the equivariant normal bundle ν of Fix(N) is equal to the pullback of ξ. Therefore if $D(\nu)$ is the unit disk bundle of ν (so that the boundary is $S(\nu)$), it follows that $(D(\nu), S(\nu))$ is homotopy equivalent to

$$(D(\xi \downarrow F), \partial D(\xi \downarrow F)) \cong (D(\xi \oplus \eta \downarrow \mathbf{CP}^n), S(\xi \oplus \eta \downarrow \mathbf{CP}^n)),$$

where $\eta \downarrow \mathbf{CP}^n$ is the canonical complex line bundle.

We claim that the signature of $D(\xi \oplus \eta)$ is nonzero; from this it will also follow that the signature of $D(\nu)$ is also nonzero. More generally, if α is an oriented vector bundle over a complex projective space $\mathbf{CP}^n$ then by the Thom isomorphism, the multiplicative structure of $H^*(\mathbf{CP}^n)$, and the basic properties of the Euler class it follows that the signature of $D(\alpha)$ is nonzero if

(*i*) dim $D(\alpha)$ is divisible by 4,
(*ii*) the middle dimensional cohomology of $(D(\alpha), S(\alpha))$ is nonzero,
(*iii*) the Euler class of α is nonzero.

Since ξ was chosen so that all three conditions are satisfied for $\xi \oplus \eta$, the signature assertion follows immediately.■

Obviously, one would like to understand these examples from a more general viewpoint, but a means for doing so is not yet available.

7. Appendix: Borderline cases of the Gap Hypothesis

As noted in Section 5, results of M. Morimoto [**Mor**] show that stepwise obstructions sometimes behave in a nonstandard fashion if $\mathbf{S}\lambda$ satisfies the Gap Hypothesis but $\mathbf{S}^2\lambda$ does not. At several points in [**Mor**] there are statements indicating that this fact had been overlooked in some earlier papers on equivariant surgery, and Morimoto raises questions about the implications for these papers. In this section we shall discuss these implications for the papers cited in [**Mor**]—specifically, [**DP2**], [**Ms1**], and [**PR**]—and we shall also discuss the implications for [**DR**].

Implications for [**DP2**]: As noted in [**Mor**, p. 466, lines 10–12], stepwise obstructions behave in the standard fashion if λ satisfies the conditions of the preceding paragraph and the underlying group G has odd order. Since [**DP2**] only deals with odd order groups, its results are unaffected.

Implications for [**PR**]: A correction is needed for Theorem 12.4 of [**PR**] when dim Sing $X = [\frac{1}{2}(\dim X - 1)]$, where $[\cdots]$ denotes the greatest integer function. Specifically, one obtains a well-defined surgery obstruction in a Bak group rather than a Wall group. No other portions of [**PR**] are affected.

Here is a more detailed explanation. Given a G-normal map $f : X \to Y$ that is adjusted over the singular set, a stepwise surgery obstruction $\sigma(f) \in L(G)$ has been described in Sections 4 and 5 (also see [**PR**, Chapter 3, Definition 10.11] or [**Dov3**, Section 3]). The construction of $\sigma(f)$ involves some choices. But, if $\sigma(f)$ vanishes, then there exists a G-normal cobordism (relative to the singular set Sing X) between (X, f) and (X', f') where f' is a G-map which is a homotopy equivalence or an equivalence in the surgery theory under consideration. The Wall group containing $\sigma(f)$ projects onto the Bak group, and $\sigma(f)$ is a lifting of the well-defined stepwise obstruction. Since the subsequent results of [**PR**] deal with the consequences of the vanishing of the obstruction $\sigma(f)$, other parts of [**PR**] are not affected by this discussion.

IMPLICATIONS FOR [**Ms1**]: Although Masuda quotes an incorrectly stated result, his proof is correct. Throughout his construction Masuda is careful to ensure that the obstruction $\sigma(f) \in L_2(\mathbb{Z}_2) \cong \mathbb{Z}_2$ vanishes. Morimoto's computation shows that the Bak group $\mathbf{W}_2(\mathbb{Z}_2, \Gamma)$ is zero, and thus the only effect on [**Ms1**] is to make some steps in the argument unnecessary.

IMPLICATIONS FOR [**DR**]: A correction is also required for [**DR**, Section 9]. This section discusses the dependence of the final stepwise obstruction $\sigma_1(f)$ on surgeries over the singular sets. If $\dim X$ is even and $\dim X^g = [\frac{1}{2}(\dim X - 1)]$, for some element $g \neq 1$ in G such that $g^2 = 1$, then the obstruction in the last paragraph on [**DR**, p. 77] is defined up to addition of an element in $\mathcal{K} = \ker(L(G) \to \mathbf{W}(G, \Gamma))$. Under the same assumptions one should substitute $\sum_i \sigma_1(\mathcal{W}_i) \in \mathcal{K}$ into Lemma 9.3, and into Theorem 9.10 (iii) one should substitute $L^i_{d_1}(\mathbb{Z}[G], \lambda) = \mathcal{K}$. No other portions of [**DR**] are affected.

References for Chapter I

[**Ba**] A. Bak, "K-Theory of Forms," Annals of Mathematics Studies No. 98, Princeton University Press, Princeton, 1981.

[**Bre1**] G. Bredon, "Equivariant Cohomology Theories," Lecture Notes in Mathematics Vol. 34, Springer, Berlin-Heidelberg-New York, 1967.

[**Bre2**] "Introduction to Compact Transformation Groups," Pure and Applied Mathematics Vol. 46, Academic Press, New York, 1972.

[**Brdr**] W. Browder, *Surgery and the theory of differentiable transformation groups*, in "Proceedings of the Conference on Transformation Groups (Tulane, 1967)," Springer, Berlin-Heidelberg-New York, 1968, pp. 3–46.

[**Brwn**] R. Brown, "Elements of Modern Topology," McGraw-Hill, New York and London, 1968.

[**BQ**] W. Browder and F. Quinn, *A surgery theory for G-manifolds and stratified sets*, in "Manifolds–Tokyo, 1973," (Conf. Proc. Univ. of Tokyo, 1973), University of Tokyo Press, Tokyo, 1975, pp. 27–36.

[**Bru**] G. Brumfiel, *Homotopy equivalences of almost smooth manifolds*, Comment. Math. Helv. **46** (1971), 381–407.

[**CS**] S. Cappell and J. Shaneson, *The codimension two placement problem and homology equivalent manifolds*, Ann. of Math. **99** (1974), 277–348.

[**CF**] P. E. Conner and E. E. Floyd, *Maps of odd period*, Ann. of Math. **84** (1966), 132–156.

[**CV**] F. Connolly and V. Vijums, *G-normal maps and equivariant homotopy theory*, preprint, University of Notre Dame, 1979.

[**CMY**] E. H. Connell, D. Montgomery, and C. T. Yang, *Compact groups in E^n*, Ann. of Math. **80** (1964), 94–103; *correction*, Ann. of Math. **81** (1965), p. 194.

[**DM**] J. F. Davis and R. J. Milgram, "A Survey of the Spherical Space Form Problem," Mathematical Reports Vol. 2 Part 2, Harwood Academic Publishers, London, 1985.

[**Dav1**] M. Davis, *Smooth G-manifolds as collections of fiber bundles*, Pac. J. Math. **77** (1978), 315–363.

[**Dav2**] ———, "Multiaxial Actions on Manifolds," Lecture Notes in Mathematics Vol. 643, Springer, Berlin-Heidelberg-New York, 1978.

[**DH**] M. Davis and W. C. Hsiang, *Concordance classes of regular $U(n)$ and $Sp(n)$ actions on homotopy spheres*, Ann. of Math. **105** (1977), 325–341.

[**tD**] T. tom Dieck, "Transformation Groups," de Gruyter Studies in Mathematics Vol. 8, W. de Gruyter, Berlin and New York, 1987.

[**Dld**] A. Dold, "Halbexakte Homotopiefunktoren," Lecture Notes in Mathematics Vol. 12, Springer, Berlin-Heidelberg-New York, 1966.

[**Dov1**] K. H. Dovermann, $\mathbb{Z}_2$ *surgery theory*, Michigan Math. J. **28** (1981), 267–287.

[**Dov2**] ——————, *Cyclic group actions on complex projective spaces*, in "Proceedings of the KIT Mathematics Workshop (Algebra and Topology, 1986)," Korea Institute of Technology, Mathematics Research Center, Taejon, South Korea, 1987, pp. 164–271.

[**Dov3**] ——————, *Almost isovariant normal maps*, Amer. J. Math. **111** (1989), 851–904.

[**DP1**] K. H. Dovermann and T. Petrie, *G-Surgery II*, Memoirs Amer. Math. Soc. **37** (1982), No. 260.

[**DP2**] ——————, *An induction theorem for equivariant surgery (G-Surgery III)*, Amer. J. Math. **105** (1983), 1369–1403.

[**DR**] K. H. Dovermann and M. Rothenberg, *Equivariant Surgery and Classification of Finite Group Actions on Manifolds*, Memoirs Amer. Math. Soc. **71** (1988), No. 379.

[**DPS**] K. H. Dovermann, T. Petrie, and R. Schultz, *Transformation groups and fixed point data*, in "*Group Actions on Manifolds (Conference Proceedings, University of Colorado, 1983)*," Contemp. Math. Vol. 36, American Mathematical Society, 1985, pp. 159–189.

[**DS**] K.H. Dovermann and R. Schultz, *Surgery on involutions with middle dimensional fixed point set*, Pac. J. Math. **130** (1988), 275–297.

[**DW**] K. H. Dovermann and L. C. Washington, *Relations between cyclotomic units and Smith equivalence of representations*, Topology **28** (1989), 81–89.

[**FH**] F. T. Farrell and W. C. Hsiang, *Rational L-groups of Bieberbach groups*, Comment. Math. Helv. **52** (1977), 89–109.

[**G+**] C. G. Gibson, K. Wirthmüller, A. A. du Plessis, and E. J. N. Looijenga, "Topological Stability of Smooth Mappings," Lecture Notes in Mathematics Vol. 552, Springer, Berlin-Heidelberg-New York, 1976.

[**GL**] A. Gleason, *Spaces with a compact Lie group of transformations*, Proc. Amer. Math. Soc. **1** (1950), 35–43.

[**Hs**] W.-C. Hsiang, *A note on free differentiable actions of S^1 and S^3 on homotopy spheres*, Ann. of Math. **83** (1966), 266–272.

[**I**] S. Illman, *Recognition of linear actions on spheres*, Trans. Amer. Math. Soc. **274** (1982), 445–478.

[**J**] K. Jänich, *On the classification of O_n-manifolds*, Math. Ann. **176** (1978), 53–76.

[**KM**] M. Kervaire and J. Milnor, *Groups of homotopy spheres*, Ann. of Math. **77** (1963), 514–537.

[**KS**] S. Kwasik and R. Schultz, *Isolated singularities of group actions on 4-manifolds*, preprint, Tulane University and Purdue University, 1989.

[Lgr] L. Lininger, *On transformation groups*, Proc. Amer. Math. Soc. **20** (1969), 191–192.

[LdM] S. S. López de Medrano, "Involutions on Manifolds," Erg. der Math. (2) Bd. 54, Springer, Berlin-Heidelberg-New York, 1971.

[Lü] W. Lück, *The equivariant degree*, in "Algebraic Topology and Transformation Groups (Proceedings, Göttingen, 1987)," Lecture Notes in Mathematics Vol. 1361, Springer, Berlin-Heidelberg-Tokyo-New York, 1988, pp. 123–166.

[LM] W. Lück and I. Madsen, *Equivariant L-theory I*, Aarhus Univ. Preprint Series (1987/1988), No. 8; [*same title*] *II*, Aarhus Univ. Preprint Series (1987/1988), No. 16 (to appear in Math. Zeitschrift).

[Ms1] M. Masuda, *$\mathbf{Z}_2$ surgery theory and smooth involutions on homotopy complex projective spaces*, in "Transformation Groups, Poznań 1985," Lecture Notes in Math. No. 1217, Springer, Berlin-Heidelberg-New York, 1986, pp. 258–289.

[Ms2] ———, *Equivariant inertia groups for involutions*, preprint, Osaka City University, 1989.

[MSc] M. Masuda and R. Schultz, *Equivariant inertia groups and rational invariants of nonfree actions*, in preparation.

[Mat] J. N. Mather, *Stratifications and mappings*, in "Dynamical Systems (Proc. Sympos., Univ. Bahia, Salvador, Brazil, 1971)," Academic Press, New York, 1973, pp. 195–232.

[Mor] M. Morimoto, *Bak groups and equivariant surgery*, *K*-Theory **2** (1989), 456–483.

[P1] T. Petrie, *Pseudoequivalences of G-manifolds*, Proc. A. M. S. Sympos. Pure Math **32 Pt. 1** (1978), 169–210.

[P2] ———, *G-Surgery I–A survey*, in "Algebraic and Geometric Topology (Conference Proceedings, Santa Barbara, 1977)," Lecture Notes in Mathematics Vol. 644, Springer, Berlin-Heidelberg-New York, 1978, pp. 197–223.

[PR] T. Petrie and J. Randall, "Transformation Groups on Manifolds," Dekker Series in Pure and Applied Mathematics Vol. 82, Marcel Dekker, New York, 1984.

[RS] M. Rothenberg and J. Sondow, *Nonlinear smooth representations of compact Lie groups*, Pac. J. Math. **84** (1980), 427–444.

[RT] M. Rothenberg and G. Triantafillou, *On the classification of G-manifolds up to finite ambiguity*, preprint, University of Chicago, 1989.

[Sc1] R. Schultz, *Homotopy sphere pairs admitting semifree differentiable actions*, Amer. J. Math. **96** (1974), 308–323.

[Sc2] ———, *Differentiable group actions on homotopy spheres II: Ultrasemifree actions*, Trans. Amer. Math. Soc. **268** (1981), 255–297.

[Sc3] ———, *Pontryagin numbers and periodic diffeomorphisms of spheres*, in "Transformation Groups (Proceedings, Osaka, 1987)," Lecture Notes in Mathematics Vol. 1375, Springer, Berlin-Heidelberg-New York-London-Paris-Tokyo, 1989, pp. 307–318.

[Sc4] ———, *An infinite exact sequence in equivariant surgery*, Mathematisches Forschungsinstitut Oberwolfach Tagungsbericht 14/1985 (Surgery and L-theory), 4–5.

[Si] L. Siebenmann, *Deformations of homeomorphisms of stratified sets*, Comment. Math. Helv. **47** (1972), 123–163.

[St] S. H. Straus, "Equivariant codimension one surgery," Ph. D. Dissertation, University of California, Berkeley, 1972.

[T] R. Thom, *Ensembles et morphismes stratifiés*, Bull. Amer. Math. Soc. (1969), 240–284.

[Ve] A. Verona, "Stratified Mappings–Structure and Triangulability," Lecture Notes in Mathematics Vol. 1102, Springer, Berlin-Heidelberg-New York, 1984.

[Vj] V. Vijums, "Characteristic subspheres of differentiable group actions on homotopy spheres," Ph.D. Thesis, Rutgers University, 1977 *(Available from University Microfilms, Ann Arbor, Mich.: Order Number 7805138.)*—Summarized in Dissertation Abstracts International Ser. B **39** (1977/1978), 5429B–5430B..

[WL1] C. T. C. Wall, "Surgery on Compact Manifolds," London Math. Soc. Monographs No. 1, Academic Press, New York, 1970.

[WL2] ———, *Formulæfor surgery obstructions*, Topology **15** (1976), 189–210; *Correction*, Topology **16** (1977), 495–496.

[Wng] K. Wang, *Differentiable S^1 actions on homotopy spheres*, Invent. Math. **24** (1974), 51–73.

CHAPTER II

RELATIONS BETWEEN EQUIVARIANT SURGERY THEORIES

... laßt uns angenehmere anstimmen ...
(*Recitative, Beethoven's Ninth Symphony*)

Although it is clear that the various approaches to equivariant surgery are based upon related ideas, there is relatively little in the literature to describe the means for passing from one theory to another. At several points in this book we shall see that the conclusions from different versions of equivariant surgery theory shed a great deal of light on each other, and consequently an explicit description of the relationships is useful for technical as well as expository purposes.

It is often extremely helpful to view the orbit space of a manifold with group action as a manifold with singularities. In particular, the Browder-Quinn approach to equivariant surgery in **[BQ]** is based explicitly upon a canonical description of a smooth action's orbit space as a *stratified set* in the sense of Thom and Mather. Although the existence of this stratification is mentioned and applied repeatedly in the literature, an easily accessible account of the proof does not exist. The standard reference has been W. Lellmann's unpublished *Diplomarbeit* **[Le]**, and the most descriptive comments in a published work appear to be on page 21 of **[G+]**. For the sake of completeness we shall explain in Section 4 how a proof of the stratification theorem can be extracted from the standard papers and books on stratified sets.

Throughout this chapter G will denote a compact Lie group unless indicated to the contrary.

1. Browder-Quinn Theories

We shall begin by recalling two of the main points from the first section of Chapter I:

(i) A free action of G on a space X is completely determined by the orbit space X/G and some additional data (*i.e.*, the homotopy class of the classifying map $X/G \to BG$).

(ii) Frequently one can obtain useful information about smooth, nonfree G-actions by applying surgery theory to the nonsingular part of a smooth orbit space $M^* = M/G$.

There are many other situations in transformation groups where group actions are completely determined by orbit spaces together with some additional structural data. During the nineteen sixties many special cases were studied extensively. A survey article by G. Bredon from that period [**Bre**] describes several of these examples and their applications to exotic group actions on spheres. Subsequent work has shown that smooth actions of compact connected Lie groups on 3- and 4-manifolds are completely determined by so-called *structured* or *weighted* orbit space data (*e.g.*, see [**Ray**] for circle actions on 3-manifolds, and see [**OR1-2**], [**Fi1-2**], and [**Prkr**] for group actions on 4-manifolds); analogous results also exist for certain group actions on higher dimensional manifolds. Such results suggest that a suitable extension of surgery theory to orbit spaces of group actions might yield new insights and further information about nonfree actions. In [**BQ**] Browder and Quinn showed that such a theory could be constructed as a special case of a *surgery theory for stratified sets.* The purpose of this section is to describe the main features of their theory, paying particular attention to the properties that will be needed later in this book.

Browder-Quinn stratification data

The *strata* of a smooth G-manifold were discussed briefly in Section 2 of Chapter I; some of the definitions will be repeated here for the sake of convenience and clarity. If M is a smooth G-manifold and H is a closed subgroup of G, then the local linearity of smooth actions implies that

$$M_{(H)} = \{x \in M | \text{ the isotropy subgroup } G_x \text{ is conjugate to } H\}$$

is a (possibly empty!) disjoint union of smooth G-invariant submanifolds of

$$M - \cup_{K \supsetneq H} M_{(K)}$$

(where different components may have different dimensions), and the projection

$$\pi_H : M_{(H)} \to M_{(H)}/G = M^*_{(H)}$$

is a fiber bundle with fiber G/H and structure group $N(H)/H$; in fact, this is a smooth fiber bundle over each component of $M^*_{(H)}$. For each $x \in M_{(H)}$ one has a tubular neighborhood in M of the form $G \times_H V(x)$ for some H-representation $V(x)$. If W is an arbitrary linear representation of H, then

$$M_{H,W} = \{x \in M_{(H)} | \ V(x) \approx_H W\},$$
$$M^*_{H,W} = M_{H,W}/G$$

define unions of components of $M_{(H)}$ and $M^*_{(H)}$ respectively. The advantages of these smaller subsets are that each such set is a smooth manifold and the restrictions of the orbit space projection defines a smooth bundle $M_{H,W} \to M^*_{H,W}$ for every H and W. The sets $M^*_{H,W}$ are called the *strata* of M/G associated to the original group action; notice that every point of M/G belongs to exactly one stratum $M^*_{H,W}$. Frequently

$M^*_{H,W}$ is called the **stratum with isotropy subgroup type** H **and slice type** W (compare [**Dav**]). It follows immediately that the number of strata is countable if M is second countable and finite if M is compact.

For our purposes it is often convenient to consider the components of a stratum $M^*_{H,W} \subseteq M^* = M/G$ individually; a component of $M^*_{H,W}$ will be called a **connected open substratum**. The considerations of Section I.2 imply that the closure of a connected open substratum is a closed substratum.

The (open) strata of the orbit space M/G satisfy certain relationships that play an important role in [**BQ**].

THEOREM 1.0. (Compare [**BQ**, (4.2) and Example (4.3.4), p. 33]). *Let M be a compact, unbounded smooth G-manifold, let $M^*_{H,W}$ be defined as above, let $F^*_{H,W}$ be the closure of $M^*_{H,W}$ in $M^* = M/G$, and let $S^*_{H,W} = F^*_{H,W} - M^*_{H,W}$. Then $S^*_{H,W}$ is a union of strata of the form $M^*_{K,\Omega}$ such that $H \subsetneqq K$, and $M^*_{H,W}$ is diffeomorphic to the interior of a canonically defined compact manifold with boundary $V^*_{H,W}$. Furthermore, there is a normal projection map*

$$\nu_{H,W} : \partial V^*_{H,W} \to S^*_{H,W}$$

and a tubular function $\rho : M^*_{H,W} \to (0,2]$ such that the following hold:

(i) There is a smooth boundary collar $\omega : \partial V^*_{H,W} \times [0, 1+\varepsilon) \to V^*_{H,W}$ such that Image $\omega \cap M^*_{H,W} = \rho^{-1}(0, 1+\varepsilon)$ and $\rho\omega$ equals projection onto the second coordinate.

(ii) There is a homeomorphism $\varphi_{H,W}$ from the mapping cylinder $\mathrm{Cyl}\,(\nu_{H,W})$ onto a neighborhood of $S^*_{H,W}$ in $F^*_{H,W}$ such that $\varphi\rho$ equals projection onto $[0,1]$ composed with $t \to 1-t$ and $\varphi_{H,W}$ maps $\mathrm{Cyl}\,(\nu_{H,W}) - S^*_{H,W} \approx \partial V^*_{H,W} \times (0,1]$ into $M^*_{H,W}$ by $\varphi_{H,W}(x,t) = \omega(x, 1-t)$ (so that $\varphi_{H,W}$ is a diffeomorphism off $S^*_{H,W}$).

(iii) If $M^*_{K,\Omega} \subset S^*_{H,W}$, then

$$\nu^{H,W}_{K,\Omega} = \nu_{H,W} | \nu^{-1}_{H,W}(M^*_{K,\Omega}) : \nu^{-1}_{H,W}(M^*_{K,\Omega}) \to M^*_{K,\Omega}$$

is a smooth fiber bundle.

A proof of Theorem 1.0 can be extracted fairly easily from a paper by M. Davis [**Dav**]. The main difference between [**Dav**] and this paper is that the former views $M^*_{H,W}$ as the interior of a smooth manifold with corners (or *faces* in the sense of [**Dav**] and [**J**]) rather than the bounded smooth manifold $V^*_{H,W}$ (see [**Dav**, Sections IV.2 and IV.4]). Theorem 1.0 follows by combining the results of [**Dav**] with standard techniques for smoothing out corners such as the Cairns-Hirsch Theorem [**HM**].■

¿From the perspective of [**BQ**], the conclusions of Theorem 1.0 follow because the subsets $M^*_{H,W}$ form a smooth stratification of M/G in the sense of R. Thom [**Tho**] and J. Mather [**Mth**] (see [**G+**], Chapters 1–2, for a more detailed treatment). This approach to Theorem 1.0 will be discussed further in Section 4.

An analog of Theorem 1.0 holds for manifolds with boundary, but the conclusion is somewhat more complicated.

COMPLEMENT TO 1.0. *Suppose now that M is a compact bounded smooth G-manifold. Then the sets $M^*_{H,W}$ are all compact manifolds such that $\partial M^*_{H,W/\mathbf{R}} = M^*_{H,W} \cap (\partial M)/G$. In this case it is possible to find an equivariant collar C for ∂M in M such that each* *intersection $M^*_{H,W} \cap C/G$ has the form $\partial M^*_{H,W/\mathbf{R}} \times [0,1]$. Furthermore, each set $M^*_{H,W}$ can be extended to a manifold with corners $V^*_{H,W}$ such that $\partial V^*_{H,W} = \partial_1 V^*_{H,W} \cup U_{H,W/\mathbf{R}}$ where $U^*_{H,W/\mathbf{R}}$ is the bounded manifold associated to $\partial M_{H,W/\mathbf{R}}$ and $\partial_1 V^*_{H,W} \cap U^*_{H,W/\mathbf{R}} = \partial U^*_{H,W/\mathbf{R}}$. Analogs of the maps $\nu_{H,W}$ and $\rho_{H,W}$ exist, and the analogs of (i)-(iii) hold if $\partial_1 V^*_{H,W}$ replaces $\partial V^*_{H,W}$. Finally, everything in sight can be chosen to be compatible with the equivariant collar neighborhood for ∂M in M.*■

One of the main ideas of [**BQ**] is to choose projections $\nu_{H,W}$ and tubular maps $\rho_{H,W}$ as in 1.0 and to consider a class of equivariant maps of smooth G-manifolds that preserve this extra structure in an appropriate sense.

In order to motivate the formal definition of such structure preserving equivariant maps, we shall first discuss an important special case. Suppose that $G = \mathbf{Z}_p$ (p prime) acts smoothly on the manifold M and the fixed set $F(M) = M^G$ is connected; for simplicity also assume $\partial M = \emptyset$. Then the data corresponding to $\nu_{G,W}$ and $\rho_{G,W}$ can be obtained directly from a closed G-invariant tubular neighborhood $D(\xi_{F(M)})$ of $F(M)$ in M. For such examples a structure preserving map f of G-manifolds will be isovariant, and compatibility with the maps ν and ρ will mean that f maps $D(\xi_{F(M)})$ to $D(\xi_{F(N)})$ by a fiber preserving equivariant map that is nonsingular and orthogonal on each fiber.

A map f with the properties described above is a special case of a ***transverse linear*** map as defined in [**BQ**].

DEFINITION: An equivariant map of smooth G-manifolds $f : M \to N$ is said to be **transverse linear (and isovariant)** if f is isovariant, the orbit space map $f^* = f/G$ sends each stratum $M^*_{H,W}$ to a stratum $N^*_{H,U}$ such that $W/W^H \cong_H U/U^H$, and

> there are G-invariant tubular neighborhoods P and Q of $M_{H,W}$ and $N_{H,U}$ in $M - \cup M_{(K)}$ and $N - \cup N_{(K)}$ such that $f(P) = Q$ and $f|P$ is a G-linear bundle map that sends fibers to fibers bijectively.

A related concept called a ***smooth stratified map*** is discussed in [**Dav**]; if $f : M \to N$ is smooth equivariant map that is transverse linear and isovariant, then the map of orbit spaces $f/G : M/G \to N/G$ is a smooth stratified map in the sense of [**Dav**]. If we pass from the setting of [**Dav**, Section IV.4] to that of Theorem 1.0, the preceding statement about f/G can be formulated and slightly extended as follows:

PROPOSITION 1.1. *Let M and N be smooth G-manifolds, and let $f : M \to N$ be a G-manifolds, and let $f : M \to N$ be a G-isovariant transverse linear map. Let $\{\nu^M_{H,W}; \rho^M_{H,W}\}$ and $\{\nu^N_{H,W}; \rho^N_{H,W}\}$ be normal projections and tubular maps for M and N respectively. Then f is isovariantly, transverse linearly homotopic to a smooth map h such that for each $M_{H,W}$ the map h^* of orbit spaces satisfies $h\nu^M_{H,W} = \nu^N_{H,W}h^*$ on some neighborhood $U_{G,W}$ of $M^*_{G,W}$ in $M^* - \cup M^*_{(K)}$ and $h\rho^M_{H,W} = \rho^N_{H,W}h^*$ on $U_{G,W} - M_{G,W}$.*

In other words, a transverse linear isovariant map can be deformed in a nice way to a map that preserves stratification data.

PROOF: (*Sketch.*) The first step is to show that f can be transverse linearly and isovariantly deformed to a smooth map. This follows from standard considerations involving smooth approximations. The smooth approximation is a stratified map h_0 in the sense of [**Dav**, Section I.1], and the induced map of orbit spaces is a stratified map in the sense of [**Dav**, Sections II.3-4]. The existence of an approximation h that preserves the extra data now follows from the uniqueness results for ordinary and augmented $\mathcal{G}$-normal systems in Section IV.4 of [**Dav**].■

The results of [**Dav**] also imply a crucial relationship between transverse linear isovariant maps and maps of vector bundle systems (or π-bundles) as defined in Section I.3 of this book.

PROPOSITION 1.2. *Let $f : M \to N$ be as in Proposition 1.1 and assume that G is finite and the Codimension ≥ 2 Gap Hypothesis holds. Then for some smooth approximation h as above there is a map of vector bundle systems*

$$c : \pi(T(M)) \to h^*\pi(T(N))$$

such that h is given by the composite

$$\pi(T(M)) \xrightarrow{c} h^*\pi(T(N)) \xrightarrow{\overline{h}} \pi(T(N))$$

on suitable neighborhoods of the strata.

More precisely, if $M^*_{H,W}$ is a stratum of W mapping to $N^*_{H,W}$ then there are tubular neighborhoods $U^M_{H,W}$, $U^N_{H,W}$ of the inverse images $M_{H,W}$ and $N_{H,W}$ in $M - \cup M_{(K)}$ and $N - \cup N_{(K)}$ such that h corresponds to $\overline{h}c$ under the identifications of $U^M_{H,W}$ and $U^N_{H,W}$ with the vector bundles in $\pi(T(M))$ and $\pi(T(N))$ associated to $M_{H,W}$ and $N_{H,W}$ respectively.

PROOF: (*Sketch.*) The total spaces of the vector bundle systems $\pi(T(M))$ and $\pi(T(N))$ can be embedded into $T(M)$ and $T(N)$ respectively. Furthermore, one can choose smooth G-compatible riemannian metrics such that the induced map of equivariant tangent bundles $T(h) : T(M) \to T(N)$ sends the normal bundles in $\pi(T(M))$ to those in $\pi(T(N))$ by maps that are fiberwise linear isometries. Transverse linearity implies that the mapping h_* from the combined total space $\pi(T(M))$ to $\pi(T(N))$ has a factorization $h_* = \overline{h}c$ of the prescribed type. If we take riemannian tubular neighborhoods with respect to these metrics, it will follow immediately that h is given by the composite $\overline{h}c$ near each stratum.■

We shall now describe the surgery theory for transverse linear isovariant maps developed by Browder and Quinn; in fact, we shall generalize the setting of [**BQ**] to deal with Σ-adjusted maps as defined in Chapter I. Throughout the discussion we shall use the equivalence

$$\begin{bmatrix}\text{transverse linear}\\ \text{isovariant smooth}\\ \text{maps}\\ X \to Y\end{bmatrix} \sim \begin{bmatrix}\text{smooth}\\ \text{stratified maps}\\ X/G \to Y/G\end{bmatrix}$$

obtained from [**Dav**, Sections III and IV] and the smooth approximation result mentioned in Proposition 1.1 above.

It will be convenient to recall some features of stratified objects that Browder and Quinn summarize in [**BQ**, Definition 4.2, p. 33]. Suppose that the space X is decomposed into strata X_a indexed by the partially ordered set A. If for each $a \in A$ we let

$X_{\partial a}$ be the union of all X_t such that $t \lneqq a$, then there is a closed neighborhood N_a of $X_{\partial a}$ in X_a and a projection $\nu_a : \partial N_a \to X_{\partial a}$ such that

(i) ∂N_a is a manifold,

(ii) N_a is the mapping cylinder of ν_a,

(iii) if $b < a \in A$, $W_b := X - \text{Int } N_b$, and $V_{a,b} = \partial N_a \cap \nu_a^{-1}(W_b)$, then $\nu_{a,b} := \nu_a | V_{a,b}$ is a smooth submersion.

In the situation of Theorem 1.0(*iii*) the product $\nu_{a,b} \times \mathbf{id}_{\mathbf{R}}$ corresponds to the fiber bundle $\nu_{K,\omega}^{H,K}$ if we let $M_{K,\Omega}^*$ correspond to X_b and take $V_{a,b}$ to be given by the bundle $\nu_{K,\Omega}^{H,W}$.

The *bundle data* for a stratified map are given in terms of the so-called *stratified tangent bundles* as defined in [**Tho**, Section 1.F, pp. 255-256] (see also [**Dav**, Section II.2]).

EXAMPLE: If M is a compact smooth G-manifold, then the stratified tangent bundle $ST(M)$ may be viewed as a G-invariant subspace of the ordinary equivariant tangent bundle $T(M)$; specifically, a tangent vector v over $x \in M$ lies in $ST(M)$ if and only if v is tangent to the unique (open) stratum of M containing x. The bundle $ST(M/G)$ is such that $ST(M)$ is the pullback of $ST(M/G)$ under the orbit space projection $M \to M/G$.

More generally, one can define **stratified vector bundles** over a smoothly stratified space X (indexed by A) to be pairs $((\xi_a), (\Psi_{ab}))$ as follows:

(i) For each $a \in A$, ξ_a is a vector bundle over X_a.

(ii) For each $a > b \in A$ let $\nu_{ab} : V_{a,b} \to W_b$ be the smooth fiber bundle described above. Then $\Psi_{a,b} : E(\xi_a)|V_{a,b} \to E(\xi_b)|W_b$ is a fiberwise linear surjection covering $\nu_{a,b}$ with Kernel $\Psi_{a,b} = \mathbf{R} \oplus TFib(\nu_{a,b})$, where $TFib$ denotes the bundle of tangents along the fibers.

We assume that consistency conditions of the form $\Psi_{a,c} = \Psi_{b,c}\Psi_{a,b}$ hold if $a > b > c$.

EXAMPLE. Let M be a compact smooth G-manifold with the usual G-invariant stratification, let A be the partially ordered set of strata (ordering $a \leq b \Leftrightarrow \overline{a} \subset b$), and suppose that $a < b \in A$. Then $\mathbf{R} \oplus TFib(\nu_{a,b})$ is merely the pullback of the equivariant normal bundle of Closure (M_b) under the map ν_{ab}.

Clearly one can stabilize a stratified vector bundle by taking direct sums with an ordinary vector bundle over the space (*e.g.*, with a trivial bundle).

Let X be a compact smooth G-manifold. The elements of the **Browder-Quinn groups** $L_n^{BQ}(X) = L_n^{BQ,h}(X)$ are represented by data (f, b, k, φ) where $f/G : M/G \to N/G$ is a degree one stratified map of n-manifolds, b is a stable isomorphism of stratified vector bundles from $ST(M/G)$ to $(f/G)^*\xi$, where ξ is some stratified vector bundle over N/G, such that the linear maps determined by f are compatible with b on tubular neighborhoods of the strata, and k is a dimension-preserving stratified map $N/G \to X/G$ that preserves the first Stiefel-Whitney classes of the strata. If some of the strata in X are non-orientable, one must also include some additional data φ in order to avoid the difficulties with [**Wa**, Chapter 9] that are described in [**FH**, pp. 102–106]. Specifically, we need to choose liftings $\varphi_{H,W}$ of the maps $f_{H,W} : N_{H,W} \to X_{H,W}$ to the oriented double coverings that are consistent in the following sense: If $X_{H,W}$ lies in the closure of $X_{K,\Omega}$ and U is the intersection of $X_{K,\Omega}$ with a tubular neighborhood of $X_{H,W}$ (so that U fibers over $X_{H,W}$), then the pullback of $\varphi_{H,W}$ with respect to the bundle $U \to X_{H,W}$ is the restriction of $\varphi_{K,\Omega}$ to the inverse image of W. If one adopts the notion of *transports* as in [**LM**], it is possible to combine the reference map k and the system of liftings φ into a single entity. This is of course more elegant, but it is also considerably less elementary; more will be said about transports in Section 3. Following the pattern in [**Wa**] and elsewhere, we also assume that these representatives are stratified homotopy equivalences on the boundary (*i.e.*, homotopy equivalences in the category of stratified maps), and a triple represents zero if it is bordant to a stratified equivalence in an appropriate sense. There is a $\pi - \pi$ theorem for objects of this type if all dimensions are sufficiently large (compare [**BQ**, Thm. 3.1, p.31]), and it follows that $L_n^{BQ,h}(X)$ basically depends only on the normal slices and the dimensions, component structures, fundamental groups, and first Stiefel-Whitney classes of the strata $M^*_{H,W}$. Furthermore, if all strata are at least 5-dimensional there is an analog of Wall's $L^1 = L^2$ theorem [**Wa**, Thm. 9.4] which implies that every class in $L_n^{BQ,h}(X)$ contains a representative $(f : M \to N; b, k)$, where $k : N/G \to X/G$ induces bijections on components of strata and bijections on the fundamental groups of all strata (compare [**LM**, Part I, Section 3]).

Let Σ be a family of connected open substrata such that the union X^*_Σ of all substrata $X^*_{H,W} \in \Sigma$ is *closed* in X/G; if G is finite, the Codimension ≥ 2 Gap Hypothesis of Chapter I holds, and $\Sigma^\#$ is the set of all closed substrata in $\pi(X)$ mapping into Σ, then it is elementary to verify that $\Sigma^\#$ is closed and G-invariant. A stratified map $f : M \to N$ will be called Σ-*adjusted* if it defines a stratified equivalence from M^*_Σ to N^*_Σ. In these cases covered by the theory of Chapter I, a stratified map $f : M \to N$ is Σ-adjusted if and only if it is $\Sigma^\#$-adjusted in the sense of Chapter I.

The preceding theory can be extended to deal with bordism classes of Σ-adjusted maps, and if this is done one obtains the Σ-adjusted versions of the Browder-Quinn groups that will be denoted by $L_n^{BQ,h}(X;\Sigma)$. If $\Sigma \subseteq \Sigma'$ then there is the usual sort of forgetful homomorphism

$$L_n^{BQ,h}(X;\Sigma') \to L_n^{BQ,h}(X;\Sigma),$$

and using triads as in Section I.5 one can define relative groups to obtain the usual sort of exact sequence of abelian groups:

(1.3)
$$\cdots \to L_{n+1}^{BQ,h}(X;\Sigma \subseteq \Sigma') \to L_n^{BQ,h}(X;\Sigma') \to L_n^{BQ,h}(X;\Sigma) \to L_n^{BQ,h}(X;\Sigma \subseteq \Sigma') \to \ldots$$

If $\Sigma' - \Sigma$ contains exactly one open connected substratum, then (1.3) is analogous to the Orbit Sequence of Section I.5, and in this case one can again describe the group $L_n^{BQ,h}(X;\Sigma \subseteq \Sigma')$ in terms of ordinary Wall groups (compare [**BQ**, Thm. 3.3, p.32]):

PROPOSITION 1.4. *In the preceding notation, assume that* $\Sigma' - \Sigma$ *only contains the open connected substratum* X^*_α *and assume dim* $X^*_\alpha + n -$ *dim* $X \geq 5$. *Then there* *is an isomorphism σ_α from $L_n^{BQ,h}(X;\Sigma \subseteq \Sigma')$ to $L^h_{n(\alpha)}(\pi_1(X^*_\alpha), w_\alpha)$, where $n(\alpha) =$ dim $X_\alpha + n -$ dim X and w_α is given by the first Stiefel-Whitney class.*

PROOF: (*Sketch.*) The map σ_α is basically given by restricting a map $M \to N$ to a map of open connected substrata $M^*_\alpha \to N^*_\alpha$. By transverse linearity and the assumptions on adjustedness this can be extended to a map of manifolds with boundary, say $\widehat{f}_\alpha : \widehat{M}_\alpha \to \widehat{N}_\alpha$, such that the induced map of boundaries is a homotopy equivalence (in fact, on collar neighborhoods of the boundary the map $\widehat{f}_\alpha$ is a product of the boundary map $\partial \widehat{f}_\alpha$ with the identity on $[0,1]$); furthermore, the bundle data also extend. The value of σ_α on $(f,-)$ is then the relative surgery obstruction for $\widehat{f}_\alpha$. This obstruction is well-defined because the transverse linearity and isovariance conditions imply that normally cobordant maps f, h (in the appropriate, restricted sense) define normally cobordant relative surgery problems $\widehat{f}_\alpha$, $\widehat{h}_\alpha$. An inverse to σ_α may be constructed exactly as at the end of Section I.5: First take an element $u \in L^h_{n(\alpha)}(\pi_1(X^*_\alpha), w_\alpha)$ and a suitable $(n-1)$-dimensional G manifold Y. Next form a normal cobordism W_α whose lower end is Y^*_α and whose relative surgery obstruction is u. Finally, extend W_α to a normal cobordism $F : W \to Y$ such that $F|\partial_- W = 1_Y$ using the data carried by the vector bundle system $\pi(Y)$.■

COROLLARY 1.5. (Orbit Sequence) *In the setting of 1.4 there is an exact sequence:*

$$\to L^h_{n(\alpha)+1}(\pi_1(X^*_\alpha), w_\alpha) \to L_n^{BQ,h}(X;\Sigma') \to L_n^{BQ,h}(X;\Sigma) \to L^h_{n(\alpha)}(\pi_1(X^*_\alpha), w_\alpha) \to$$

PROOF: This follows immediately from (1.3) and Proposition 1.4. A somewhat different proof of this result can be obtained from [**BQ**], Prop. 4.8, p. 35.■

The final paragraph of [**BQ**] derives a periodicity theorem for the groups $L_*^{BQ,h}(X)$; the same methods yield a Σ-adjusted analog.

COROLLARY 1.6. (Periodicity) *There are isomorphisms*

$$L_n^{BQ,h}(X;\Sigma) \approx L_{n+4k}^{BQ,h}(X;\Sigma)$$

for all $k > 0$ induced geometrically by sending $(f : M \to N; -)$ to $(f \times 1 : M \times P \to N \times P; -)$ where P is a closed, oriented simply connected k-manifold with signature equal to 1.■

Simple homotopy. The groups $L_n^{BQ,h}(X)$ can be viewed as analogs of the homotopy Wall groups, say $L_n^h(\pi_1(X), w)$. In [**BQ**] Browder and Quinn also describe analogs $L_n^{BQ,s}(X)$ of the simple Wall groups $L_n^s(\pi_1(X), w)$. The entire discussion for $L_n^{BQ,s}(X)$ is parallel to the discussion of $L_n^{BQ,h}(X)$, the only significant difference being the replacement of transverse linear isovariant homotopy equivalences by *simple* transverse linear isovariant homotopy equivalences. For the present it suffices to say that simple equivalences are characterized by the fact that for each map of open connected substrata $f^*_\alpha : A^*_\alpha \to B^*_\alpha$ the canonical extension to compact bounded manifolds $\widehat{f_\alpha} : \widehat{A_\alpha} \to \widehat{B_\alpha}$ has trivial Whitehead torsion in the usual sense. More will be said about equivariant Whitehead torsion in Section 3.

Equivariant surgery exact sequences. As noted in [**BQ**], the main result of that paper is a generalization of the Sullivan-Wall surgery sequence that applies to smooth G-manifolds (see Thm. 4.6, p. 34). For this theory the analog of a homotopy equivalence is a transverse linear isovariant homotopy equivalence, the analog of a simple homotopy equivalence is the concept described in the preceding paragraph, and the analogs of the Wall groups $L_*^c(\pi_1(X), w_X)$ ($c = h$ or s) are the groups $L_*^{BQ,c}(X)$.

In this section we have only discussed those portions of [**BQ**] dealing with equivariant surgery groups, and we have not attempted to explain the relevance of these groups to the Stratified Surgery Exact Sequence of [**BQ**, Thms. 2.2 and 4.6, pp. 29–30, 34]. Further information on the construction and applications of this sequence can be found in the original paper of Browder and Quinn [**BQ**] and also in subsequent work on multiaxial actions by M. Davis and W. C. Hsiang [**DH**].

Homotopically stratified surgery

In [**Q2**] Quinn considers a notion of *homotopically stratified space* that generalizes the concept of smoothly stratified space but still has many useful properties. For example, modulo difficulties with low-dimensional strata and low-dimensional gaps between strata, the group of stratum-preserving self-homeomorphisms acts transitively on every stratum [**Q2**, Cor. 1.3, p. 444]. In [**Wb**] S. Weinberger develops a stratified surgery theory for homotopy stratified spaces that is in many respects analogous to the Browder-Quinn theory for smooth stratifications. As noted in [**Wb**, Subsection 9E] this theory also provides an enlightening setting for certain other approaches to equivariant surgery.

2. Theories with Gap Hypotheses

The methods of Section 1 develop a version of equivariant surgery theory with weak restrictions on the objects but strong restrictions on the mappings (*i.e.*, transverse linearity and isovariance). In contrast, the methods of Chapter I develop a version of equivariant surgery theory with significant restrictions on objects (*i.e.*, the Standard Gap Hypothesis) but weak restrictions on mappings. There are in fact several closely related versions of equivariant surgery theory with Gap Hypotheses. One reason for this is that somewhat different methods are needed for each of the three basic categories of G-manifolds (smooth, locally linear PL, and locally linear topological). Another reason for multiple versions of equivariant surgery is that several different concepts of equivariant equivalence have proved to be useful in transformation groups. In this section we shall describe the main types of equivariant surgery theories with Standard Gap Hypotheses that have been constructed.

Formal properties of equivariant surgery theories

One purpose of this section is to note that different versions of equivariant surgery theory all have certain formal properties. Before listing the various theories we shall describe these properties and note that they hold for Browder-Quinn theories $L^{BQ,h}$ and $L^{BQ,s}$ of Section 1 as well as the theory $I^{ht,DIFF}$ of Chapter I.

We shall assume that G is finite and all group actions satisfy the Codimension ≥ 2 Gap Hypothesis of Chapter I. An equivariant surgery theory begins with some family $\mathcal{F}^a$ of compact G-manifolds in one of three categories:

$DIFF_G$: smooth G-manifolds.
PL_G^{LL}: piecewise linear manifolds with piecewise linear, locally linear G-actions.
TOP_G^{LL}: topological manifolds with (continuous) locally linear G-actions.

Each theory also specifies a family of subgroups $\mathcal{H}^a$ that is closed under conjugation and passage to subgroups. Usually $\mathcal{H}^a$ will consist of all subgroups of G, but in Chapter V we shall consider cases where $\mathcal{H}^a$ is all subgroups whose orders are powers of certain primes; other choices are also possible but have not been considered in the literature. If X is an object in $\mathcal{F}^a$ and $\mathcal{H}^a$ is the distinguished class of subgroups we consider the subset

$$\pi(X, \mathcal{H}^a)$$

of the geometric poset $\pi(X)$ consisting of all closed substrata M_α that are components of the fixed point sets of subgroups in $\mathcal{H}^a$.

We are interested in constructions that define abelian groups $I^a(G; X; \Sigma \subseteq \Sigma')$ for each object X in $\mathcal{F}^a$ and each pair of closed G-invariant subsets $\Sigma \subseteq \Sigma'$ in $\pi(X; \mathcal{H}^a)$ such that Σ contains all closed substrata of dimension ≤ 4.

EXAMPLES: 1. Take $\mathcal{F}^a$ to be all compact smooth G-manifolds satisfying the Codimension ≥ 2 Gap Hypothesis, let $\mathcal{H}^a =$ all subgroups, and take $I^a(G; X; \Sigma \subseteq \Sigma')$ to be the Browder-Quinn groups $L_n^{BQ,c}(X; \Sigma \subseteq \Sigma')$ defined in Section 1, where $c = h$ or s.

2. Take $\mathcal{F}^a$ to be all compact smooth G-manifolds such that each closed substratum is simply connected and at least 5-dimensional, let $\mathcal{H}^a$ be all subgroups, and choose $I^a(G; X; \Sigma \subseteq \Sigma')$ to be the group $I^{ht,DIFF}(G; X; \Sigma \subseteq \Sigma')$ defined in Section I.5.

The results of Section 1 (of this chapter) and Section I.5 imply that both of these examples satisfy the following important property:

(2.0) EXACTNESS CONDITIONS. *(i) If $\Sigma \subseteq \Sigma'$ in $\pi(X, \mathcal{H}^a)$ and Σ contains all closed substrata of dimension ≤ 4, then there are abelian group homomorphisms f, g such that the sequence*

$$I^a(G; X; \Sigma') \xrightarrow{f} I^a(G; X; \Sigma) \xrightarrow{g} I^a(G; X; \Sigma \subseteq \Sigma')$$

is exact.

(ii) If in addition $X \times I$ is an object in $\mathcal{F}^a$ then there is an abelian group homomorphism

$$I^a(G; X \times I; \Sigma \subseteq \Sigma') \xrightarrow{\partial} I^a(G; X; \Sigma')$$

such that

$$I^a(G; X \times I; \Sigma) \xrightarrow{g} I^a(G; X \times I, \Sigma \subseteq \Sigma') \xrightarrow{\partial} I^a(G; X; \Sigma') \xrightarrow{f} I^a(G; X; \Sigma)$$

is also exact.

If $\Sigma' - \Sigma$ consists of a single G-orbit, then one generally has an isomorphism from $I^a(G; X; \Sigma \subseteq \Sigma')$ to some ordinary surgery obstruction group in the sense of Wall or

Cappell-Shaneson; for example this is the case if $I^a = L_n^{BQ,c}$ or $I^{ht,DIFF}$. Since the usefulness of (2.0) depends upon such identifications, the existence of such isomorphisms can be viewed as an additional axiom for a theory of equivariant surgery obstruction groups. Unfortunately, there is no fully satisfactory way of formalizing this, but in each case it is not difficult to show that there is an isomorphism of the desired type.

Other typical features of equivariant surgery theories.

An equivariant surgery theory I^a generally has an associated notion of *normal map* (more precisely, *of type* I^a) given by ($f : M \to N$, other data) and a distinguished subcollection of normal maps known as *equivalences* for the theory I^a or more succinctly as I^a-equivalences; the G-manifolds M and N are assumed to belong to the family $\mathcal{F}^a$, and the map f may be required to satisfy certain technical conditions. For the theory $I^{ht,DIFF}$ of Section I.5 the appropriate concept of normal map appears in Section I.4, and the associated notion of equivalence is ordinary homotopy equivalence. Furthermore, for each closed and G-invariant subset $\Sigma \subset \mathcal{H}^a$ there is an associated notion of Σ-adjusted maps that satisfy the conditions for an I^a-equivalence on the closed substrata in Σ. In each of the theories we consider the elements of the surgery obstruction groups $I^a(G;X;\Sigma)$ are essentially represented by Σ-adjusted normal maps modulo the equivalence relation generated by identifying two normal maps that are normally cobordant by a Σ-adjusted cobordism and setting equivalences equal to zero. This is entirely analogous to the definition of the groups for $I^{ht,DIFF}$ in Section I.5. Finally, the relative groups $I^a(G;X;\Sigma \subset \Sigma')$ can also be defined by analogy with the definition of the relative groups $I^{ht,DIFF}(G;X;\Sigma \subset \Sigma')$ in Section I.5.

The basic examples

We begin with analogs of $I^{ht,DIFF}$ that are discussed in [**DR1**] and also in [**MR**].

(2.1) *The theories* $I^{ht,CAT}$, *where* $CAT = PL$ *or* TOP.

For these theories one takes $\mathcal{F}^{ht,CAT}$ to be the class of all PL or topological locally linear G-manifolds, whose closed substrata are simply connected and such that the Standard Gap Hypothesis holds. Extension of the methods of Chapter I to the PL category is entirely straightforward (compare [**LM**] or [**MR**]), but the extension to the topological category requires additional work to overcome some complications. First of all, a compact locally linear G-manifold need not have the G-homotopy type of a finite G-CW complex (compare [**DR1**, Section 12]; there are also other classes of examples); one must restrict to degree 1 maps between objects having similar equivariant finiteness obstructions. The machinery for dealing with this is presented in Section 2 of [**DR1**]. A second problem involves the definition of bundle data. The appropriate analogs of stable G-vector bundles are stable locally linear $G - \mathbf{R}^n$ bundles in the sense of [**LaR**]; these objects satisfy many of the same properties as stable G-vector bundles, but unstable $G - \mathbf{R}^n$ bundles are far less well-behaved than stable G-vector bundles. In order to circumvent this problem, it is useful to deal with unstable bundle data that are not defined everywhere; specifically, one would like to allow bundle data that are defined on the complement of some nice configuration of low-dimensional submanifolds (see [**DR1**,

p. 21]). As indicated by the results of [**DR1**], the existence of such bundle data suffices for equivariant surgery theory.

If $\Sigma' - \Sigma = G\{\alpha\}$ the groups $I^{h,CAT}(G; X; \Sigma \subseteq \Sigma')$ are the same regardless of whether CAT equals $DIFF$, PL, or TOP; specifically, the group in question is the ordinary Wall group $L^h_{\dim\alpha}(\mathbb{Z}[W_\alpha], w_\alpha)$ where $\dim\alpha = \dim X_\alpha$ and $W_\alpha \subseteq N(H_\alpha)/H_\alpha$ is the subgroup sending X_α to itself (here H_α is the generic isotropy subgroup for X_α, and $W_\alpha = N_\alpha(H_\alpha)/H_\alpha$, where $N_\alpha(H_\alpha)$ is defined in Section I.4, and w_α is a first Stiefel-Whitney class).

(2.2) *The theory $I^{s,DIFF}$ for surgery up to equivariant simple homotopy equivalence.*

This theory is also discussed at length in [**DR1**]. In this case $\mathcal{F}^{s,DIFF} = \mathcal{F}^{ht,DIFF}$, $\mathcal{H}^{s,DIFF}$ = all subgroups, and the objective is to perform equivariant surgery to obtain an *equivariant simple homotopy equivalence* in the sense of Illman [**I1**] or Rothenberg [**Ro**]. The relation between Poincaré duality and equivariant Whitehead torsion turns out to be quite delicate, and methods for dealing with this are developed in Sections 5c and 6 of [**DR1**]. In this case, if $\Sigma' - \Sigma = G\{\alpha\}$ the construction of stepwise obstructions in I.5.3 generalizes to yield an isomorphism

$$I^{s,DIFF}(G; X; \Sigma \subseteq \Sigma') \cong L^s_{\dim\alpha}(\mathbb{Z}[W_\alpha], w_\alpha);$$

as in (2.1) the notation means $\dim\alpha = \dim X_\alpha$ and W_α is the subgroup of $N(H_\alpha)/H_\alpha$ that sends X_α to itself.

It is also possible to construct a theory $I^{s,PL}$; the methods used in the smooth category go through with relatively few changes. Specific information on this theory can be found in [**MR**]. On the other hand, it is well known that equivariant Whitehead torsion is not topologically invariant (*e.g.*, see [**Ro**] or [**St**]), and therefore it is not possible to construct a theory $I^{s,TOP}$ by a direct extension of the techniques used in the smooth and PL locally linear categories. In a series of papers M. Steinberger and J. West have constructed an equivariant topological Whitehead group that is appropriate to the category of locally linear topological manifolds (one reference is [**St**]). As noted in the introduction to [**DR1**], it is very conceivable that the methods of [**DR1**] could be combined with those of Steinberger and West to produce a viable theory $I^{s,TOP}$.

(2.3) *Analogous theories $I^{ht(R),CAT}$ for equivariant surgery with coefficients in $R \subseteq \mathbb{Q}$.*

Settings of this type have been considered in [**DP1**, 9.3], Appendix A of [**DP1**], and Section 10 of [**DR1**] (as well as further unpublished work of the first author). In this case $\mathcal{F}$ and $\mathcal{H}$ are the same as before, and the objective is to obtain an isogeneric map that induces isomorphisms of homology with coefficients in R *over every closed substratum.* The extension of the methods of (2.1) and Section I.5 to cover this theory is fairly straightforward. If we let Σ' equal $\Sigma - G\{\alpha\}$ as usual then we have that $I^{ht(R),CAT}(G; X; \Sigma \subseteq \Sigma')$ is isomorphic to $L^h_{\dim\alpha}(R[W_\alpha], w_\alpha)$, where the notation is the same as in (2.1).

(2.4) *Analogous theories with nonsimply connected closed substrata.*

It is possible to remove the simple connectivity condition on closed substrata in (2.1)-(2.3); as elsewhere in topology, this passage from simply connected to nonsimply connected subjects requires a more complicated setting and more sophisticated techniques.

Results of W. Lück and I. Madsen [**LM**] provide all the tools that are needed. Following their notation, for each G-manifold X of the prescribed type there is a *simple reference object* R_X containing data including the information carried by λ_X from Chapter I, and there are equivariant surgery obstruction groups

$$\mathcal{L}_n^{c,CAT}(G; R_X; \Sigma \subseteq \Sigma')$$

$$(n = \dim X,\ c = h \text{ or } s,\ CAT = DIFF,\ PL, \text{ or } TOP \text{ with } s, TOP \text{ excluded})$$

such that the exactness condition (2.0) is satisfied (see [**LM**, Part II, sequence (1.1)]). It should be noted that the Gap Hypothesis is not needed to define the Lück-Madsen groups $\mathcal{L}_n^{c,CAT}(G; R_X; \Sigma \subseteq \Sigma')$ or to derive the exact sequence condition (2.0); in particular, the Lück-Madsen sequence extends indefinitely to the left. However, one needs weak gap conditions (codimension ≥ 3 gaps, all closed substrata of dimension ≥ 4) in order to obtain an analog of Wall's $L^1 = L^2$ theorem (see [**LM**, Thm. 3.2]). Specifically, one wants every class in the surgery obstruction group to have a representative with the same reference data as R_X (in the nonequivariant case, this means the fundamental group maps bijectively to $\pi_1(X)$ and this bijection preserves the first Stiefel-Whitney class).

As in (2.1) and (2.2) there are forgetful transformations $\mathcal{L}^{c,DIFF} \to \mathcal{L}^{c,PL}$, $\mathcal{L}^{h,PL} \to \mathcal{L}^{h,TOP}$, and $\mathcal{L}^{s,CAT} \to \mathcal{L}^{h,CAT}$. The maps $\mathcal{L}^{c,DIFF} \to \mathcal{L}^{c,PL}$ turn out to be isomorphisms, but the exact relationship between $\mathcal{L}^{h,PL}$ and $\mathcal{L}^{h,TOP}$ is not fully understood. Frequently we shall denote $\mathcal{L}^{c,DIFF} \cong \mathcal{L}^{c,PL}$ simply by $\mathcal{L}^c$.

In order to go further and describe the groups $\mathcal{L}_n^c(G; R_X; \Sigma \subseteq \Sigma')$ when $\Sigma' = \Sigma \cup G\{\alpha\}$ and α is minimal in $\Sigma' - \Sigma$, it is necessary to assume a version of the Standard Gap Hypothesis. As in Chapter I, the appropriate condition is that $\mathbf{S}^2 R_X$ satisfy the Gap Hypothesis. Under these conditions, by [**LM**, Part II, Thm. 1.4] there is an isomorphism

$$\sigma_\alpha : \mathcal{L}_n^c(G; R_X; \Sigma \subseteq \Sigma') \cong L^c_{\dim\,\alpha}(\mathbf{Z}[E_\alpha], w_\alpha)$$

where E_α is the same group $\pi_1(X_\alpha^*)$ that appears for the Browder-Quinn stepwise surgery obstruction (see Proposition 1.4) and w_α is the first Stiefel-Whitney class; if the Codimension ≥ 3 Gap Hypothesis holds, the group E_α is simply an extension of $\pi_1(X_\alpha)$ by by W_α. The map σ_α is a stepwise obstruction homomorphism analogous to the maps in Chapter I. There is also an analog of Orbit Sequence I.5.4.

As suggested in the Appendix to [**LM**, Part I], the groups $\mathcal{L}_n^c(G; R_X; \Sigma \subseteq \Sigma')$ and the associated exact sequences reduce to those of [**DR1**] and Chapter I of this paper if all closed substrata are simply connected and a suitable form of the Gap Hypothesis holds. We shall explain this further in Section 3.

(2.5) *Theories I^h and $I^{h(R)}$ for equivariant surgery up to pseudoequivalence, where* $R \subseteq \mathbf{Q}$ (note the difference between I^h and the symbol I^{ht} that appears in Chapter I and (2.1) and (2.3) above).

These theories were studied in [**DP1**] and [**DP2**]. The goal of the theory I^h is to convert an equivariant degree 1 map $M \to X$ into an equivariant map that is an homotopy equivalence (but not necessarily a G-homotopy equivalence); for the theories $I^{h(R)}$

one wants an equivariant map that is an R-local homotopy equivalence. Since [**DP1**] and [**DP2**] are mainly interested in maps of simply connected smooth G-manifolds, localization can be understood in any of the usual senses. In this case one must be fairly restrictive on the choices of groups, manifolds, and maps; in order to avoid certain technical problems, *we shall assume henceforth that G is nilpotent* (compare [**DP1**, Thm. 2.11, pp. 19–20] and the accompanying discussion). The appropriate families of subgroups $\mathcal{H}^h$ and $\mathcal{H}^{h(R)}$ consist of all subgroups whose orders are prime powers (possibly 1) in the first case, and all subgroups whose orders are powers of primes p such that $p^{-1} \notin R$ in the second case. Since many of the formal properties that we want are not in [**DP1**] or [**DP2**], we shall discuss such properties further in Chapter V of this book. The appropriate notion of indexing data is considerably more involved than for any of the other theories. For the present it will suffice to note that a suitable degree 1 map $f : M \to X$ has indexing data λ containing the G-poset $\pi(X; \mathcal{H}^a)$.

If $\Sigma \subseteq \Sigma'$ in $\pi(X, \mathcal{H}^a)$ the methods of [**DP1**], [**DP2**] and [**LM**] yield surgery obstruction groups $I^c(G; \lambda; \Sigma \subseteq \Sigma')$ provided all closed substrata in $\pi(X, \mathcal{H}^a)$ are simply connected and have dimension ≥ 5. Partial results on exactness and the structure of these groups when $\Sigma' - \Sigma = G\{\alpha\}$ appear in [**DP1**] and [**DP2**]; more specific statements will be given in Chapter V. For the sake of simplicity we assume at this point that G has odd order (and is nilpotent as before). The results of [**DP2**] then yield the following description of $I^a(G; \lambda; \Sigma \subseteq \Sigma')$ if $\Sigma' - \Sigma = G\{\alpha\}$:

(2.6a) *If α is not the maximal element in $\pi(X; \mathcal{H}^a)$ then*

$$I^a(G; \lambda; \Sigma \subseteq \Sigma') \cong L^h_{\dim \alpha}(R_\alpha[W_\alpha]),$$

where $\dim \alpha$ and W_α are given as before and R_α is the ring R localized at the order of H_α (if $a = h(R)$; if $a = h$ take $R = \mathbb{Z}$).

(2.6b). *If α is the maximal element and $a = h$, then*

$$I^a(G; \lambda; \Sigma \subseteq \Sigma') \cong L^{D(G)}_{\dim \alpha}(\mathbb{Z}[G]),$$

where $D(G)$ is the kernel of the map

$$\widetilde{K}_0(\mathbb{Z}[G]) \to \widetilde{K}_0(\mathcal{M} = \text{maximal order})$$

induced by sending a module A to $\mathcal{M} \otimes_{\mathbb{Z}[G]} A$.

In some contexts it is useful to work with bundle data that are weaker than the data of [**DP1–2**]. One example is the notion of Smith bundle data in [**PR2**]. Almost everything in [**DP1–2**] goes through for this type of bundle data.

Variations on the preceding themes

As usual, it is possible to modify the theories presented above in many different ways. We shall limit ourselves to three directions in which one might proceed.

Actions of positive-dimensional compact Lie groups. Versions of the theory in (2.5) for such groups have been considered in work of Petrie [**Pe2-3**].

Theories with coefficients in Cappell-Shaneson groups rather than Wall groups. Some ideas along these lines will be presented in Section IV.6. It is also possible to define variants of the theory in (2.5) such that the groups corresponding to (2.6a) are Cappell-Shaneson groups rather than Wall groups. This theory is discussed further at the end of Section V.8 where it is called $I^{h\Gamma}$. One advantage of $I^{h\Gamma}$ is that the simple connectivity assumption on the nonmaximal substrata X_α $(\alpha \in \pi(X;\mathcal{H}^h))$ can be eliminated. A related theory of this type is considered in **[DS]**.

Surgery up to high connectivity. In **[Kr]** M. Kreck introduces an approach to surgery theory whose objective is to construct a map that is an homotopy equivalence through some reasonably high dimension. An equivariant analog of this theory might be just what one needs for certain applications.

3. Passage from one theory to another

One of the most basic features of topology is the introduction of related but distinct definitions for the same object or similar objects. This is particularly apparent in homology and cohomology theory, where one has various definitions—each with its own advantages and disadvantages—and natural transformations between the different versions such that some of the transformations induce isomorphisms under favorable conditions. The preceding sections and Chapter I suggest that similar things happen in equivariant surgery; the goal of this section is to describe the various transformations between the theories in Chapter I and the first two sections of this paper, and to indicate when these maps define isomorphisms.

As indicated in Section 2, equivariant surgery theories have exactness properties similar to those of homology/cohomology theories or half exact functors. Similarly, transformations of equivariant surgery theories will be expected to behave like natural transformations of homology/cohomology theories or half exact functors. Specifically, we want families of homomorphisms

$$\varphi(X;\Sigma\subseteq\Sigma') : I^a(G,X;\Sigma\subseteq\Sigma') \to I^b(G;X;\Sigma\subseteq\Sigma')$$

such that the following diagrams are commutative if all objects in sight can be defined:

$$\text{(3.1.A)}\qquad \begin{array}{ccccc} I^a(G;X;\Sigma') & \longrightarrow & I^a(G;X;\Sigma) & \longrightarrow & I^a(G;X;\Sigma\subseteq\Sigma') \\ \downarrow & & \downarrow & & \downarrow \\ I^b(G;X;\Sigma') & \longrightarrow & I^b(G;X;\Sigma) & \longrightarrow & I^b(G;X;\Sigma\subseteq\Sigma') \end{array}$$

$$\text{(3.1.B)}\qquad \begin{array}{ccc} I^a(G;X\times I;\Sigma\subseteq\Sigma') & \longrightarrow & I^a(G;X;\Sigma') \\ \downarrow & & \downarrow \\ I^b(G;X\times I;\Sigma\subseteq\Sigma') & \longrightarrow & I^b(G;X;\Sigma') \end{array}$$

When $\Sigma' = \Sigma \cup G\{\alpha\}$, the discussion of Section 2 indicates that the relative groups $I^a(G;X;\Sigma \subsetneq \Sigma')$ and $I^b(G;X;\Sigma \subsetneq \Sigma')$ should be given by algebraic L- or Γ-groups of appropriate types, and that in these cases the transformations $\varphi(G;X;\Sigma \subsetneq \Sigma')$ should have reasonable interpretations as maps of surgery obstruction groups.

We shall consider five basic types of transformations on equivariant surgery theories:

(3.2.A) *Change-of-category morphisms.*

(3.2.B) *Change of K-theory morphisms.*

(3.2.C) *Change-of-coefficients morphisms.*

(3.2.D) *Morphisms to Lück-Madsen groups.*

(3.2.E) *Morphisms from Browder-Quinn groups.*

(3.2.F) *Restriction-to-subgroup morphisms.*

There are transformations that can be placed in more than one of these classes; in such cases we shall choose one possibility and not try to enumerate the others.

3A. Change of categories

In nonequivariant topology one can pass from the category of smooth manifolds to the category of PL manifolds by taking smooth triangulations (compare Part II of [**Mun**]). Similarly, one can pass from PL manifolds to topological manifolds by forgetting the underlying PL structure. Similar considerations hold for G-manifolds. A smooth G-manifold (G finite) determines a PL locally linear G-manifold by the local linearity properties of smooth G-manifolds ([**Bre2**], Cor VI.2.4, p. 308) and the smooth equivariant triangulation theorem of S. Illman [**I2**], and one can pass from PL locally linear G-manifolds to the corresponding topological objects by forgetting the PL structure. Furthermore, G-vector bundles determine PL locally linear $G-\mathbf{R}^n$ bundles in the sense of Lashof and Rothenberg [**LaR**], and one can pass to topological locally linear $G-\mathbf{R}^n$ bundles simply by forgetting the PL structure. It follows that the transformations

smooth G-objects $\to$ PL locally linear G-objects
PL locally linear G-objects $\to$ topological locally linear G-objects

define maps of equivariant surgery groups

$$\beta : I^{ht,DIFF}(G;\lambda,\Sigma \subsetneq \Sigma') \to I^{ht,PL}(G;\lambda;\Sigma \subsetneq \Sigma')$$
$$\gamma : I^{ht,PL}(G;\lambda;\Sigma \subsetneq \Sigma') \to I^{ht,TOP}(G;\lambda;\Sigma \subsetneq \Sigma')$$

where $I^{ht,CAT}$ is given as in (2.1) and Chapter I and $\lambda = \lambda_X$ denotes the indexing data for the G-manifold X. Since all abelian group operations are given by the constructions in [**Do1**], [**Do2**], and [**DR1**], it follows that β and γ are homomorphisms.

THEOREM 3.3. *Suppose that* $S^{-1}\lambda_X = \lambda_Y$ *for some* λ_Y *such that the obstruction set* $I^{ht,CAT}(G; \mathbf{S}^{-1}\lambda_X; \Sigma \subseteq \Sigma')$ *satisfies the conditions for constructing an abelian group operation as above, and suppose that* $\mathbf{S}^2\lambda_X$ *satisfies the Gap Hypothesis. Then the maps* $\beta(\lambda_X; \Sigma)$ *and* $\gamma(\lambda_X; \Sigma)$ *are isomorphisms and the maps* $\beta(\mathbf{S}^{-1}\lambda_X; \Sigma)$ *and* $\gamma(\mathbf{S}^{-1}\lambda_X; \Sigma)$ *are monomorphisms.*

REMARK: Since the possible choices of indexing data for topological locally linear G-actions on compact manifolds are the same as the choices for smooth G-actions [**Ktz**], Theorem 3.3 implies that the groups $I^{ht,CAT}(G; \lambda; \Sigma \subseteq \Sigma')$ do not depend on CAT if all closed substrata have dimension ≥ 7.

PROOF: The method closely resembles the standard uniqueness proof for homology theories on finite cell complexes; one shows first that the natural transformations in question are isomorphisms (or monomorphisms or epimorphisms) for "atomic" pairs and then obtains the general case by induction using exact sequences and the five lemma.

Suppose that $\Sigma' = \Sigma \cup G\{\alpha\}$ where α is minimal in $\pi(X) - \Sigma$. If $\mathbf{S}^2\lambda_X$ satisfies the Gap Hypothesis, then the stepwise surgery obstruction σ_α^{CAT} defines an isomorphism

$$I^{ht,CAT}(G; \lambda_X; \Sigma \subseteq \Sigma') \to L^h_{\dim \alpha}(\mathbb{Z}[W_\alpha]; w_\alpha)$$

by Proposition I.5.3 if $CAT = DIFF$ and the corresponding statements in this section if $CAT = PL$ or TOP. Furthermore, from the construction it follows that $\sigma_\alpha^{PL}\beta = \sigma_\alpha^{DIFF}$ and $\sigma_\alpha^{TOP}\gamma = \sigma_\alpha^{PL}$. Since the maps σ_α^{CAT} are isomorphisms, the same must be true for β and γ.

We shall also need information on β and γ when $\Sigma' = \Sigma \cup G\{\alpha\}$ as above but we only know that $\mathbf{S}\lambda_X$ satisfies the Gap Hypothesis. In this case Proposition 5.4 and the corresponding statements in this section imply that the stepwise surgery obstructions σ_α^{CAT} define isomorphisms to quotient groups $L^h_{\dim \alpha}(\mathbb{Z}[W_\alpha]; w_\alpha)/B_{CAT}$. By construction these maps are compatible with β and γ as before, and it follows that β and γ are epimorphisms in this case.

The preceding paragraphs provide the necessary information about the "atomic" case $\Sigma' = \Sigma \cup G\{\alpha\}$. The general case is a consequence of the atomic cases and the following result:

THEOREM 3.4. (Comparison Theorem) *Let* I^a *and* I^b *be abelian group valued constructions satisfying the exactness condition in* (2.0), *and let* $\varphi : I^a \to I^b$ *define a family of homomorphisms* $\varphi(G; X; \Sigma \subseteq \Sigma') : I^a(G; X; \Sigma \subseteq \Sigma') \to I^b(G; X; \Sigma \subseteq \Sigma')$ *that satisfy the commutativity relations of* (3.1.A)–(3.1.B). *Assume that* $\pi(X; \mathcal{H}^a) = \pi(X, \mathcal{H}^b)$ *and that* $I^c(G; X; \pi(X, \mathcal{H}^c)) = 0$ *for all* X *in* $\mathcal{F}^c$, *where* $c = a$ *or* b. *Furthermore, suppose that* I^a *and* I^b *can be defined for each of the manifolds* $D^k \times Y$, *where* $0 \leq k \leq 3$, *and that the maps* $\varphi(G; D^k \times Y; \Sigma \subseteq \Sigma \cup G\{\alpha\})$ *are isomorphisms if* $k = 0, 1$ *and epimorphisms if* $k = 2$. *Then for arbitrary* $\Sigma \subseteq \pi(X; \mathcal{H}^a)$ *the maps* $\varphi(G; D^k \times Y; \Sigma)$ *are isomorphisms if* $k = 1$ *and monomorphisms if* $k = 0$.

REMARK: 1. In many cases the map $\varphi(G; D^0 \times Y = Y; \Sigma)$ is in fact an isomorphism. For example, this follows from the theorem if $Y = D^k \times Y_0$ where $k \geq 1$ and I^a, I^b can be defined for Y_0. The main difficulty in showing that $\varphi(G; Y; \Sigma)$ is an isomorphism arises from closed substrata of Y that are 6-dimensional, so that the various $\pi - \pi$

theorems do not apply in the next lower dimension. In Chapter III we shall discuss some additional conditions under which $\varphi(G;Y;\Sigma)$ is an isomorphism.

2. The triviality conditions $0 = I^{ht,CAT}(G;X;\pi(X))$ follow directly from the definitions, and therefore Theorem 3.4 applies to the theories $I^{ht,CAT}$.

PROOF: The triviality condition implies that the domain and codomain of the homomorphism $\varphi(X;\pi(X,\mathcal{H}^c))$ are both zero, and therefore $\varphi(X;\Sigma)$ is an isomorphism in the extreme case $\pi(X;\mathcal{H}^c)$. Proceed by downward induction on Σ; specifically, assume that the maps $\varphi(D^k \times Y;\Sigma_0)$ satisfy the conclusion of the theorem when Σ_0 properly contains Σ. Let $\Sigma_0 = \Sigma \cup G\{\alpha\}$ where α is minimal in $\Sigma_0 - \Sigma$. We then have the following commutative diagram with exact rows (the group G is omitted from the notation to save space):

$$\begin{array}{ccccccc}
I^a(D^2 \times Y;\Sigma \subseteq \Sigma') & \longrightarrow & I^a(D^1 \times Y;\Sigma') & \longrightarrow & I^a(D^1 \times Y;\Sigma) & \longrightarrow & \cdots \\
{\scriptstyle C_2}\downarrow & & {\scriptstyle A_1}\downarrow & & {\scriptstyle B_1}\downarrow & & {\scriptstyle C_1}\downarrow \\
I^b(D^2 \times Y;\Sigma \subseteq \Sigma') & \longrightarrow & I^b(D^1 \times Y;\Sigma') & \longrightarrow & I^b(D^1 \times Y;\Sigma) & \longrightarrow & \cdots
\end{array}$$

$$\begin{array}{ccccccc}
I^a(D^1 \times Y;\Sigma \subseteq \Sigma') & \longrightarrow & I^a(Y;\Sigma') & \longrightarrow & I^a(Y;\Sigma) & \longrightarrow & I^a(Y;\Sigma \subseteq \Sigma') \\
{\scriptstyle C_1}\downarrow & & {\scriptstyle A_0}\downarrow & & {\scriptstyle B_0}\downarrow & & {\scriptstyle C_0}\downarrow \\
I^b(D^1 \times Y;\Sigma \subseteq \Sigma') & \longrightarrow & I^b(Y,\Sigma') & \longrightarrow & I^a(Y;\Sigma) & \longrightarrow & I^a(Y;\Sigma \subseteq \Sigma')
\end{array}$$

Our general hypotheses imply that C_2 is an epimorphism and C_0 and C_1 are isomorphisms. Furthermore, the induction assumption implies that A_1 is an isomorphism and A_0 is a monomorphism. The refined version of the five lemma (compare [**McL**, Chapter I, Lemma 3.3]). now implies that B_1 is an isomorphism and B_0 is a monomorphism. This completes the proof of the inductive step, and it follows that $\varphi(X;\Sigma)$ is an isomorphism for all choices of Σ.■

The Comparison Theorem will be used repeatedly in this book to show that several different approaches to equivariant surgery theory yield the same equivariant surgery groups and also to prove the periodicity theorems in Chapter III-V.

Comparison theorems for relative groups. In most settings where Theorem 3.4 applies, the maps

$$\varphi(G;X;\Sigma \subseteq \Sigma') : I^a(G;X;\Sigma \subseteq \Sigma') \to I^b(G;X;\Sigma \subseteq \Sigma')$$

are isomorphisms for all pairs $\Sigma \subseteq \Sigma'$; such a statement contains the result in 3.4 because $I^c(G;X;\Sigma) = I^c(G;X;\Sigma \subseteq \pi(X;\mathcal{H}^c))$ by construction (as usual, $c = a$ or b). All that is needed for this generalization are extensions of the exactness and naturality properties to triples $\Sigma \subseteq \Sigma' \subseteq \Sigma''$; the appropriate statements can be obtained by replacing $I^c(G;X;\Sigma)$ with $I^c(G;X;\Sigma \subseteq \Sigma'')$ and making corresponding change at other places where absolute objects appear (recall the previous remark about the equivalence of Σ^* and the pair $\Sigma^* \subseteq \pi(X;\mathcal{H}^c)$, where $\Sigma^* = \Sigma$ or Σ' and $c = a$ or b).

3B. Rothenberg sequences in equivariant surgery

If CAT is either the category of smooth ($DIFF$) or PL manifolds, then there are obvious forgetful maps from the simple equivariant surgery groups $I^{s,CAT}(G;\lambda;\Sigma\subseteq\Sigma')$ to the corresponding equivariant surgery groups $I^{ht,CAT}(G;X;\Sigma\subseteq\Sigma')$ because a simple G-homotopy equivalence is a special type of G-homotopy equivalence. Furthermore, these forgetful maps are abelian group homomorphisms, and one has a commutative diagram involving boundary homomorphisms as in (3.1.A) and (3.1.B) with $a=(s,CAT)$ and $b=(ht,CAT)$.

PROPOSITION 3.5. *In the setting above, assume that $\Sigma'=\Sigma\cup G\{\alpha\}$ and $\lambda=\mathbf{S}\lambda'$ for some $\lambda'=\lambda_Y$. Then the stepwise surgery obstruction map*

$$I^{s,CAT}(G;\lambda;\Sigma\subseteq\Sigma\cup G\{\alpha\})\to L^s_{\dim\alpha}(\mathbf{Z}[W_\alpha],w_\alpha)$$

is an isomorphism of abelian groups if $\mathbf{S}^2\lambda$ satisfies the Gap Hypothesis and an epimorphism if $\mathbf{S}\lambda$ does. Furthermore, the change-of-category maps $I^{s,DIFF}\to I^{s,PL}$ are isomorphisms if all closed substrata have dimension ≥ 7 and monomorphisms if all closed substrata have dimension ≥ 6.■

The proofs are very similar to the preceding arguments with ht replacing s.

PROPOSITION 3.6. *Assume the same setting. Then the forgetful homomorphisms from $I^{s,CAT}(G;X;\Sigma\subseteq\Sigma\cup G\{\alpha\})$ to $I^{ht,CAT}(G;X;\Sigma\subseteq\Sigma\cup G\{\alpha\})$ correspond to the canonical maps $L^s_{\dim\alpha}(\mathbf{Z}[W_\alpha],w_\alpha)\to L^h_{\dim\alpha}(\mathbf{Z}[W_\alpha],w_\alpha)$ under the stepwise surgery obstruction. Furthermore, the forgetful homomorphisms commute with those in the exact sequences as in (2.0).*■

The proof of Proposition 3.6 is straightforward.

COROLLARY 3.7. *The kernels and cokernels of the forgetful maps*

$$I^{s,CAT}(G;X;\Sigma)\to I^{ht,CAT}(G;X;\Sigma)$$

are finite 2-groups.

PROOF: The forgetful maps $L^s_m(\pi,w)\to L^h_m(\pi,w)$ fit into a long exact sequence due to M. Rothenberg (compare [**Sh**]); the third terms of this sequence are given by $\widehat{H}_m(\mathbf{Z}_2;Wh(\pi))$, where $\mathbf{Z}_2$ acts on the Whitehead group $Wh(\pi)$ by a map obtained from conjugation on $\mathbf{Z}[\pi]$. Therefore the kernel and cokernel of the forgetful maps $I^{s,CAT}(*)\to I^{ht,CAT}(*)$ are finite with exponent two when $(*)$ is given by $(G;X;\Sigma\subseteq\Sigma\cup G\{\alpha\})$. Furthermore, if $\Sigma'=\Sigma\cup G\{\alpha\}=\pi(X)$ then the map $I^{s,CAT}(G;X;\Sigma')\to I^{ht,CAT}(G;X;\Sigma')$ goes from 0 to 0 and therefore is an isomorphism. The general result proceeds by induction on the size of $\pi-\Sigma$. There is a commutative diagram relating $I^{s,CAT}$ to $I^{ht,CAT}$ as in the hypotheses of Theorem 3.4, and one can modify the proof of that result to show that the kernel and cokernel of $f(\Sigma): I^{s,CAT}(\Sigma)\to I^{s,CAT}(\Sigma)$ are possibly trivial finite 2-groups if the same is true for $f(\Sigma\cup G\{\alpha\})$. If we combine this with the preceding remarks, we see that the kernel and cokernel of $f(\Sigma)$ and $f(\Sigma')$ are always finite 2-groups as claimed.■

In ordinary surgery theory one has the Rothenberg exact sequence containing the forgetful maps $L^s \to L^h$; it is natural to ask about the existence of similar sequences in equivariant surgery theory. The existence of a long exact sequence can be established by purely formal considerations. A basic construction known as **spacification** shows that the Wall groups $L_*^c(\pi, w)$ can be realized as the homotopy groups of some spectrum $\mathbb{L}_0^c(\mathbb{Z}[\pi], \varepsilon_w)$ that reflects the geometric definition of $L_*^c(\pi, w)$ in [**Wa**, Chapter 9]; the specifics of this process appear in the original paper by Quinn on the subject [**Q**] and also in subsequent works by Taylor and Williams [**TW**], Nicas [**Ni**, in Section 2.2], Ranicki [**Ran**], and Weinberger [**Wb**, Section 2]. In this setting the forgetful maps correspond to morphisms of spectra $f : \mathbb{L}^s \to \mathbb{L}^h$, and the Rothenberg sequence is just the homotopy exact sequence of the inclusion $\mathbb{L}^s \to$ Mapping Cylinder(f). A similar construction works in equivariant surgery; in this case the spectra have the form $\mathbb{I}_0^{c,CAT}(G; \lambda_X; \Sigma \subseteq \Sigma')$; however, there are isomorphisms $\pi_k(\mathbb{I}_0^{c,CAT}(G; \lambda_X; \Sigma \subseteq \Sigma')) \cong I^{c,CAT}(G; S^k\lambda_X; \Sigma \subseteq \Sigma')$ only if $S^{k+\varepsilon}\lambda_X$ satisfies the Gap Hypothesis for a suitably chosen $\varepsilon \geq 0$. There are also forgetful maps $\Phi : \mathbb{I}^{s,CAT} \to \mathbb{I}^{c,CAT}$ such that the forgetful map $I^s \to I^{ht}$ equals the map on homotopy groups associated to Φ. It follows that the forgetful maps $I^s \to I^{ht}$ can be embedded into long exact sequences that are equal to the homotopy exact sequences of suitably defined pairs of spectra.

In analogy with the Rothenberg sequence, the third terms of this long exact sequence can be described in terms of generalized equivariant Whitehead torsion as defined in [**DR3**], [**DR4**] or [**I3**]. For example, the results of [**Do3**] describe the cokernel of the forgetful maps $I^s \to I^{ht}$ in terms of generalized torsion invariants; applications of these torsion formulas to group actions on spheres and homotopy complex projective spaces are noted in the introduction to [**Do3**]. It is also possible to derive equivariant analogs of other features of the Rothenberg sequence. However, complications in the relationship between Poincaré duality and equivariant torsion lead to substantial difficulties in formulating correct statements and proofs. Further information can be found in [**DR2**] and [**CL**] (also see [**LM**, Part II, Section 3]).

Analogs for Browder-Quinn theory. There are also forgetful homomorphisms $L^{BQ,s} \to L^{BQ,h}$, and there are analogs for Proposition 3.5-3.6, Corollary 3.7 and the equivariant Rothenberg sequence described above. In fact, everything works better in this case because the Gap Hypothesis is unnecessary; for example, the Browder-Quinn groups $L_*^{BQ,c}$ are always isomorphic to the homotopy groups of the spacifications $\mathbb{L}_0^{BQ,c}$.

3C. Change of coefficients

The groups $I^{ht(R),CAT}$ were defined to be generalizations of the groups $I^{ht,DIFF}$ where the appropriate notion of equivalence is a map $f : X \to Y$ such that f is isogeneric and for each α the map $f_\alpha : X_\alpha \to Y_\alpha$ is an R-local homotopy equivalence, where $\mathbb{Z} \subseteq R \subseteq \mathbb{Q}$. Since we are assuming that all closed substrata are 1-connected, this is equivalent to either (a) each map f_α is an R-local homology equivalence or (b) the map f is an R-equivariant localization as described in [**MMT**]. It follows that there are homomorphisms

$$I^{ht,CAT}(-; \Sigma \subseteq \Sigma') \to I^{ht(R),CAT}(-; \Sigma \subseteq \Sigma') \to I^{ht(\mathbb{Q}),CAT}(-; \Sigma \subseteq \Sigma')$$

and if $\Sigma' = \Sigma \cup G\{\alpha\}$ the groups and morphisms have the form

$$L^h_{\dim\alpha}(\mathbb{Z}[W_\alpha], w_\alpha) \xrightarrow{\otimes R} L^h_{\dim\alpha}(R[W_\alpha], w_\alpha) \xrightarrow{\otimes \mathbb{Q}} L^h_{\dim\alpha}(\mathbb{Q}[W_\alpha], w_\alpha).$$

As in the preceding subsections, the groups $I^{ht(R),CAT}$ are essentially independent of the choice of category CAT. Furthermore, if $\mathbb{Z} \subseteq R \subseteq A \subseteq \mathbb{Q}$ then there are obvious forgetful maps $I^{ht(R),CAT} \to I^{ht(A),CAT}$ as above, and these maps embed into exact localization sequences generalizing the usual exact localization sequences (see **[CM]**, **[Pa1]**, **[Pa2]**), at least if a suitable version of the Gap Hypothesis holds. Since the kernels and cokernels of the localization maps

$$L^h_*(R[\pi], w) \to L^h_*(A[\pi], w)$$

are torsion groups by the results of the given references, the following is an immediate consequence.

PROPOSITION 3.8. *If all closed substrata have dimension* ≥ 7, *then the kernel and cokernel of* $I^{ht(R),CAT} \to I^{ht(A),CAT}$ *are torsion groups.*■

REMARK: In general the kernel and cokernel are not finitely generated; *e.g.*, the cokernel of $L^h_0(\mathbb{Z}) \to L^h_0(\mathbb{Q})$ is $\mathbb{Z}_8 \oplus (\mathbb{Z}_4)^\infty \oplus (\mathbb{Z}_2)^\infty$ by standard results on Witt groups.

3D. Change to pseudoequivalence

Modulo some technicalities, a G-pseudoequivalence is a G-equivariant map that is an ordinary (*i.e.*, nonequivariant) homotopy equivalence; it is immediate that every G-homotopy equivalence is a pseudoequivalence, and standard examples as in **[Pe1]** show that pseudoequivalence is a much weaker notion than G-homotopy equivalence. It is a routine exercise to verify that every $(ht, DIFF)$ equivariant surgery problem (as defined in Chapter I) determines an $(h, DIFF)$ equivariant surgery problem (as defined in **[DP1]** and mentioned in Section 2 of this chapter). Furthermore, in both cases the abelian group operation is defined by the methods of **[Dov1]** and **[Dov2]**, and it follows that there are homomorphisms

$$I^{ht,DIFF}(G; \lambda_X; \Sigma \subseteq \Sigma') \to I^{h,DIFF}(G; \widehat{\lambda_X}; \Sigma \subseteq \Sigma')$$

for $\Sigma \subseteq \Sigma'$ in $\pi(X; \mathcal{H}^h)$; here $\widehat{\lambda_X}$ denotes the indexing data for the identity map of X in the theory $I^{h,DIFF}$ (see Chapter V for further information). If α is not the maximal element in $\pi(X; \mathcal{H}^h)$ and $\Sigma \subseteq \Sigma \cup G\{\alpha\} = \Sigma'$ in $\pi(X; \mathcal{H}^h)$ then the map of stepwise obstructions

$$I^{ht,DIFF}(G; \lambda_X; \Sigma \subseteq \Sigma') \to I^{h,DIFF}(G; \widehat{\lambda_X}, \Sigma \subseteq \Sigma')$$

corresponds to the localization map

$$L^h_{\dim\alpha}(\mathbb{Z}[W_\alpha], w_\alpha) \to L^h_{\dim\alpha}(R_\alpha[W_\alpha], w_\alpha)$$

where R_α is defined as in Section 2. If α is maximal in $\pi(X; \mathcal{H}^h)$ then the map of stepwise obstructions corresponds to the forgetful map

$$L^h_{\dim X}(\mathbb{Z}[G], w_G) \to L^{D(G)}_{\dim X}(\mathbb{Z}[G], w_G)$$

where the codomain is the projective Wall group with respect to the defect subgroup $D(G) \subseteq \widetilde{K}_0(\mathbb{Z}[G])$ described in Section 2.

Similar results can be formulated for the groups $I^{h(R),DIFF}$ and the forgetful maps E, F suggested by the following diagram:

$$\begin{array}{ccc} I^{ht,DIFF} & \xrightarrow{B} & I^{h,DIFF} \\ \downarrow C & & \downarrow F \\ I^{ht(R),DIFF} & \xrightarrow{E} & I^{h(R),DIFF} \end{array}$$

Since no new complications arise, a more detailed discussion will not be given.

3E. Passage to Lück-Madsen groups

Working in a more categorical setting, W. Lück and I. Madsen have presented another approach to equivariant surgery with improved formal properties [**LM**]. One obvious feature of their work is the removal of the simple connectivity assumption on closed substrata. A comparison of [**LM**] with [**DR1**] and various remarks in [**LM**] suggest that both approaches should yield the same equivariant surgery groups in most if not all cases. However, there are enough differences to justify an explicit comparison of the two formulations. We shall assume this dimension restriction and the Gap Hypothesis throughout the discussion that follows.

The most obvious difference between [**DR1**] and [**LM**] is the formulation of the indexing data. It is not enough to specify the geometric poset $\pi(X)$; one needs an object ${}_1\pi(X)$ that also incorporates the fundamental groups of the open substrata

$$(X_\alpha - \bigcup_{\beta<\alpha} X_\beta)/W_\alpha$$

and certain homomorphisms between these groups that arise indirectly from inclusions of closed substrata. All of this information is contained in an equivariant generalization of the fundamental groupoid of a space; there is a very clear description of this material in Section I.10 of tom Dieck's book on transformation groups [**tD**] (see pp. 74–77 in particular). The dimension, slice, and orientation data are carried by a *fiber transport*

$$t_X : \pi^G(X) \to \mathsf{B}_n$$

where $\pi^G(X)$ is the equivariant fundamental groupoid of X, B_n is the equivariant fundamental groupoid of the classifying space $B(G,n)$ for n-dimensional G-vector bundles or $CAT\ G-\mathbb{R}^n$ bundles (if $CAT \neq DIFF$), and t_X is induced by the classifying map of the G-equivariant tangent bundle (if $CAT \neq DIFF$ this is defined as in [**LaR**]). The morphism t_X can be viewed as a refined first Stiefel-Whitney class for all the bundles that arise. A **Lück-Madsen equivariant surgery problem** is defined to be an isogeneric equivariant map $f : M \to X$ that has degree 1 in a strong sense [**Lü**] together with an $\mathbb{R}$-stable G-bundle map $\widehat{f} : TM \to (\xi \downarrow X)$ covering f; here TM denotes the G-tangent bundle, defined as in [**LR**] if $CAT \neq DIFF$, and $\mathbb{R}$-stable means one only allows stablization by trivial representations. In this situation we also assume that the transports t_X and t_{TX} induce the same map ${}_1\pi(X) \to {}_1\pi(B(G, n+k))$ for a suitable $k \geq 0$.

In the Appendix to Part I of [LM], Lück and Madsen show that bundle data in the sense of [**DP1**] or [**DR1**] yield bundle data in the sense of [**LM**].

The Lück-Madsen indexing data associated to X are essentially the pair $R_X = ({}_1\pi(X), t_{X*})$, where t_{X*} is an induced map from ${}_1\pi(X)$ to ${}_1\pi(B(G, n+k))$ as before. Since this is simpler than the definition of geometric reference in [**LM**, Definitions I.2.3 and I.3.1], some explanation is needed. The Lück-Madsen notion of reference includes an object like ${}_1\pi$ and *two* maps t_{0*}, t_{1*} induced by fiber transport together with a weak compatibility condition relating t_{0*} and t_{1*}; the latter is given by an equivalence called τ in [**LM**, Definition I.2.3]; the equivalence τ must be simple in an appropriate sense in order to discuss simple surgery for the reference. In our situation $t_{0*} = t_{1*}$ and τ is taken to be the identity.

If we let R_X denote the geometric reference of X, it follows that one has a canonical map from $I^{c,CAT}(G; \lambda_X)$ to the Lück-Madsen group $\mathcal{L}^{c,CAT}(G; R_X)$ if $CAT = DIFF$ or PL.

We now assume that the methods of both [**DR1**] and [**LM**] apply to everything under consideration. If Σ and Σ' are closed, G-invariant subsets of $\pi(X)$ with $\Sigma \subseteq \Sigma'$, then one can define $(\Sigma \subseteq \Sigma')$-adjusted Lück-Madsen groups as in [**LM**, Part II, Section 1]; our notion of $(\Sigma \subseteq \Sigma')$-adjusted is identical to their notion of $(\Sigma \subseteq \Sigma')$-**restricted**.

The group operation in the Lück-Madsen groups $\mathcal{L}_n^{c,CAT}(G; R; \Sigma)$ is given by disjoint union, and the group operation in the Dovermann-Rothenberg groups $I^{c,CAT}(G; \lambda; \Sigma)$ is given by the construction of [**Do2**, Section 5]. Despite this difference the group operations are compatible:

PROPOSITION 3.9. *The canonical maps*

$$\varphi : I^{c,CAT}(G; \lambda_X; \Sigma \subseteq \Sigma') \to \mathcal{L}_N^{c,CAT}(G; R_X; \Sigma \subseteq \Sigma')$$

are homomorphisms.

PROOF: (*Sketch.*) Let $(f_1; -)$ and $(f_2; -)$ represent classes u_1, u_2 in $I^{c,CAT}(G; \lambda_X; \Sigma \subseteq \Sigma')$. The construction of [**Do2**, Section 5] describes a representative $(f_0; -)$ for $u_1 + u_2$ such that there is a cobordism $(F; -)$ such that $\partial_+(F; -) = (f_1; -)\amalg(f_2; -)$; $\partial_-(F; -) = (f_0; -)$, and $(F; -)$ has certain additional properties (compare Lemma 2.2 in Chapter III of this book). In fact, the bordism $(F; -)$ is a bordism of G-surgery problems in the sense of [LM], and from this it follows that the classes of these problems in $\mathcal{L}^{c,CAT}(G; R_X; \Sigma \subseteq \Sigma')$ also satisfy

$$-\ \text{class}(f_0; -) + \text{class}(f_1; -) + \text{class}(f_2; -) = 0.$$

Since the classes in question are $\varphi(u_1 + u_2)$, $\varphi(u_1)$, and $\varphi(u_2)$ respectively, it follows that $\varphi(u_1 + u_2) = \varphi(u_1) + \varphi(u_2)$.■

As noted in Section I.5 and Section 2 of this chapter, the theories $I^{c,CAT}$ and $\mathcal{L}_n^{s,CAT}$ all satisfy the exactness conditions of (2.0). In fact, if $\varphi : I^{c,CAT}(-) \to \mathcal{L}_n^{c,CAT}(-)$ is the transformation of equivariant surgery groups described above, φ is compatible with these exact sequences:

PROPOSITION 3.10. *(i) Suppose that $\Sigma \subseteq \Sigma' \subseteq \pi(X)$ and $T \subseteq T' \subseteq \pi(X)$ define closed G-invariant pairs such that $T \subseteq \Sigma$ and $T' \subseteq \Sigma'$. Then the following diagram commutes:*

$$\begin{array}{ccc} I^{c,CAT}(G;\lambda_X;\Sigma \subseteq \Sigma') & \xrightarrow{\text{forgetful map}} & I^{c,CAT}(G;\lambda_X;T \subseteq T') \\ \varphi \downarrow & & \varphi \downarrow \\ \mathcal{L}_n^{c,CAT}(G;R_X;\Sigma \subseteq \Sigma') & \xrightarrow{\text{forgetful map}} & \mathcal{L}_n^{c,CAT}(G;R_X;T \subseteq T') \end{array}$$

(ii) The following diagram also commutes:

$$\begin{array}{ccc} I^{c,CAT}(G;\mathbf{S}\lambda_X;\Sigma \subseteq \Sigma') & \longrightarrow & I^{c,CAT}(G;\lambda_X;\Sigma') \\ \downarrow & & \downarrow \\ \mathcal{L}_{n+1}^{c,CAT}(G;\mathbf{S}R_X;\Sigma \subseteq \Sigma') & \longrightarrow & \mathcal{L}_n^{c,CAT}(G;R_X;\Sigma') \end{array}$$

These follow directly from the geometric definitions of the various equivariant surgery groups.

The next step in comparing $I^{c,CAT}$ to $\mathcal{L}^{c,CAT}$ is to show that the previously described transformation φ is compatible with the stepwise obstructions that were described in Proposition I.5.3 and (2.1) for $I^{ht,CAT}$, in (2.2) for $I^{s,CAT}$, and in (2.4) for $\mathcal{L}^c$.

PROPOSITION 3.11. *(i) Assume that $\mathbf{S}^2\lambda_X$ satisfies the Gap Hypothesis, let $\Sigma' = \Sigma \cup G\{\alpha\}$ as usual, and denote the stepwise surgery obstruction maps in the theories $I^{c,CAT}$ and $\mathcal{L}^{c,CAT}$ by $\sigma_\alpha(I)$ and $\sigma_\alpha(\mathcal{L})$ respectively. Then $\sigma_\alpha(I) = \sigma_\alpha(\mathcal{L})\varphi$. Furthermore, φ is an isomorphism.*

(ii) Assume that $\mathbf{S}\lambda_X$ satisfies the Gap Hypothesis, take $\Sigma \subseteq \Sigma'$ and $\sigma_\alpha(I)$ as above, and write the codomain as $L^c_{\dim \alpha}(\mathbb{Z}[W_\alpha], w_\alpha)/B(I)$. If the stepwise surgery obstruction isomorphism $\sigma_\alpha(\mathcal{L})$ for $\mathcal{L}^{c,CAT}$ has range given by $L^c_{\dim \alpha}(\mathbb{Z}[W_\alpha]; w_\alpha)/B(\mathcal{L})$, then $B(I) \subseteq B(\mathcal{L})$ and the following diagram commutes:

$$\begin{array}{ccc} I^{c,CAT}(-;\Sigma \subseteq \Sigma') & \xrightarrow{\sigma_\alpha(I)} & L^c_{\dim \alpha}(\mathbb{Z}[W_\alpha], w_\alpha)/B(I) \\ \varphi \downarrow & & \downarrow \text{projection} \\ \mathcal{L}^c(-;\Sigma \subseteq \Sigma') & \xrightarrow{\sigma_\alpha(\mathcal{L})} & L^c_{\dim \alpha}(\mathbb{Z}[W_\alpha], w_\alpha)/B(\mathcal{L}). \end{array}$$

Furthermore, the map φ is an epimorphism.

PROOF: (i) Consider the inverses to $\sigma_\alpha(I)$ and $\sigma_\alpha(\mathcal{L})$. In both cases the inverse γ_α is given by choosing Y such that $Y \times I$ and X have the same indexing or reference data, adding handles to $Y_\alpha \times I$ away from the strictly smaller closed substrata to realize a prescribed element of $L^c_{\dim \alpha}(\mathbb{Z}[W_\alpha], w_\alpha)$ to obtain a cobordism V_α with $\partial_0 V_\alpha = Y_\alpha$, and thickening $Y \times I \cup V_\alpha$ into an equivariant cobordism V whose bottom end is Y and whose top end is G-homotopy equivalent to Y (and in fact G-simple equivalent if $c = s$); there is also a corresponding construction of a map from W to $Y \times I$ whose restriction to the bottom end is the identity. From this description it is clear that $\sigma_\alpha(\mathcal{L})\varphi\gamma_\alpha(I)$ is the identity; therefore we also have $\sigma_\alpha(\mathcal{L})\varphi = \gamma_\alpha(I)^{-1}$; and (i) follows because $\gamma_\alpha^{-1} = \sigma_\alpha$. Since $\sigma_\alpha(\mathcal{L})$ and $\sigma_\alpha(I)$ are isomorphisms, the same is true for φ.□

(ii) If $\mathbf{S}\lambda_X$ satisfies the Gap Hypothesis but $\mathbf{S}^2\lambda_X$ does not, then a stepwise surgery obstruction map

$$\sigma_\alpha(I) : I^{c,CAT}(G; \lambda_X; \Sigma \subseteq \Sigma') \to L^c_{\dim \alpha}(\mathbb{Z}[W_\alpha], w_\alpha)/B(I)$$

is described in Proposition 5.5 of Chapter I. In particular, this map is a surjective homomorphism and there is a homomorphism

$$\gamma_\alpha(I) : L^c_{\dim \alpha}(\mathbb{Z}[W_\alpha], w_\alpha) \to I^{c,CAT}(G; \lambda_X; \Sigma \subseteq \Sigma')$$

such that $\sigma_\alpha\gamma_\alpha$ is reduction *mod* $B(I)$. Similar constructions and arguments apply if $I^{c,CAT}$ is replaced by $\mathcal{L}^{c,CAT}$; yielding homomorphisms $\gamma_\alpha(\mathcal{L})$ and $\sigma_\alpha(\mathcal{L})$; this requires a systematic check of both [**DR1**, Section 9] and [**LM**, Part II, Section 1]. By construction it follows that $\sigma_\alpha(\mathcal{L})\varphi\gamma_\alpha(I)$ is reduction mod $B(\mathcal{L})$. The surjectivity of φ and the commutativity of the diagram in (ii) follows immediately from this.■

We are now prepared to prove a comparison theorem for $I^{c,CAT}$ and $\mathcal{L}^{c,CAT}$.

THEOREM 3.12. *Suppose that the indexing and reference data for X have the form $\mathbf{S}\lambda_Y$ and $\mathbf{S}R_Y$ for some Y such that $I^{c,CAT}$ and $\mathcal{L}^{c,CAT}_{n-1}$ are both defined for Y. Furthermore, assume that $D^2 \times X$ satisfies the Gap Hypothesis. Then the maps $\varphi(X;\Sigma)$ are isomorphisms and the maps $\varphi(Y;\Sigma)$ are monomorphisms.*

This follows directly from Theorem 3.4 (the general comparison theorem) and Propositions 3.9–3.11.

Rothenberg sequences. As in other similar situations, there is a canonical map from the simple equivariant surgery groups $\mathcal{L}^s_*(G; R_X; \Sigma \subseteq \Sigma')$ to the corresponding homotopy surgery groups $\mathcal{L}^h_*(G; R_X; \Sigma \subseteq \Sigma')$, and therefore one expects to have a Rothenberg-type exact sequence containing the maps $\mathcal{L}^s_* \to \mathcal{L}^h_*$. A sequence of this type is constructed explicitly in [**LM**, Part II, Section 3]. The construction of this sequence does not require the Gap Hypothesis, but the weak gap conditions (codimension ≥ 3 gaps, closed substrata $\geq$ 4-dimensional) are needed to describe the third terms of the sequence in a reasonable algebraic fashion (compare [**LM**, Part II, Cor. 3.8]). If the weak gap conditions hold, then the third terms of this sequence have the form $\widehat{H}^{*+1}(\mathbb{Z}/2; \mathcal{W}(X))$ where $\widehat{H}^q$ denotes Tate cohomology and $\mathcal{W}(X)$ is a generalized equivariant Whitehead group with a conjugation involution defined as in [**CL**].

3F. Passage from Browder-Quinn groups

The Browder-Quinn groups $L^{BQ,c}_*$ ($c = h$ or s) were chronologically the first family of equivariant surgery groups to be constructed, and they have also proved to be initial objects in an informal categorical sense. Specifically, given an equivariant surgery theory I^a it is almost always possible to find a theory $L^{BQ,T(a)}$ of Browder-Quinn type that maps into I^a in the sense of (3.1.A) and (3.1.B). Furthermore, in many cases the map from $L^{BQ,T(a)}_*$ to I^a is an isomorphism. Such relationships often provide a means for using one approach to equivariant surgery to shed light on another. In Section 5 of Chapter III we shall discuss some specific examples.

In order to relate the Browder-Quinn theories to other approaches, it is necessary to know that G-normal maps in the sense of [**BQ**] determine G-normal maps both in the sense of [**LM**] and in the sense of [**DR1**] and Section I.4 of this book. As indicated in the preceding subsection, a G-normal map in the sense of [**DR1**] or Section I.4 determines a G-normal map in the sense of [**LM**]; specifically, this is done in the Appendix to [**LM**, Part I]). Therefore it is only necessary to show how G-normal maps in the sense of [**BQ**] determine G-normal maps in the sense of [**DR1**] and Section I.4 of this book. This has been understood by many users of equivariant surgery theory for some time (compare [**DMS**, Section 2]), but a definitive reference does not seem to exist in the literature.

Recall from Section 1 that a degree one Browder-Quinn G-normal map consists of a G-isovariant transverse linear map $f : M \to X$ with degree 1 and a stable isomorphism of stratified vector bundles $b : ST(M/G) \to ((f/G)^*, \xi)$, where $ST(M/G)$ is the stratified tangent bundle of M/G as defined in Section 1 of this chapter.

REMARK: The formal definition of normal map in [**BQ**, page 34, lines 16–24] may not explicitly mention the need for bundle data, but it is clear from the context (especially lines 11–15 in the preceding citation) that a condition of this sort is necessary and presumably intended.

PROPOSITION 3.13. *Let $(f : M \to X, b)$ be a Browder-Quinn G-normal map (not necessarily of degree 1). Then f can be approximated by an isovariant transverse linear map h such that there is a system of equivariant bundle data for h in the sense of Section I.4.*

PROOF: (*Sketch.*). Let ξ_0 be a stratified vector bundle over X/G such that b induces a stable isomorphism from $ST(M/G)$ to $f^*\xi_0$.

Take $\Pi(T(M))$ and $\Pi(T(X))$ to be the vector bundle systems defined by the equivariant normal bundles of closed substrata as in Section I.3. By Proposition 2.1 (of this chapter) there is an approximation h to f such that h corresponds to a map of vector bundle systems

$$c : \pi(T(M)) \to h^*\pi(T(X))$$

on a neighborhood of the singular strata; as usual, the equivariant normal bundle of a closed substratum is identified with a neighborhood of itself by the equivariant tubular neighborhood theorem. By construction h is isovariantly and transverse linearly homotopic to f.

A covering homotopy argument shows that the isovariant transverse linear homotopy from f to h can be covered by a homotopy of stable stratified bundle isomorphisms from b to some map b_1 covering h/G. The stable isomorphism b_1 lifts to an equivariant stable isomorphism of stratified bundles

$$\widetilde{b}_1 : ST(M) \to h^*\xi_1$$

where ξ_1 is the pullback of ξ_0 under the orbit space projection $X \to X/G$.

The crucial step is to form a stable G-vector bundle isomorphism

$$b_2 : T(M) \to h^*\xi_2$$

(where ξ_2 is some G-vector bundle over X) by taking the sum of $\widetilde{b}_1$ and the map c in a suitable manner. Choose a sequence of closed G-invariant subsets of $\pi(X)$

$$\pi(X) = \Sigma_0 \supseteq \Sigma_1 \supseteq \cdots \supseteq \Sigma_{r+1} = \varnothing$$

such that $\Sigma_i = \Sigma_{i+1} \cup G\{\alpha(i)\}$, where $\alpha(i)$ is minimal in $\pi(X) - \Sigma_{i+1}$, and let

$$U_n = X - \bigcup_{\alpha \in \Sigma_{i+1}} X_\alpha$$

so that $X - \text{Sing } X = U_0 \subseteq U_1 \subseteq \cdots \subseteq U_r = X$. Define corresponding open subsets of M by taking $U_i' = h^{-1}(U_i)$. We shall inductively construct b_2 over each set U_i'. If $i = 0$ this is easy because c is only defined on the singular set Sing X; we simply take $b_2 = \widetilde{b}_1 | \text{Nonsing } M$. Assume now that b_2 and ξ_2 are defined over U_i' and U_i respectively. More precisely, assume there is a vector bundle $\xi_2(i)$ over U_i whose restriction to each intersection $U_{j+1} - U_j = G \cdot X_{\alpha(j)}$ or U_{j+1} $(j < i)$ is the direct sum

$$(\xi_1 | G \cdot X_{\alpha(j)} \cap U_{j+1}) \oplus (\pi(T(X)) | G \cdot X_{\alpha(j)} \cap U_{j+1})$$

and a stable G-vector bundle isomorphism

$$b_2(i) : T(M)|U_i' \to h^* \xi_2(i)$$

such that $b_2(i)$ is a fiberwise direct sum of $\widetilde{b}_1$ and c with respect to the decomposition given above. Let V be an equivariant tubular neighborhood of $G \cdot X_{\alpha(i)} \cap U_{i+1}$ in U_{i+1}, and let

$$V' = V \cap U_i = V - G \cdot X_{\alpha(i)}.$$

The compatibility conditions for stratified bundles and vector bundle systems imply that the restriction $\xi_2(i)|V'$ is the pullback of

$$(\star) \qquad (\xi_1 | G \cdot X_{\alpha(i)} \cap U_{i+1}) \oplus (\pi(T(X)) | G \cdot X_{\alpha(i)} \cap U_{i+1})$$

to V' under the projection from V' to $G \cdot X_{\alpha(i)} \cap U_{i+1}$. Therefore if we define $\xi_2(i+1)$ over V to be the pullback of $(\star)$ under the projection from V to $G \cdot X_{\alpha(i)} \cap U_{i+1}$, it follows that $\xi_2(i)$ and $\xi_2(i+1)$ agree over $V \cap U_I$. This yields the desired extension of $\xi_2(i) \downarrow U_i$ to $\xi_2(i+1) \downarrow U_{i+1}$.

The next order of business is to construct an extension of $b_2(i)$ to the stabilization of $T(M)|U_{i+1}'$. Isovariance considerations imply that $U_{i+1}' - U_i'$ is a union of open substrata with isotropy subgroup $G_{\alpha(i)}$. Let N be an equivariant tubular neighborhood of $U_{i+1}' - U_i'$ in U_{i+1}', and let $N' = N \cap U_i'$. Then the compatibility properties of stratified bundle maps and maps of vector bundle systems imply that $b_2(i)|N$ splits as a direct sum of $\widetilde{b}_1$ and c with respect to the direct sum decomposition of $(\star)$ and the corresponding decomposition for the stabilization of $T(M)|N'$. Thus if we define $b_2(i+1)$ over N by $\widetilde{b}_1 \oplus c$, it follows that $b_2(i+1)$ and $b_2(i)$ agree over $N' = N \cap U_{i+1}$. It follows that $b_2(i)$ and $b_2(i+1)$ can be combined to form the desired extension of $b_2(i)$ to the stabilization of $T(M)|U_{i+1}'$.

In order to complete the system of bundle data it is necessary to specify a compatibility (stable) isomorphism φ relating stabilizations of the vector bundle systems $\Pi(b_2)$ and c; since $\Pi(b_2) = c$ by construction, one can do this trivially by taking φ to be the identity.■

One additional fact is needed for the construction of transformations from Browder-Quinn groups.

LEMMA 3.14. *Let $f : M \to N$ be a transverse linear isovariant homotopy equivalence of smooth G-manifolds. Then f is a simple transverse linear isovariant homotopy equivalence if and only if the generalized Whitehead torsion of f (in the sense of* [**I3**] *or* [**DR2**]) *is trivial.*

This follows immediately from the usual splittings of generalized Whitehead groups as direct sums of ordinary Whitehead groups (compare [**I3**] or [**DR2**]).■

Let X be a compact smooth G-manifold whose indexing and reference data have the form $\lambda_X = \mathbf{S}\lambda_Y$ and $R_X = \mathbf{S}R_Y$ for some compact smooth G-manifold Y, and let $\Sigma \subseteq \Sigma'$ be closed and G-invariant in $\pi(X)$. The main comparison properties of Browder-Quinn groups with index space X may be summarized as follows:

THEOREM 3.15. *(i) Let $c = s$ or h. Then there are homomorphisms*

$$\theta^c_{X,\Sigma\subseteq\Sigma'} : L^{BQ,c}_*(G; X; \Sigma \subseteq \Sigma') \to \mathcal{L}^c_*(G; R_X; \Sigma \subseteq \Sigma')$$

from the Browder-Quinn groups to the Lück-Madsen groups satisfying the compatibility relations (3.1.A)–(3.1.B). *If $\mathbf{S}^2R_X$ satisfies the Gap Hypothesis and all closed substrata of X have dimension ≥ 7, then the maps $\theta^c_{X,\Sigma}$ are isomorphisms. Furthermore, if Y is such that $\mathbf{S}R_Y = R_X$, then the maps $\theta^c_{Y,\Sigma}$ are monomorphisms.*

(ii) Let $c = ht$ or s, and assume X and Y satisfy the Gap Hypothesis with all closed substrata simply connected of dimension ≥ 6. Then there are homomorphisms

$$\rho^c_{X,\Sigma\subseteq\Sigma'} : L^{BQ,c}(G; X; \Sigma \subseteq \Sigma') \to I^{c,DIFF}(G; \lambda; \Sigma \subseteq \Sigma')$$

of equivariant surgery groups satisfying the compatibility relations (3.1.A)–(3.1.B). *If $\mathbf{S}^2\lambda_X$ satisfies the Gap Hypothesis and all closed substrata of X have dimension ≥ 7, then the maps $\rho^c_{X,\Sigma}$ are isomorphisms. Furthermore, in this case the maps $\rho^c_{Y,\Sigma}$ are monomorphisms.*

PROOF: We have described canonical methods for constructing bundle data in the sense of Section I.4 or [**LM**] from Browder-Quinn bundle data. It follows immediately that one has maps $L^{BQ,c}_*(G; X; \Sigma \subseteq \Sigma') \to \mathcal{L}^c_*(G; R_X; \Sigma \subseteq \Sigma')$ satisfying (3.1.A)-(3.1.B) and that these maps are homomorphisms. The construction of homomorphisms

$$\rho^c_{X,\Sigma\subseteq\Sigma'} : L^{BQ,c}_n(G; X; \Sigma \subseteq \Sigma') \to I^{c,DIFF}(G; \lambda_X; \Sigma \subseteq \Sigma')$$

requires a little more work. The $L^1 = L^2$ principle for $L^{BQ,c}_*$ implies that every class in $L^{BQ}_n(G; X; \Sigma \subseteq \Sigma')$ contains a "good" representative $f : M \to X$ where f induces an isomorphism of indexing data and all closed substrata of M are 1-connected. Such a class determines an element of $I^{c,DIFF}(G; X; \Sigma \subseteq \Sigma')$; in fact, a direct extension of the $L^1 = L^2$ argument shows that this class does not depend upon the choice of good representative. Additivity may be shown as follows: Take good representatives $f_i : M_i \to X_i$ for two classes u_1, u_2 in the Browder-Quinn group, and let $F : W \to Y$ be the cobordism construction of [**Do2**], Section 5, such that $\partial_+ F = f_1 \amalg f_2$ and $-\partial_- F$ represents $\rho(u_1) + \rho(u_2)$. A direct examination of the construction shows that F is a transverse linear isovariant map. It follows that $-\partial F$ represents a class in the Browder-Quinn group, and this equals the class of $f_1 \amalg f_2$. Therefore we have

$$\rho(u_1 + u_2) = \rho(f_1 \amalg f_2) = \rho(-\partial f) = \rho(f_1) + \rho(f_2),$$

and it follows that ρ is a homomorphism. By the standard comparison principle of Theorem 3.4, the isomorphism and monomorphism statements will follow if we can show that (i) θ and ρ are isomorphisms when $\mathbf{S}^2 R_X$ or $\mathbf{S}^2\lambda_X$ satisfies the Gap Hypothesis, and $\Sigma' = \Sigma \cup G\{\alpha\}$, with α minimal in $\pi(X)-\Sigma$ and $\dim\alpha \geq 6$, (ii) θ and ρ are epimorphisms if $\mathbf{S}^2 R_X$ and $\mathbf{S}^2\lambda_X$ are replaced by $\mathbf{S}R_X$ and $\mathbf{S}\lambda_X$ in the hypotheses. This can be proved by the sort of argument employed in Proposition 3.11. For each of the theories one has stepwise surgery obstructions with values in a quotient $L^c_{\dim\alpha}(\mathbf{Z}[E_\alpha], w_\alpha)/B$; these obstructions are described in Proposition 1.4 for the Browder-Quinn theories and in (2.4) for the Lück-Madsen theories. Furthermore, if $I^a \to I^b$ is either transformation of theories then one has the following diagram, in which the horizontal composites are the natural quotient projections:

$$\begin{array}{ccccc} L^c_{\dim\alpha}(\mathbf{Z}[E_\alpha], w_\alpha) & \xrightarrow{\gamma_\alpha(a)} & I^a(-;\Sigma\subseteq\Sigma') & \xrightarrow{\sigma_\alpha(a)} & L^c_{\dim\alpha}(\mathbf{Z}[E_\alpha], w_\alpha)/B(a) \\ & & \downarrow \varphi & & \\ L^c_{\dim\alpha}(\mathbf{Z}[E_\alpha], w_\alpha) & \xrightarrow{\gamma_\alpha(b)} & I^b(-;\Sigma\subseteq\Sigma') & \xrightarrow{\sigma_\alpha(b)} & L^c_{\dim\alpha}(\mathbf{Z}[E_\alpha], w_\alpha)/B(b). \end{array}$$

In each case the map γ_α is formed by a geometric process like that described in Proposition 3.11, and the same geometric considerations as in that proof imply the relation

$$\sigma_\alpha(b)\varphi\gamma_\alpha(a) = \sigma_\alpha(b)\gamma_\alpha(b) = \text{reduction mod } B(b).$$

Thus φ is an epimorphism if $\mathbf{S}R_X$ or $\mathbf{S}\lambda_X$ satisfies the Gap Hypothesis and the dimension condition holds. Furthermore, if $\mathbf{S}^2R_X$ or $\mathbf{S}^2\lambda_X$ satisfies the Gap Hypothesis, then $B(a) = B(b) = 0$. Therefore φ is an isomorphism in this case.■

Localized equivariant surgery. At the beginning of this subsection we mentioned that Browder-Quinn type groups can be associated to many constructions for equivariant surgery obstruction groups. In order to illustrate this, we shall describe a class of examples beyond $L^{BQ,h}$ and $L^{BQ,s}$. If R is a subring of the rationals, it is possible to define Browder-Quinn type groups for surgery up to R-local equivariant homotopy equivalence in analogy with the groups $I^{ht(R),DIFF}$ described in (2.3). The existence of such variants of the Browder-Quinn groups has been well known to workers in the area for a long time (in particular, Browder described some generalizations along these lines in his lectures at the 1976 A.M.S. Summer Symposium at Stanford). In order to keep the discussion brief we shall only consider manifolds whose closed substrata are 1-connected. Elements of the relevant groups $L_n^{BQ,h(R)}(G;X)$ are transverse linear isovariant normal maps $M \to X$ (with Browder-Quinn bundle data) such that the degrees of the associated stratum maps are units in R, and maps that are R-local equivariant homotopy equivalences represent zero. It is also possible to define adjusted groups $L_n^{BQ,h(R)}(G;X;\Sigma\subseteq\Sigma')$, and if $\Sigma' = \Sigma\cup G\{\alpha\}$ then this group is isomorphic to $L^h_{\dim\alpha}(R[W_\alpha], w_\alpha)$ if $D^2 \times X$ satisfies the Gap Hypothesis and a quotient of the latter group if we only know that $D^1 \times X$ satisfies the Gap Hypothesis. All of this can be done if all closed substrata have dimension ≥ 5; it is not necessary to assume the Gap Hypothesis (however, we are assuming the Codimension ≥ 2 Gap Hypothesis throughout this section). If X does satisfy the Gap Hypothesis, then the methods of 3.13 and 3.15 yield homomorphisms from these groups to the groups of (2.3),

$$\rho^{h(R)}_{X,\Sigma\subseteq\Sigma'} : L^{BQ,h(R)}_n(G;X;\Sigma\subseteq\Sigma') \to I^{ht(R),DIFF}(G;\lambda_X;\Sigma\subseteq\Sigma'),$$

and these homomorphisms are isomorphisms if all closed substrata are at least 6-dimensional.

There are also other variants of Browder-Quinn groups for equivariant surgery with coefficients. In **[DHM]** Davis, Hsiang, and Morgan deal with a setting in which the fixed coefficient ring $R \subset \mathbf{Q}$ is replaced by a judiciously chosen family of subrings R_α indexed by the closed substrata. It is also possible to define Browder-Quinn type groups that are equivariant analogs of the homology surgery obstruction groups $\Gamma^c_n(\varphi : \mathbb{Z}[\pi] \to A)$ of Cappell and Shaneson [**CS**]; we shall consider such Browder-Quinn Γ-groups further in Section IV.6.

Dynamic geometric references and indexing data.

There are some situations in equivariant surgery where one does not have associated Browder-Quinn groups; since phenomena of this sort will become unavoidable in Chapter V, we shall indicate how examples arise in some of the theories discussed above. It is often desirable to have groups of type $I^{ht,CAT}$ or $\mathcal{L}^h$ for maps $f : M \to N$ such that the indexing data or references for M and N are not isomorphic. For example, it might be appropriate to study cases where $\check{f} : \pi(M) \to \pi(N)$ is bijective but the slice data s_M and s_N involve linearly inequivalent representations; particularly important examples of this type are discussed in [**PR**, Chapter IV]. In such cases the proper notion of indexing data or geometric reference will depend upon f and will include the appropriate information for both M and N, the map $\check{f}$, and some further compatibility conditions. One can think of such information as **dynamic** reference data that specializes to the previous **static** reference data if f is taken to be an identity map. The dynamic analog of the indexing data λ_X is called the G-**Poset pair** $\lambda(f)$ in **[DR1]** (the footnote following this paragraph discusses the terminology), and the dynamic version of the geometric reference is given by a triple (R_M, R_N, Φ) where Φ includes an isomorphism ${}_1\check{f}$ relating the objects ${}_1\pi(M)$ and ${}_1\pi(N)$ mentioned in Subsection E and a homotopy compatibility relation φ between the fiber transports t_M and t_N (recall that the latter carry the dimension, slice, and orientation data); more precise information in the latter case is given in Definitions I.2.3 and I.3.1 of **[LM]**. In each case it is also possible to define dynamic reference information λ or (R_1, R_2, Φ) directly without specifying a map $f : M \to N$; in fact, this level of generality is needed for many applications, including those of [**PR**, Chapter IV]. Nonstatic indexing data can also be studied for other theories besides $I^{ht,CAT}$ and $\mathcal{L}^h$; two particularly important cases are the theories $I^{ht(R),CAT}$ for equivariant surgery with coefficients and the theories I^h, $I^{h(R)}$ for equivariant surgery up to pseudoequivalence discussed in (2.5). For all of these theories the constructions and results involving static reference information extend to situations with dynamic reference information. The generalization for $I^{ht(R),CAT}$ is entirely straightforward, but there are technical difficulties to carring out the generalization to pseudoequivalence theories; the problems in the latter case are analyzed in Chapter V. In contrast, the transverse linearity condition implies that Browder-Quinn theories are definable only with static reference information.

(*Footnote on terminology:* The *capitalized* G-Posets of [**DR1**] are not the same as the *uncapitalized* G-posets of [**DR1**] and this book; the capitalized G-Poset coincides with this book's notion of indexing data.)

3G. *Restrictions to subgroups*

Up to this point we have dealt with transformations of equivariant surgery theories where the group G is fixed. However, for some purposes it is useful to consider transformations of equivariant surgery theories for different groups. The simplest examples are maps induced by restricting actions of a group G to a subgroup H (compare [**DP2**]). Although these restriction maps have interesting formal properties and important applications (see [**Pe3**]), we shall bypass this aspect of the subject because it is not needed subsequently in this book. The basic references for restriction homomorphisms are [**DP2**] and Section II.4 of [**LM**].

4. Appendix: Stratifications of smooth G-manifolds

Many familiar smooth manifolds have natural descriptions as nonsingular real algebraic varieties. Furthermore, each irreducible nonsingular algebraic variety over the real or complex numbers is a smooth manifold (compare [**Mum**, Section 1B, pp. 9–12]). It follows that smooth manifolds can be viewed as generalizations of nonsingular varieties (in fact, results of J. Nash [**Na**] and A. Tognoli [**To**] show that every compact smooth manifold comes from a real algebraic variety, but in general there may be many nonsingular real algebraic varieties that have the same underlying smooth manifold). There are many situations in topology, geometry, and analysis where one needs a corresponding generalization of *singular* varieties. In particular, if a compact Lie group G acts smoothly on a manifold M, then topologists have often found it useful to view the orbit space M/G as some sort of smooth manifold with (smooth) singularities. Between the late nineteen fifties and early seventies H. Whitney [**Wh1-2**], R. Thom [**Tho**], and J. Mather [**Mth**] developed a notion of *smooth stratification* that is general enough to include smooth manifolds, real algebraic varieties, and many other objects including orbit spaces of smooth actions of compact Lie groups. In fact, such orbit spaces have canonical smooth stratifications, and this result underlies the Browder-Quinn surgery theory for G-manifolds. There are also many other places in the literature where this stratification appears; for example, the triangulation theorem for orbit spaces in [**V3**, p. 128] follows from the existence of stratifications for orbit spaces and the triangulation theorem for smooth stratified sets in [**V3**]. However, there is very little information in the literature on the proof of the stratification theorem for orbit spaces (the main ideas are mentioned briefly in [**G+**, p. 21], but generally the result is merely stated as something that can be proved). The most detailed reference for the proof seems to be Lellmann's *Diplom* thesis [**Le**], and this is basically unavailable except from some topologists' private files. Since we deal extensively with the Browder-Quinn surgery theory and the latter depends upon the stratification theorem, we shall explain in this section how a proof of the stratification theorem can be extracted from standard papers and books on stratified objects. We shall mainly follow Lellmann's approach in [**Le**], but we shall use subsequently published works when these are helpful; in particular, at points where Lellmann refers to an unpublished manuscript by Mather [**Mth0**] we shall substitute references to books by Gibson-Du Plessis-Wirthmüller-Looijenga [**G+**] and Verona [**V3**].

Various types of prestratifications

There are actually three different definitions of stratifications, each of which us useful in its own way. The first approach to defining stratifications was due to Whitney [**Wh1-2**]. His definition is the most concrete, but it requires a fair amount of preliminary discussion.

DEFINITION: Let X and Y be smooth submanifolds of the unbounded smooth n-manifold manifold M^n, let q denote the codimension of X in M, and let $y \in Y$. We shall say that the pair (X, Y) *satisfies* **Condition A** *at the point* y if for each sequence $\{x_k\}$ in X such that

(i) $\lim_k x_k = y$,
(ii) the sequence of tangent planes $\{T(x_k; X)\}$ has a limit in the Grassmann bundle $G_{n,q}(M)$ of $(n-q)$-planes in the tangent space TM,

we have

$$T_\infty := \lim_k T(x_k; X) \supset T(y; Y).$$

If u and v are distinct points in $\mathbf{R}^n$ let $\overrightarrow{(u,v)}$ denote the shifted secant line that passes through the origin and contains $u - v$ (and also $v - u$). It is immediate that $\overrightarrow{(\ldots)}$ defines a smooth map

$$\mathbf{R}^n \times \mathbf{R}^n - \text{Diagonal} \longrightarrow \mathbf{RP}^{n-1}.$$

Following standard terminology we shall say that $\varphi : V \to M$ is a *smooth chart at* Y if φ is a diffeomorphism from an open subset $V \subset \mathbf{R}^m$ to an open neighborhood of y in M.

DEFINITION: The pair (X, Y) is said to *satisfy* **Condition B** *at the point* y if there is a smooth chart $\varphi : V \to M$ at y such that for every pair of sequences $\{x_k\}$, $\{y_k\}$ in $\varphi^{-1}(X)$ satisfying

(i) $x_k \neq y_k$ for all k,
(ii) $\lim_k x_k = \lim_k y_k = \varphi^{-1}(y)$,
(iii) $\lim_k \overrightarrow{(x_k, y_k)} := L_\infty$ exists in $\mathbf{RP}^{n-1}$,
(iv) $\lim_k \varphi_*^{-1}(T(x_k; X)) := T_\infty$ exists in the Grassmann manifold $G_{n,c}(\mathbf{R}^n)$,

one has $L_\infty \subset T_\infty$.

This definition turns out to be independent of the choice of smooth chart (*e.g.*, see the paragraph following the statement of Condition B on page 203 of [**Mth**]).

DEFINITION: Let M be as above, and let S be a subspace of M. A **prestratification** Σ of S is a covering of S by pairwise disjoint smooth submanifolds of M, where each submanifold is contained in Σ. The elements of Σ are called the *strata* of the prestratification. We shall say Σ is *locally finite* if each point of M has a neighborhood that only intersects finitely many strata nontrivially and that Σ satisfies the **Frontier Condition** if the following holds:

(‡) *If X and Y are strata such that $\overline{X} \cap Y \neq \emptyset$, then $\overline{X} \supset Y$.*

We shall say that Σ is a **Whitney prestratification** if Σ is locally finite, satisfies the Frontier Condition, and also satisfies Condition B for all $X, Y \in \Sigma$ and all $y \in Y$.

If M is a smooth G-manifold and every stratum of a Whitney prestratification Σ is a G-invariant submanifold we shall call Σ a *G-Whitney prestratification*.

In [**Tho**] Thom defined a more abstract concept of stratification that includes Whitney prestratifications, and in [**Mth**] (see also [**Mth0**]) Mather reformulated Thom's definition in terms that are more convenient for many purposes. Both of these approaches turn out to be useful in studying the stratification properties of smooth G-manifolds. For technical reasons we shall consider Mather's definition first.

DEFINITION: Let A be a locally compact, separable, and metrizable topological space. An **abstract prestratification** on A (in the sense of Mather) is a triple $(A, \Sigma, \mathbf{T})$ such that the following hold:

(1) Σ is a locally finite decomposition of A into pairwise disjoint, locally closed subsets (the *strata*) such that each stratum $X \in \Sigma$ carries the structure of a smooth manifold and the previous Frontier Condition (‡) holds.
(2) $\mathbf{T}$ is a family $\{T_X, \pi_X, \rho_X\}$ indexed by Σ such that T_X is an open neighborhood of X in A (the associated *tubular neighborhood* of X), $\pi_X : T_X \to X$ is a continuous retraction (the associated *local retraction of T_X*), and $\rho : T_X \to [0, \infty)$ is a continuous map (the associated *tubular function of T_X*) whose zero set is precisely the stratum X.
(3) If $T_{X,Y} := T_X \cap Y$ and the restrictions of π_X and ρ_X to $T_{X,Y}$ are $\pi_{X,Y}$ and $\rho_{X,Y}$ respectively, then the associated map

$$(\pi_{X,Y}, \rho_{X,Y}) : T_{X,Y} \to X \times (0, \infty)$$

is a proper smooth submersion.
(4) If $X, Y, Z \in \Sigma$ and $a \in T_{Y,Z} \cap T_{X,Z} \cap (\pi_{Y,Z})^{-1}(T_{X,Y})$ then there are compatibility relations

$$\pi_{X,Y}\pi_{Y,Z}(a) = \pi_{X,Z}(a),$$
$$\rho_{X,Y}\rho_{Y,Z}(a) = \rho_{X,Z}(a).$$

If a compact Lie group G acts smoothly on A such that the strata are smooth G-manifolds and all the relevant subsets and mappings are invariant and equivariant respectively, then we shall say that $(A, \Sigma, \mathbf{T})$ is an *abstract G-prestratification.*

DEFINITION: Two abstract G-prestratifications $(A, \Sigma, \mathbf{T})$ and $(A', \Sigma', \mathbf{T}')$ are said to be *germ equivalent* if $A = A'$, $\Sigma = \Sigma'$, the smooth structures on the strata agree (equivariantly) and for each $X \in \Sigma$ there exists a G-invariant neighborhood $T_X^* \subseteq T_X \cap T'_X$ such that $\pi_X | T_X^* = \pi'_X | T_X^*$ and $\rho_X | T_X^* = \rho'_X | T_X^*$.

REMARK: Every abstract G-prestratification is germ equivalent to one with the following additional properties for all $X, Y \in \Sigma$:
(i) $T_X \cap T_Y \neq \emptyset \Rightarrow X \subset \overline{Y}$ or $Y \subset \overline{X}$.
(ii) $T_X \cap Y \neq \emptyset \Rightarrow X \subset \overline{Y}$.
We shall assume these conditions henceforth on all abstract G-prestratifications that we consider.

The basic relationship between Whitney G-prestratifications and abstract G-prestratifications is contained in the following result:

THEOREM 4.1. *If $A \subseteq M^n$ has a Whitney G-prestratification Σ, then there exists a family of data $\mathbf{T} = \{T_X, \pi_X, \rho_X\}_{X \in \Sigma}$ such that $(A, \Sigma, \mathbf{T})$ is an abstract G-prestratification.*

COMMENTS ON THE PROOF: The nonequivariant version of this result is due to Mather and can be found in [**Mth**, Sections 5–6, pp. 212–216] and [**G+**, Section II.2] (also see [**Mth0**] beginning with page 33). The same arguments work in the equivariant category with no significant changes (this is discussed explicitly in [**Le**, Sätze II.1.3–4, pp. 23–26, and Beispiel, p. 27]).■

Mather's notion of an abstract prestratification is in fact equivalent to Thom's notion of *ensemble stratifié* in [**Tho**]; a proof of this is described in a paper of A. Verona [**V1**, Footnote 2, p. 1116]. For our purposes it suffices to know that Mather's definition implies some of the structural properties that appear in Thom's definition. The two principal results are straightforward equivariant analogs of [**V1**, Prop. 2.6 and Thm. 2.7, pp. 1113–1117]; these correspond to Satz II.2.3 and Satz II.2.4 in [**Le**, pp. 30–31].

THEOREM 4.2. *Let $(A, \Sigma, \mathbf{T})$ be an abstract prestratified G-set of dimension n. Let $A' \in \Sigma$ be n-dimensional such that A' is dense in A and $A - A' \neq \emptyset$. Then there exists an unbounded G-manifold W of dimension $n-1$ and a proper, continuous, equivariant map $F : W \times [0, r] \to A$ with the following properties:*
(i) $F | W \times [0, r]$ is a smooth embedding into A'.
(ii) $F(W \times \{0\}) = A - A'$.
(iii) $F(W \times [0, r])$ is a neighborhood of $A - A'$ in A.
(iv) If $w \in W$ satisfies $F(w, 0) = x \in X \in \Sigma$, then there exists a stratum $Y \subset \overline{X}$ with $F(w, t) \in T_Y$ and $\pi_Y(F(w, t)) = \pi_Y(x)$ for all $t \in [0, r]$.■

REMARK: The proof of [**V1**, Prop. 2.6] relies heavily on the existence of a *controlled vector field* on a stratified set. There are no difficulties in constructing such vector fields equivariantly if the stratification is G-invariant; as noted in [**Le**, Satz II.2.1, p. 29], an equivariantly controlled vector field can be obtained by averaging an arbitrary such vector field over G using the latter's Haar measure.

THEOREM 4.3. [**Le**, Satz II.2.4, p. 31] *Let $(A, \Sigma, \mathbf{T})$ be a prestratified G-set with finite dimension n. For each stratum $X \in \Sigma$ there is a smooth G-manifold W_X (possibly with boundary) and a proper, continuous, equivariant mapping $F_X : W_X \to A$ with the following properties:*

(i) $F_X|(W_X - \partial W_X) : W_X - \partial W_X \to X$ is an equivariant diffeomorphism.

(ii) $F_X(W_X) = \overline{X}$ and $F_X(\partial W_X) = \overline{X} - X$.

(iii) There is an equivariant collar $c : \partial W_X \times [0,1] \to W_X$ such that for all $v \in \partial W_X$ there is a stratum Y with $F_X c(v,t) \in T_Y$ and $\pi_Y(F_X c(v,t)) = \pi_Y(F_X(v))$ for all $t \in [0,1]$. ■

REMARK: If $(A, \Sigma, \mathbf{T})$ is obtained from a Whitney G-prestratification via Theorem 4.1, then the mapping F_X can be constructed to be smooth of class C^1 (see [**V2**, Prop.1.3 and Thm.1.4]).

Proof of the stratification theorem

Theorem 4.1 is the central step in proving that orbit spaces of smooth G-manifolds have stratifications. The other steps in the proof can be described as follows:

INITIAL STEP. (compare [**G+**, p. 21]) *Use the differentiable slice theorem to show that a smooth G-manifold M^n has a Whitney G-prestratification whose strata are the sets $M_{(H,V)}$ described in Section* II.1.

By Theorem 4.1 it will follow that M^n has an abstract G-prestratification in the sense of Mather.

FINAL STEP. *Push the G-stratification on M down to a prestratification of M/G whose strata are the sets $M^*_{(H,V)} = M_{(H,V)}/G$.*

We shall deal with the Initial and Final Steps in order.

THEOREM 4.4. [**Le**, Satz II.1.1, pp. 21–22] *Let G be a compact Lie group, let M be a smooth G-manifold, and let Σ be the sets $M_{(H,V)}$ of all points x whose isotropy subgroups G_x are conjugate to H and whose slice representations are equivalent to V. Then Σ determines a Whitney G-prestratification of M (viewing M as a subspace of itself).*

PROOF: The sets $M_{(H,V)}$ are pairwise disjoint submanifolds of M that are G-invariant and cover M. Furthermore the family of all such sets is locally finite by a refinement of the usual local finiteness theorems for orbit types (see [**Bre**, Chapter IV]).

Suppose that $\overline{M}_{(H,V)} \cap M_{(H',V')} \neq \varnothing$. Then there is a point $y \in M_{(H',V')}$ so that $y \in \overline{M}_{(H,V)}$, and it follows that the slice type (H, V) appears in $G \times_{H'} V'$. Since a tubular neighborhood of each orbit in $M_{(H',V')}$ is G-diffeomorphic to $G \times_{H'} V'$, every neighborhood of a point in $M_{(H',V')}$ also contains points with slice type (H,V), and from this we have $M_{(H',V')} \subset \overline{M}_{(H,V)}$.

Let $y \in M_{(H,V)}$; without loss of generality we may assume $G_y = H$ and $V_y \approx V$. For each (H',V') such that $M_{(H,V)} \subset \overline{M}_{(H',V')}$ we shall show that the pair $(M_{(H',V')}, M_{(H,V)})$ satisfies the Whitney Condition B at y. By the differentiable slice theorem it suffices to establish Condition B at the point CLASS$[1,0] \in G \times_H V$ with respect to the pair $(G \times_H V', G \times_H V)$. Since the projection $G \to G/H$ is a principal H-bundle projection, there is a submanifold $F \subset G$ containing the identity and projecting diffeomorphically to a neighborhood of the image of 1 in G/H. In fact, one can choose F so that a neighborhood of CLASS$[1,0] \in G \times_H V$ is diffeomorphic to $F \times V$ and $M_{(H,V)}$ corresponds to $F \times V_{(H,V)}$. Furthermore, the points with slice type (H',V') correspond to $F \times V_{(H',V')}$, and since the slice types at $[g,v]$ and $[1,v]$ agree for all $g \in F$ it follows that we need only establish Condition B for the pair $(V_{(H',V')}, V_{(H,V)})$ at $0 \in V$.

But if (H_Z, σ_Z) denotes the isotropy subgroup and slice type at a point z, then for $x' \in V_{(H',V')}$ and $y' \in V_{(H,V)}$ we have

$$(H_{tx'+(1-t)y'}, \sigma_{tx'+(1-t)y'}) = (H_{x'}, \sigma_{x'}) = (H', V')$$

for all $t \neq 0$ because y' is a fixed point of the linear action on V, so that $H_{tx'+(1-t)y'} = H_{x'}$ for $t \neq 0$; the slice types agree because there is only one slice type on each connected component of $V_{(H')}$.

Therefore if $\{x_k\}$ and $\{y_k\}$ are sequences as in the definition of Condition B, then for sufficiently large values of k one in fact has

$$\overrightarrow{(x_k, y_k)} \in T_{x_k}(V_{(H',V')})$$

and in particular

$$\lim_k \overrightarrow{(x_k, y_k)} \subset \lim_k T_{x_k}(V_{(H',V')}). \blacksquare$$

If M is a smooth G-manifold and Σ is the set of submanifolds $\{M_{(H,V)}\}$, then the preceding results yield G-equivariant control data $\mathbf{T} = \{T_X, \pi_X, \rho_X\}_{X \in \Sigma}$. Let $\Sigma' = \{X/G | X \in \Sigma\}$, and let $p : M \to M/G$ be the orbit space projection. For each $X' := X/G \in \Sigma'$ we define a triple $(T_{X'}, \pi_{X'}, \rho_{X'})$ as follows:

(1) $T_{X'} = p(T_X) \subset M/G$.
(2) $\pi_{X'} : T_{X'} \to X'$ satisfies $\pi_{X'}(v) = p\pi_X(v^*)$, where $p(v^*) = v$.
(3) $\rho_{X'} : T_{X'} \to [0, \infty)$ satisfies $\rho_{X'}(v) = \rho_X(v^*)$, where $p(v^*) = v$.

The maps $\pi_{X'}$ and $\rho_{X'}$ are well-defined continuous mappings since π_X and ρ_X are equivariant. Set $\mathbf{T}' = \{T_{X'}, \pi_{X'}, \rho_{X'}\}_{X' \in \Sigma'}$.

THEOREM 4.5. [**Le**, Satz III.1.1, pp. 37–38] *The triple $(M/G, \Sigma', \mathbf{T}')$ defines an abstract prestratified set. Furthermore, the orbit space projection p commutes with the appropriate local retractions and tubular functions, and the restriction of p to each stratum is a proper submersion onto the image stratum.*

This is the stratification theorem for smooth orbit spaces.

PROOF: It is immediate that M/G and Σ' satisfy the basic conditions in the definition of an abstract prestratification. All of the required properties of $\mathbf{T}'$ except the submersion property follow immediately from the definitions.

To prove the remaining property, let $X' = M_{(H,V)}/G$ and $Y' = M_{(H',V')}/G$ where H' is contained in a conjugate of H and there is an open equivariant embedding of $G \times_{H'} V'$ into $G \times_H V$; we need to show that $(\pi_{X',Y'}, \rho_{X',Y'}) : T_{X'} \cap Y' \to X' \times (0, \infty)$ is a submersion. Take an arbitrary point $y \in T_X \cap Y$, and set $y' = p(y)$ and $x = \pi_X(y)$. The differentiable slice theorem implies we can change coordinates equivariantly such that near the orbit $G \cdot y$ the map $(\pi_{X,Y}, \rho_{X,Y})$ takes the form

$$G/G_y \times W_y \xrightarrow{h} G/G_x \times (W_x \oplus \mathbf{R}),$$

where we can assume $y = (G_y, 0)$ and $x = (G_x, 0)$. Since $(\pi_{X,Y}, \rho_{X,Y})$ is an equivariant submersion and $G/G_y \to G/G_x$ is a submersion, it follows that the composite

$$W_y \cong \{G_y\} \times W_y \subset G/G_y \times W_y \xrightarrow{h} G/G_x \times (W_x \oplus \mathbf{R}) \xrightarrow{\text{proj}} W_x \oplus \mathbf{R}$$

is also a submersion at $0 \in W_y$. However, this composite represents the mapping (π'_X, ρ'_X) on local charts for Y' and $X' \times \mathbf{R}$, and therefore it follows that $(\pi_{X'}, \rho_{X'})$ is a submersion as required.■

COROLLARY 4.6. *Let M be a compact G-manifold and let (H, V) be a slice type for M. Then there exists a (possibly bounded) G-manifold $W_{(H,V)}$ with a single slice type $(H$, trivial$)$ and a mapping pair (f, f') consisting of an equivariant proper C^1-map $f : W_{(H,V)} \to M$ and a proper continuous mapping $f' : W_{(H,V)}/G \to M/G$ such that the following hold:*

(i) The following diagram commutes:

$$\begin{array}{ccc} W_{(H,V)} & \xrightarrow{f} & M \\ {\scriptstyle p}\downarrow & & \downarrow{\scriptstyle p} \\ W_{(H,V)}/G & \xrightarrow{f'} & M/G \end{array}$$

(ii) f maps the interior Int *$W_{(H,V)}$ diffeomorphically into $M_{(H,V)}$ and f' maps the interior* Int *$W_{H,V}/G$ diffeomorphically into $M_{(H,V)}/G$.*

(iii) If Fr *$M_{(H,V)} = \overline{M_{(H,V)}} - M_{(H,V)}$ and similarly for $M_{(H,V)}/G$, then $f(\partial W_{(H,V)})$ is equal to* Fr *$M_{(H,V)}$ and $f'(\partial W_{(H,V)}/G)$ is equal to* Fr *$M_{(H,V)}/G$.*

PROOF: This follows immediately from Theorems 4.5 and 4.3.■

Orbifolds

If G is a finite group acting locally linearly on a manifold M^n, then the slice theorem implies that every point of M/G has a neighborhood homeomorphic to an orbit space V/H where V is an n-dimensional representation of a subgroup $H \subset G$. Furthermore, these identifications of neighborhoods with quotients V/H have good differentiability properties if G acts smoothly on M. More generally, one can consider abstract spaces X such that every point has a neighborhood of the form V/H where V is a representation of some finite group H. Such spaces were originally studied by I. Satake [**Sat**] during the nineteen fifties, and in recent years they have received considerable attention, largely in connection with W. Thurston's geometrization program in 3-dimensional topology (see [**Thu**, Chapter 11] or [**DavMrg**, Section 1]). Following [**Thu**], such spaces are now generally called **orbifolds**. A survey of work on this topic is beyond the scope of this book. However, we should note that the methods of this section also apply to orbifolds that have suitably defined structures of differentiable varieties with singularities.

References for Chapter II

[Bre] G. Bredon, *Exotic actions on spheres*, in "Proceedings of the Conference on Transformation Groups (Tulane, 1967)," Springer, Berlin-Heidelberg-New York, 1968, pp. 47–76.

[Bre2] ________, "Introduction to Compact Transformation Groups," Pure and Applied Mathematics Vol. 46, Academic Press, New York, 1972.

[BQ] W. Browder and F. Quinn, *A surgery theory for G-manifolds and stratified sets*, in "Manifolds–Tokyo, 1973," (Conf. Proc. Univ. of Tokyo, 1973), University of Tokyo Press, Tokyo, 1975, pp. 27–36.

[CS] S. Cappell and J. Shaneson, *The codimension two placement problem and homology equivalent manifolds*, Ann. of Math. **99** (1974), 277–348.

[CM] G. Carlsson and R. J. Milgram, *Some exact sequences in the theory of Hermitian forms*, J. Pure Applied Algebra **18** (1980), 233–252.

[CL] F. Connolly and W. Lück, *The involution on the equivariant Whitehead group*, K-Theory **3** (1989), 123–140.

[Dav] M. Davis, *Smooth G-manifolds as collections of fiber bundles*, Pac. J. Math. **77** (1978), 315–363.

[DH] M. Davis and W. C. Hsiang, *Concordance classes of regular $U(n)$ and $Sp(n)$ actions on homotopy spheres*, Ann. of Math. **105** (1977), 325–341.

[DHM] M. Davis, W. C. Hsiang, and J. Morgan, *Concordance of regular $O(n)$-actions on homotopy spheres*, Acta Math. **144** (1980), 153–221.

[DavMrg] M. Davis and J. Morgan, *Finite group actions on homotopy 3-spheres*, in "The Smith Conjecture (Symposium, New York, 1979)," Pure and Applied Mathematics Vol. 112, Academic Press, Orlando, Florida, 1984.

[tD] T. tom Dieck, "Transformation Groups," de Gruyter Studies in Mathematics Vol. 8, W. de Gruyter, Berlin and New York, 1987.

[Do1] K. H. Dovermann, "Addition of equivariant surgery obstructions," Ph.D. Thesis, Rutgers University, 1978 *(Available from University Microfilms, Ann Arbor, Mich.: Order Number DEL79-10380.)*—Summarized in Dissertation Abstracts International **39** (1978/1979), B5406.

[Do2] ____________, *Addition of equivariant surgery obstructions*, in "Algebraic Topology, Waterloo 1978 (Conference Proceedings)," Lecture Notes in Mathematics Vol. 741, Springer, Berlin-Heidelberg-New York, 1979, pp. 244-271.

[Do3] ____________, *Almost isovariant normal maps*, Amer. J. Math. **111** (1989), 851–904.

[DMS] K. H. Dovermann, M. Masuda, and R. Schultz, *Conjugation involutions on homotopy complex projective spaces*, Japanese J. Math. **12** (1986), 1–34.

[DP1] K. H. Dovermann and T. Petrie, *G–Surgery II*, Memoirs Amer. Math. Soc. **37** (1982), No. 260.

[DP2] ————————, *An induction theorem for equivariant surgery (G-Surgery III)*, Amer. J. Math. **105** (1983), 1369–1403.

[DR1] K. H. Dovermann and M. Rothenberg, *Equivariant Surgery and Classification of Finite Group Actions on Manifolds*, Memoirs Amer. Math. Soc. **71** (1988), No. 379.

[DR2] ————————, *Poincaré duality and generalized Whitehead torsion*, preprint, University of Hawaii and University of Chicago, 1986.

[DR3] ————————, *An algebraic approach to the generalized Whitehead group*, in "Transformation Groups (Proceedings, Poznań, 1985)," Lecture Notes in Mathematics Vol. 1217, Springer, Berlin-Heidelberg-New York, 1986, pp. 92–114.

[DR4] ————————, *The generalized Whitehead torsion of a G-fiber homotopy equivalence*, in "Transformation Groups (Proceedings, Osaka, 1987)," Lecture Notes in Mathematics Vol. 1375, Springer, Berlin-Heidelberg-New York-London-Paris-Tokyo, 1989, pp. 60–88.

[DS] K.H. Dovermann and R. Schultz, *Surgery on involutions with middle dimensional fixed point set*, Pac. J. Math. **130** (1988), 275–297.

[FH] F. T. Farrell and W. C. Hsiang, *Rational L-groups of Bieberbach groups*, Comment. Math. Helv. **52** (1977), 89–109.

[Fi1] R. Fintushel, *Circle actions on simply connected 4-manifolds*, Trans. Amer. Math. Soc. **230** (1977), 147-171.

[Fi2] ————————, *Classification of circle actions on 4-manifolds*, Trans. Amer. Math. Soc. **242** (1978), 377-390.

[G+] C. G. Gibson, K. Wirthmüller, A. A. du Plessis, and E. J. N. Looijenga, "Topological Stability of Smooth Mappings," Lecture Notes in Mathematics Vol. 552, Springer, Berlin-Heidelberg-New York, 1976.

[HM] M. Hirsch and B. Mazur, "Smoothings of Piecewise Linear Manifolds," Annals of Mathematics Studies No. 80, Princeton University Press, Princeton, 1974.

[I1] S. Illman, *Whitehead torsion and group actions*, Ann. Acad. Sci. Fenn. Ser. A I **588** (1974).

[I2] ————————, *Smooth triangulations of G-manifolds for G a finite group*, Math. Ann. **233** (1978), 199–220.

[I3] ————————, *Equivariant Whitehead torsion and actions of compact Lie groups*, in "*Group Actions on Manifolds (Conference Proceedings, University of Colorado, 1983)*," Contemp. Math. Vol. 36, American Mathematical Society, 1985, pp. 91–106.

[Ka] G. Katz, *Normal combinatorics of G-actions on manifolds*, in "Transformation Groups (Proceedings, Poznań, 1985)," Lecture Notes in Mathematics Vol. 1217, Springer, Berlin-Heidelberg-New York, 1986, pp. 167–182.

[Kr] M. Kreck, *Surgery and duality—an extension of results of Browder, Novikov, and Wall about surgery on compact manifolds*, preprint, Universität Mainz, 1985.

[**LaR**] R. Lashof and M. Rothenberg, *G-smoothing theory*, Proc. A. M. S. Sympos. Pure Math. **32 Pt.1**, 211–266.

[**Le**] W. Lellmann, *Orbiträume von G-Mannigfaltigkeiten und stratifizierte Mengen*, Diplomarbeit, Universität Bonn, 1975.

[**Lü**] W. Lück, *The equivariant degree*, in "Algebraic Topology and Transformation Groups (Proceedings, Göttingen, 1987)," Lecture Notes in Mathematics Vol. 1361, Springer, Berlin-Heidelberg-New York, 1988, pp. 123–166.

[**LM**] W. Lück and I. Madsen, *Equivariant L-theory I*, Aarhus Univ. Preprint Series (1987/1988), No. 8; [same title] *II*, Aarhus Univ. Preprint Series (1987/1988), No. 16 (to appear in Math. Zeitschrift).

[**McL**] S. MacLane, "Homology," Grundlehren der math. Wissenschaften Bd. 114, Springer, Berlin-Göttingen-Heidelberg, 1963.

[**MR**] I. Madsen and M. Rothenberg, *On the classification of G-spheres*, in "Proceedings of the Northwestern Homotopy Conference (Conference Proceedings, Northwestern University, 1982)," Contemporary Math. 19, American Mathematical Society, 1983, pp. 193–226; *Detailed version:* [same title] *I: Equivariant transversality*, Acta Math. **160** (1988), 65–104,, pp.; [same title] *II: PL automorphism groups*, Math. Scand. **64** (1989), 161–218,, pp.; [same title] *III: Topological automorphism groups*, Aarhus Univ. Preprint Series (1985/1986), No. 14.

[**MR2**] ———————, *On the homotopy groups of equivariant automorphism groups*, Invent. Math. **94** (1988), 623–637.

[**Mth**] J. N. Mather, *Stratifications and mappings*, in "Dynamical Systems (Proc. Sympos., Univ. Bahia, Salvador, Brazil, 1971)," Academic Press, New York, 1973, pp. 195–232.

[**Mth0**] ———————, *Notes on topological stability*, lecture notes, Harvard University, 1970.

[**MMT**] J. P. May, J. McClure, and G. Triantafillou, *Equivariant localization*, Bull. London Math. Soc. **14** (1982), 223–230.

[**Mum**] D. Mumford, "Algebraic Geometry I: Complex Projective Varieties," Grundlehren der math. Wiss. Bd. 221, Springer, Berlin-Heidelberg-New York, 1976.

[**Mun**] J. R. Munkres, "Elementary Differential Topology (Revised Edition)," Annals of Mathematics Studies Vol. 54, Princeton University Press, Princeton, N. J., 1966.

[**Na**] J. Nash, *Real algebraic manifolds*, Ann. of Math. **56** (1952), 405–421.

[**Ni**] A. Nicas, *Induction Theorems for Groups of Homotopy Manifold Structures*, Memoirs Amer. Math. Soc. **39** (1982), No. 267..

[**OR1**] P. Orlik and F. Raymond, *Actions of the torus on 4-manifolds I*, Trans. Amer. Math. Soc. **152** (1971), 531–559.

[**OR2**] ———————, *Actions of the torus on 4-manifolds II*, Topology **13** (1974), 89–112.

[**Prdn1**] W. Pardon, *Local Surgery and the Exact Sequence of a Localization*, Memoirs Amer. Math. Soc. **12** (1977), No. 196.

[**Prdn2**] ———, *The exact sequence of a localization for Witt groups II*, Pac. J. Math. **102** (1982), 283–305.

[**Prkr**] J. Parker, *Four-dimensional G-manifolds with 3-dimensional orbits*, Pac. J. Math. **125** (1986), 187–204.

[**Pe1**] T. Petrie, *Exotic S^1-actions on* $\mathbf{CP}^3$ *and related topics*, Invent. Math. **17** (1972), 317-327.

[**P2**] ———, *Pseudoequivalences of G-manifolds*, Proc. A. M. S. Sympos. Pure Math **32 Pt. 1** (1978), 169–210.

[**Pe3**] ———, *One fixed point actions on homotopy spheres I–II*, Adv. in Math. **46** (1982), 3–70.

[**PR**] T. Petrie and J. Randall, "Transformation Groups on Manifolds," Dekker Series in Pure and Applied Mathematics Vol. 82, Marcel Dekker, New York, 1984.

[**PR2**] ————, *Spherical isotropy representations*, I. H. E. S. Publ. Math. **62** (1985), 5–40.

[**Q**] F. Quinn, *A geometric formulation of surgery*, in "Topology of Manifolds (Conference Procedings, University of Georgia, 1969)," Markham, Chicago, 1970, pp. 500-511. *(Available from University Microfilms, Ann Arbor, Mich.: Order Number 7023630.)*—Summarized in Dissertation Abstracts International **31** (1970/1971), p. B3564.

[**Q2**] ———, *Homotopically stratified sets*, J. Amer. Math. Soc. **1** (1987), 441–499.

[**Ran**] A. Ranicki, *Algebraic L-theory assembly*, preprint, University of Edinburgh, 1989.

[**Ray**] F. Raymond, *Classification of the actions of the circle on 3-manifolds*, Trans. Amer. Math. Soc. **131** (1968), 51–78.

[**Ro**] M. Rothenberg, *Torsion invariants and finite transformation groups*, Proc. A. M. S. Sympos. Pure Math. **32 Pt. 1** (1978), 267–311.

[**Sat**] I. Satake, *On a generalization of the notion of a manifold*, Proc. Nat. Acad. Sci. U. S. A. **42** (1956), 359–363.

[**Sh**] J. Shaneson, *Wall's surgery obstruction groups for* $G \times \mathbf{Z}$, Ann. of Math. **90** (1969), 296–334.

[**St**] M. Steinberger, *The equivariant topological s-cobordism theorem*, Invent. Math. **91** (1988), 61–104.

[**TW**] L. Taylor and B. Williams, *Surgery spaces: Formulæ and structure*, in "Algebraic Topology, Waterloo 1978 (Proceedings)," Lecture Notes in Math. Vol. 741, Springer, Berlin-Heidelberg-New York, 1979, pp. 334–354.

[**Tho**] R. Thom, *Ensembles et morphismes stratifiés*, Bull. Amer. Math. Soc. **75** (1969), 240–284.

[Thu] W. Thurston, *The geometry and topology of 3-manifolds*, lecture notes, Princeton University, 1978–1979.

[To] A. Tognoli, *Su una congettura di Nash*, Ann. Sc. Norm. Sup. Pisa **27** (1973), 167–185.

[V1] A. Verona, *Homological properties of abstract prestratifications*, Rev. Roum. Math. Pures et Appl. **17** (1972), 1109–1121.

[V2] ———, *Integration on Whitney prestratifications*, Rev. Roum. Math. Pures et Appl. **17** (1972), 1473–1480.

[V3] ———, "Stratified Mappings–Structure and Triangulability," Lecture Notes in Mathematics Vol. 1102, Springer, Berlin-Heidelberg-New York, 1984.

[Wa] C. T. C. Wall, "Surgery on Compact Manifolds," London Math. Soc. Monographs Vol. 1, Academic Press, London and New York, 1970.

[Wb] S. Weinberger, *The topological classification of stratified spaces*, preprint, University of Chicago, 1989.

[Wh1] H. Whitney, *Elementary structure of real algebraic varieties*, Ann. of Math. **66** (1957), 545-566.

[Wh2] ———, *Local properties of analytic varieties*, in "Differential and Combinatorial Topology (A Symposium in Honor of Marston Morse)," Princteon Mathematical Series Vol. 27, Princeton University Press, Princeton, N. J., 1965, pp. 205–244.

CHAPTER III

PERIODICITY THEOREMS IN EQUIVARIANT SURGERY

C. T. C. Wall's formulation of surgery theory [**WL**] contains both algebraic and geometric descriptions of the surgery obstruction groups $L_n^s(\pi, w)$ associated to a group π and a homomorphism $\pi \to \mathbf{Z}_2$. From the algebraic definition it is immediate that $L_n(\pi, w)$ really depends not on n itself but only on n mod 4. This algebraic periodicity also has a simple geometrical interpretation: namely, the surgery obstructions of f and $f \times \mathbf{id}(\mathbf{CP}^2) : M \times \mathbf{CP}^2 \to X \times \mathbf{CP}^2$ correspond under the algebraic isomorphism from L_n to L_{n+4}. In the simply connected case this is due to D. Sullivan and A. Casson [**Su**] (compare [**Bro3**, Section III.5]), and the general case is due to Wall (see [**WL**], Thm. 9.9). The passage from algebra to geometry depends upon two basic properties of $\mathbf{CP}^2$; namely, it is a closed simply connected manifold and an integral valued invariant called its *signature* (see [**ACH**] or [**MS**]) is equal to 1.

Similar periodicity results are also true for many generalizations of Wall's groups, including the homology surgery obstruction groups of S. Cappell and J. Shaneson [**CS**] and the transverse linear G-isovariant surgery obstruction groups of W. Browder and F. Quinn [**BQ**]. In each case one has a fourfold periodicity induced by taking products with $\mathbf{CP}^2$. In this chapter we shall establish periodicity theorems in other settings for equivariant surgery. Some of these statements are straightforward extensions of the periodicity theorems in ordinary surgery theory. However, equivariant surgery theories have much richer periodicity properties than their nonequivariant counterparts. In the next few paragraphs we shall discus an example that has a curious implication regarding the role of the Gap Hypothesis: For groups of **odd order** it is possible to construct a *periodic stabilization* $I^{a,\infty}$ of an equivariant surgery theory I^a such that

(*i*) the original theory maps isomorphically to the periodic stabilization if the Gap Hypothesis and standard dimension restrictions hold,

(*ii*) the periodic stabilization provides a framework for studying first order approximations to equivariant surgery obstructions even if the Gap Hypothesis does not hold.

In fact, for groups of odd order periodic stabilization provides a method for splitting an equivariant surgery problem into two parts, one of which has the good formal properties that follow from the Gap Hypothesis. We shall discuss some consequences and special cases in Section 5.

Background information

We shall begin by describing some results that W. Browder presented in lectures at the 1976 A. M. S. Summer Institute on Algebraic and Geometric Topology at Stanford

University. Assume that G is a group of odd order, and denote its order by $|G|$. If X is a topological space define a G-space

$$X\!\uparrow\! G = X^{|G|} \ (= |G| - \text{fold cartesian product})$$

with G acting on $X\!\uparrow\! G$ by permuting the coordinates. If X is a smooth G-manifold, then $X\!\uparrow\! G$ is a smooth G-manifold by general considerations involving multiplicative induction construction (compare tom Dieck [tD]). In his lectures Browder made the following two observations:

(B1) Crossing with $\mathbf{CP}^2\!\uparrow\! G$ preserves the Atiyah-Singer G-signature as defined as in [**AS**, Section 6] (recall this is a generalization of the ordinary signature to G-manifolds and takes values in the complex representation ring $R(G)$).

(B2) For every G-manifold M, there is some integer $r_M > 0$ such that product manifold $M \times (\mathbf{CP}^2\!\uparrow\! G)^r$ satisfies the Gap Hypothesis if $r \geq r_M$.

Therefore, given a surgery problem $f : M \to X$ that does not satisfy the Gap Hypothesis, one can obtain a problem that does by considering $f \times \mathbf{id}(\mathbf{CP}^2\!\uparrow\! G)^r$ for r sufficiently large. If this problem has a reasonable equivariant surgery obstruction, then we have obtained a halfway reasonable candidate for an equivariant surgery invariant of the original problem. Strictly speaking, we should say "candidates" rather than "candidate" because it is not immediately obvious that the obstruction for

$$f \times \mathbf{id}((\mathbf{CP}^2\!\uparrow\! G)^r)$$

is somehow independent of r if $r \geq r_M$. The periodicity theorem of this paper will imply that this surgery obstruction is in fact independent of r if $r \geq r_M$.

Here is a very crude formulation of our periodicity result:

PERIODICITY PRINCIPLE. *Let Y be a closed smooth G-manifold (G of odd order) satisfying some regularity hypotheses for equivariant surgery such as the Gap Hypothesis and no strata in troublesome low dimensions. Let P be a suitable periodicity manifold such as $\mathbf{CP}^2\!\uparrow\! G$ (or in some cases $\mathbf{CP}^2$). Then the product map*

$$\mu_P : \begin{bmatrix} \textit{obstruction group} \\ \textit{for surgery problems} \\ \textit{over } Y \end{bmatrix} \to \begin{bmatrix} \textit{obstruction group} \\ \textit{for surgery problems} \\ \textit{over } Y \times P \end{bmatrix}$$

defined by $\mu_P(f) = f \times \mathbf{id}(P)$ is an isomorphism of abelian groups.

<u>Special case:</u> Suppose that G acts freely on Y and orientation-preservingly on P. Then the product map μ_P corresponds to the twisted product map

$$\mu_P : L^s_{\dim Y}(\pi_1(Y/G), w_{Y/G}) \to L^s_{\dim Y + \dim P}(\pi_1(Y/G), w_{Y/G})$$

considered by T. Yoshida [**Yo**]. In this case one needs Yoshida's results on twisted product formulas plus an auxiliary calculation (*i.e.*, Prop. 3.1(*iii*)) to show that μ_P is bijective.

Incidentally, both (B1) and (B2) are false for groups of even order; further information on these cases appears at the end of Section 3 (see Proposition 3.7 in particular).

Organization of this chapter

For most equivariant surgery theories it is almost a triviality to define product maps taking an equivariant surgery problem $(f : M \to X, -)$ to $(f \times \mathbf{id}_Y : M \times Y \to X \times Y, -)$. In Section 1 we verify that this construction has good formal properties such as additivity. Statements of the main theorems of this chapter appear in Chapter 2. These include both abstract and specific forms of the periodicity theorem (Theorems 2.7 and 2.8–9 respectively) and the properties of the manifolds $\mathbf{CP}^2 \uparrow G$ that are relevant to the periodicity theorems. Proofs of the results on $\mathbf{CP}^2 \uparrow G$ are given in Sections 2 and 3. In Section 4 we show that product maps on equivariant surgery groups are compatible with the sorts of exact sequences described in (II.2.0); precise statements of these elementary but important relationships are given in Theorem 4.1. We also interpret the equivariant product maps using stepwise surgery obstructions in ordinary surgery obstruction groups. Specifically, Theorem 4.2 says that equivariant products on stepwise surgery obstructions correspond to the twisted products of ordinary obstructions that were studied by T. Yoshida [**Yo**]. Section 4 concludes by using these observations and the results of Section 3 to prove the main periodicity theorems. In Section 5 we use the periodicity manifolds $\mathbf{CP}^2 \uparrow G$ to study equivariant surgery obstructions outside the Gap Hypothesis range by taking products with $(\mathbf{CP}^2 \uparrow G)^r$ for some large integer r; this is the process we call periodic stabilization. As noted earlier in this chapter, one consequence is a splitting of equivariant surgery obstruction problems for odd order groups into stable and unstable portions.

Recently M. Yan has obtained periodicity theorems similar to ours for the stratified isovariant surgery theory of S. Weinberger (see [**Ya**] and [**Wb**]). Relationships between his work and ours will be discussed at appropriate points in Sections 2 and 3.

In subsequent papers we shall study further topics related to the results of this chapter. The periodicity theorems may be viewed as a first step towards understanding more general product formulas in equivariant surgery, and in the future we hope to consider questions of this sort. Separate results of Conner-Floyd [**CF**] and the first author [**Dov3**] imply that periodic stabilization cannot be extended to even order groups (see Proposition 3.7 and [**Dov3**, Example 4.13, p. 281]); however, one can still draw some positive conclusions. A second topic for further study will be the applications of periodic stabilization to study the role of the Gap Hypothesis in equivariant surgery for actions of odd order groups. Further work of the second author [**Sc3**] shows that periodic stabilization yields an infinite extension of the long-but-finite exact sequence of equivariant surgery due to the first author and M. Rothenberg [**DR**] if the group under consideration has odd order. The meaning of this surgery sequence for semilinear differentiable actions on spheres seems to be a topic that merits further study. In the semifree case it is enlightening to compare the information carried by the surgery exact sequence with the standard Browder-Petrie and Rothenberg-Sondow classifications (*e.g.*, see [**Sc1**]). This comparison should lead to a much better understanding of semilinear actions for groups such as $\mathbf{Z}_p \times \mathbf{Z}_p$ (where p is an odd prime).

Acknowledgment. Our work on the topics in this chapter began in the summer of 1982. As noted above, the initial stimulus for this work came from earlier lectures by Bill Browder at the 1976 A. M. S. Summer Institute on Algebraic and Geometric Topology at Stanford. The contents of these lectures were neither published nor circulated in preprint form. We appreciate the good will he has shown in discussing the contents of his lecture and answering questions based upon the second author's notes from these lectures.

1. Products in equivariant surgery

In this paper we shall only consider actions of finite groups. Furthermore, we shall assume the Codimension ≥ 2 Gap Hypothesis of Chapter I (*i.e.*, if $D_1 \subsetneqq D_2$ are closed substrata then the difference in dimensions is at least 2).

The periodicity theorems of equivariant surgery are basically statements about certain product constructions. Consequently, the proofs of these theorems should begin with some observations about product operations in equivariant surgery theories. Product formulas have been studied extensively in ordinary surgery theory, both geometrically (see **[Mrg]**, **[Sh2]**, and **[Wi]**) and algebraically (compare **[Ra1]**–**[Ra3]**), but the literature of equivariant surgery contains very little information about products. It is essentially a routine exercise to verify the existence of well behaved products in most versions of equivariant surgery theory. However, for the sake of completeness it is necessary to state explicitly the basic properties of such products. In this section we shall restrict attention to the following three types of equivariant surgery theories:

I: The theories $I^{c,CAT}$ of **[DR1]** and **[MR]**, where $c = ht$ (surgery up to equivariant homotopy equivalence), $ht(R)$ (surgery up to R-local equivariant homotopy equivalence), or s (surgery up to simple equivariant homotopy equivalence) and $CAT = DIFF$, PL, or TOP (but no theory is defined for $c = s$ and $CAT = TOP$). The theory $I^{ht,DIFF}$ is discussed in Section I.5, and the remaining theories are discussed in Section II.2.

II: The theories $\mathcal{L}^c$ of **[LM]**, where $c = h = ht$ or s. These theories are discussed in (II.2.2) and Subsections II.3E–F.

III: The theories $L^{BQ,c}$ of **[BQ]**, where $c = h = ht$ or s. These theories are discussed in Section II.1 and Subsection II.3F.

Products in the theories for equivariant surgery up to pseudoequivalence (*i.e.* the theories I^h and $I^{h(R)}$) will be considered in Section V.2; in these cases there are a few technical complications.

The obstruction theories for equivariant surgery groups generally depend upon some simple numerical or representation-theoretic invariants of the target manifolds. For the theories of Type I, the necessary information corresponding to the target manifold X is given by the *indexing data* λ_X described in Section I.2, and for the theories of Type II the information is given by the *geometric reference* R_X, the main features of which are described in Section II.3 (see Part I of [**LM**] for more precise information). The geometric reference also contains the necessary invariants for the theories of Type III; this follows from [**BQ**, Thm. 3.1, p. 31].

NOTATION: We shall use I^a to denote an equivariant surgery theory of one of the types described above. If X is an admissible G-manifold for the theory I^a we shall let Λ_X denote the indexing data (for Type I) or geometric reference (for Types II or III). The I^a equivariant surgery group for X will be denoted by $I^a(G;\Lambda_X)$. An I^a equivariant surgery problem will mean a representative of a class in $I^a(G;\Lambda_X)$.

The definition of products is nonequivariant surgery theory is relatively straightforward. Given a surgery problem

$$(f : M \to X,\ b : \nu_M \to f^*\xi)$$

and a manifold Y, the product with Y is simply the pair

$$(f \times \mathbf{id}(Y),\ b \times \mathbf{id}(\nu_Y)),$$

and it is elementary to verify that this induces a map $\mathbf{P}_Y$ from (bordism classes of) surgery problems into X to (bordism classes of) surgery problems into $X \times Y$. Specifically, in terms of the notation at the end of Section 1, we need to know that crossing with a suitable closed smooth G-manifold Y will induce a map of surgery obstruction sets

$$\mathbf{P}_Y : I^a(G;\Lambda_X) \to I^a(G;\Lambda_X \times \mathrm{data}\,(Y)).$$

where I^a denotes one of the theories described above. In each case it is a routine exercise to verify that the product of an I^a equivariant surgery problem with the identity map of and I^a admissible G-manifold Y will yield a new I^a equivariant surgery problem. The analogous statements hold for admissible bordisms between I^a equivariant surgery problems. Our first result states that product maps are additive if group structures can be defined as in [**Dov2**] or [**DR**].

PROPOSITION 1.1. *Let I^a be one of the equivariant surgery theories described above, let Y be an I^a admissible G-manifold, and let $\Lambda = \Lambda_X$ be either the indexing data or the geometric reference associated to X. Suppose that $\mathbf{S}\Lambda$ and $\mathbf{S}\Lambda \times \mathrm{data}\,(Y)$ satisfy the basic conditions for defining I^a equivariant surgery obstruction groups, and for Type I theories also assume that $\mathbf{S}\Lambda$ and $\mathbf{S}\Lambda \times \mathrm{data}\,(Y)$ satisfy the Gap Hypothesis. Then the product map*

$$\mathbf{P}_Y : I^a(G;\Lambda_X) \to I^a(G;\Lambda_X \times \mathrm{data}\,(Y))$$

is a homomorphism of abelian groups.

REMARK: There are examples where the Gap Hypothesis holds for both, either, and neither of Λ_X and $\Lambda_X \times \mathrm{data}\,(Y)$. In particular, if G acts trivially on Y and $\dim Y > \dim X$, then $\Lambda_X \times \mathrm{data}\,(Y)$ will not satisfy the Gap Hypothesis even if Λ_X does. On the other hand, Theorem 2.1 gives examples where $\Lambda_X \times \mathrm{data}\,(Y)$ satisfies the Gap Hypothesis but Λ_X does not.

PROOF: In the Lück-Madsen and Browder-Quinn theories (Types II and III) the group operation is given by disjoint union, and in these cases the additivity of products follows from the distributive law

$$(A \amalg B) \times Y = (A \times Y) \amalg (B \times Y).$$

On the other hand, for theories of Type I the addition is given by the construction of [**Dov1**] and [**Dov2**], and some additional work is needed to verify that $\mathbf{P}_Y$ is additive. In this case the proof depends upon the following recognition principle for sums that is implicit in [**Dov2**].

LEMMA 1.2. *Let I^a be one of the theories $I^{c,CAT}$. Suppose we are given representatives $(f_1 : M_1 \to X_1, -)$ and $(f_2 : M_2 \to X_2, -)$ for classes in I^a and a map of cobordisms $(F^* : W \to Z, -)$ such that*

(i) $\partial_+ F^* = f_1 \amalg f_2$,

(ii) $\partial_- F^*$ *and the corresponding data determine a representative for an element of* I^a,

(iii) for $i = 1, 2$ the inclusion maps $M_i \subset W$, $X_i \subset Z$, $\partial_- Z \subset Z$ all induce isomorphisms of G-posets,

(iv) the object $(F^*, -)$ is an I^a normal map but not necessarily an I^a normal cobordism.

Then $(-\partial_- F^*, -)$ represents $(f_1, -) + (f_2, -)$.

The reason for condition *(iv)* is that a normal cobordism is usually defined to be a cobordism of normal maps with certain additional properties (compare [**DR**, Definition 3.10, p. 22]).

PROOF OF 1.2: Let $(F, -)$ be the sum cobordism constructed as in [**Dov2**, Section 5]; the top end of $(F, -)$ is given by $(f_1, -) \amalg (f_2, -)$ and the bottom end is given by the **negative** of the sum $(f_{12}, -)$. Attach $-F^*$ to F along the upper boundaries. This almost yields a normal cobordism between $-\partial_- F^*$ and f_{12} (with the appropriate extra data); the most basic problem is that the cobordisms of closed substrata are not simply connected. However, it is possible to proceed as in the first two paragraphs of [**Dov2**, page 267], and modify $F \cup -F^*$ on the interior so that a genuine normal cobordism is obtained.■

PROOF OF 1.1 FOR TYPE I THEORIES: Let $(f_1, 1)$ and $(f_2, -)$ represent classes in $I^c(G; \lambda)$, and construct the sum cobordism $(F; f_1 \amalg f_2, \partial_- F)$ as in [**Dov2**]. The most important features are that $\lambda(\partial_- F) \cong \lambda(F)$, for by the preceding lemma these relations determine the sum uniquely. Consider the cobordism

$$(F \times \mathbf{id}_Y; f_1 \times \mathbf{id}_Y \amalg f_2 \times \mathbf{id}_Y,\ \partial_- F \times \mathbf{id}_Y).$$

By the characterization of sums in 1.2 it follows that

$$[-\partial_- F \times \mathbf{id}_Y] = [f_1 \times \mathbf{id}_Y] + [f_2 \times \mathbf{id}_Y],$$

which translates to

$$\mathbf{P}_Y[-\partial_- F] = \mathbf{P}_Y[f_1] + \mathbf{P}_Y[f_2].$$

But $-\partial_- F$ represents $[f_1] + [f_2]$, and therefore the preceding equation reduces to

$$\mathbf{P}_Y([f_1] + [f_2]) = \mathbf{P}_Y[f_1] + \mathbf{P}_Y[f_2],$$

so that $\mathbf{P}_Y$ is in fact a homomorphism.■

2. Statements of periodicity theorems

Most of our periodicity theorems involve pairs of smooth G-manifolds X, Y such that both X and $X \times Y$ satisfy the Gap Hypothesis; however, we are also interested in situations where X does not satisfy the Gap Hypothesis but $X \times Y$ does. In order to show that the periodicity theorems apply in many situations it is necessary to describe systematic choices for Y. The following examples of W. Browder [Bro2] will play an important role in this chapter:

Theorem 2.1. *Assume G has odd order and let $Y = (\mathbb{CP}^2 \uparrow G)$ as in the introduction. Then given any λ there is an $r > 0$ so that $\lambda \times \mathrm{data}\,(Y^{r+k})$ satisfies the Gap Hypothesis for all $k \geq 0$.*

We shall prove this in Section 3.

REMARK. The information contained in the indexing data λ_X is also contained in the geometric reference R_X, so it is meaningful to talk about λ_X in all cases of interest to us.

It will be useful to have periodicity theorems for $\mathbf{P}_Y$ with several different choices of Y. We shall axiomatize what we need using the *Witt invariants* as defined in [ACH]. If π is a finite group, then $Witt_+(\mathbb{Z}[\pi])$ will denote the symmetric Witt group of the group ring $\mathbb{Z}[\pi]$ in the sense of [ACH] (where the group is called $W_0(\mathbb{Z}[\pi])$; we have changed notation to avoid conflict with the terminology for Bak groups in Chapters I and II).

Definition. A smooth G-manifold Y is said to be a **periodicity manifold** if the following hold:

(i) For every subgroup $K \subseteq G$, the fixed point set Y^K is a closed, 1-connected, oriented manifold whose dimension is divisible by 4.

(ii) Each integral cohomology group $H^j(X^K; \mathbb{Z})$ is $N(K)/K$-equivariantly isomorphic to a direct sum of permutation modules of the form $\mathbb{Z}[N(K)/L_\alpha]$, where $\{L_\alpha\}$ is some set of subgroups of $N(K)$ containing K (as usual, $N(K)$ denotes the normalizer of K in G).

(iii) For every K, the **Witt invariant**

$$Witt_{N(K)/K}(Y^K) \in Witt_+(\mathbb{Z}[N(K)/K])$$

(see [**ACH**]) is that of the standard multiplication form

$$\text{mult.} : \mathbb{Z} \otimes \mathbb{Z} \to \mathbb{Z}.$$

We do *not* assume the action of G is effective. Therefore, it follows immediately from the definition that

(2.2) $\mathbb{C}P^{2n}$ *with the trivial G-action is a periodicity manifold.*

Here is the example in which we are particularly interested:

THEOREM 2.3. *If G has odd order, then $\mathbb{C}P^{2n}\uparrow G$ is a periodicity manifold.*

In his Stanford lectures Browder stated a version of this result involving the G-signature rather than the Witt invariant, and by the results of [**Yo**] it seems fair to say that Theorem 2.3 was essentially known to Browder. A simultaneous generalization of (2.2) and Theorem 2.3 is given in [**Ya**]; namely, if G has odd order and S is a finite G-set, then the analogous product $\mathbb{C}P^2\uparrow S$ ($:= |S|$ copies of $\mathbb{C}P^2$ with G acting by permuting the coordinates according to the action on S) is a periodicity manifold.

We shall verify Theorem 2.3 in Section 3.

In order to formulate a reasonable periodicity theorem for $\mathbf{P}_Y$ we need to make an assumption that $\pi(X)$ and $\pi(X \times Y)$ are essentially the same. We shall say that λ *is* Y-**preperiodic** if the projection onto X induces isomorphisms of posets

$$\pi(X \times Y) \cong \pi(X).$$

By abuse of language we shall sometimes say that X or $\pi(X)$ is Y-preperiodic or that the component structure (of X) is (Y-pre)periodic.

If Y has a trivial G-action and is connected, then every G-manifold is Y preperiodic. On the other hand, if $G = \mathbb{Z}_{p^r} (r \geq 2)$ and G acts semifreely on (say) X with $X^G \neq \varnothing$, then $\pi(X \times (\mathbb{C}P^2\uparrow G))$ and $\pi(X)$ are not isomorphic since $X \times (\mathbb{C}P^2\uparrow G)$ has more orbit types than X. Fortunately there are some simple sufficient conditions that guarantee $\mathbb{C}P^2\uparrow G$-preperiodicity.

DEFINITION. A G-manifold V has **strongly saturated orbit structure** if given a point $x \in V$, a neighborhood U of x, and a subgroup H of G_x, there is a point $y \in U$ with $G_y = H$. In the notation of [**DP1**, (1.16)], this means that the poset $\pi(V)$ is **complete**.

Before discussing the significance of strong saturation for $\mathbb{C}P^2\uparrow G$-preperiodicity, we note that strong saturation implies a condition introduced in Section I.1.

PROPOSITION 2.4. *Let $f : M \to X$, and suppose that X has strongly saturated orbit structure. Then f is* **isogeneric** *in the sense of Section* I.1; *i.e., the generic isotropy subgroups of the closed substrata satisfy $G(\check{f}(\alpha)) = G(\alpha)$ for all $\alpha \in \pi(M)$.*

PROOF: We shall prove the contrapositive statement. Assume that $G(\check{f}(\alpha))$ strictly contains $G(\alpha)$. By construction $X_{\check{f}(\alpha)}$ is a component of the fixed point set of $G(\alpha)$, and therefore there is a neighborhood U of $X_{\check{f}(\alpha)}$ such that no point of $U - X_{\check{f}(\alpha)}$ is fixed by $G(\alpha)$. Since $G(\check{f}(\alpha))$ is strictly larger than $G(\alpha)$, every point on $X_{\check{f}(\alpha)}$ has an isotropy subgroup which properly contains $G(\alpha)$. Therefore no point of U has isotropy subgroup exactly $G(\alpha)$; since $G(\alpha) \subseteq G_x$ for every $x \in X_{\check{f}(\alpha)}$, this implies that X does not have strongly saturated orbit structure.■

We shall need the following elementary observation in the proof of Theorem 2.5:

SUBLEMMA 2.5A. *If V is a locally smoothable G-manifold with strongly saturated orbit structure with H, x, y, U as above, then there is a smooth curve $\gamma : [0,1] \to U$ such that $\gamma(0) = x, \gamma(1) = y$ and $G_{\gamma(t)} = H$ for $t > 0$.*■

The proof is an easy exercise in local linearity at fixed points.

THEOREM 2.5. *(i) If Y is a 1-connected closed manifold then $Y{\uparrow}G$ has strongly saturated orbit structure.*

(ii) If V has strongly saturated orbit structure and V' is a G-manifold, then $V' \times V$ has strongly saturated orbit structure.

(iii) If V is locally smoothable and has strongly saturated orbit structure then the projection $\pi : V \times \mathbf{CP}^{2n}{\uparrow}G \to V$ induces an isomorphism $\check{p}$ from $\pi(V \times \mathbf{CP}^2{\uparrow}G)$ to $\pi(V)$.

This has a very useful consequence.

COROLLARY 2.6. *Given $f : M \to X$ as before, the data sets $\lambda_X \times \mathrm{data}(\mathbf{CP}^{2n}{\uparrow}G)^r$ are $(\mathbf{CP}^{2n}{\uparrow}G)$-preperiodic for all $r \geq 1$.*■

PROOF OF THEOREM 2.5(*i*): Consider first the case of a fixed point $(y, \ldots, y) \in \mathrm{Diagonal}(Y{\uparrow}G)$. Recall that the G-action is given by

$$g(x_t) = (x_{tg^{-1}})$$

where t ranges over all the elements of G. Express G as a union of H-cosets

$$G = Hg_1 \cup \cdots \cup Hg_k,$$

and choose distinct points $z_1, \ldots, z_k$ that are arbitrarily close to y. For $t \in G$ let

$$y'_t = z_j \text{ if } t^{-1} \in Hg_j.$$

It follows that $y' = (y'_t)$ is left fixed under H; furthermore, by appropriate choice of the z_j's one can take y' arbitrarily close to $(y, \ldots, y)$. If the isotropy subgroup H' at y' were strictly larger than H, then one could find $|H'| > |H|$ points with equal coordinates. By construction, this does not happen.

The general case is proved similarly. Notice that in this case a point (y_t) with isotropy subgroup K has the property that $y_t = y_u$ if $t^{-1}u \in K$, and thus the coordinates of (y_t) involve $|G/K|$ distinct points, each repeated $|K|$ times,

PROOF OF (*ii*): Let $(x, y) \in V' \times V$. Then $G_{(x,y)} = G_x \cap G_y$. Let $H \subseteq G_{(x,y)}$. By strong saturation there is a point $z \in V$ with $G_z = H \subseteq G_{(x,y)} \subseteq G_y$ such that z is arbitrarily close to y. Then (x, z) is arbitrarily close to (x, y) and $G_{(x,z)} = H$.

PROOF OF (*iii*): If V is locally smoothable and has strongly saturated orbit structure, then

$$\pi(V) \cong \coprod_{H \in \mathcal{F}} \pi_0(X^H),$$

where $\mathcal{F}$ is some family of subgroups that is closed under taking subgroups. By (ii), a similar statement applied to $V \times (\mathbb{CP}^{2n}\uparrow G)$.

Suppose now that we have

$$(x, y) \in V \times (\mathbb{CP}^{2n}\uparrow G)$$

with $H = G_{x,y} = G_x \cap G_y$. Then we may approximate x and y by x' and y' with $G_{x'} = G_{y'} = H$. Furthermore, by the sublemma we have that (x, y) and (x', y') are joined by a continuous, H-fixed curve. Therefore (x, y) lies in a component Q of $V \times (\mathbb{CP}^{2n}\uparrow G)$ such that

$$Q \supseteq (\text{ component of } V^H) \times (\text{ component of } \mathbb{CP}^{2n}\uparrow G)^H.$$

On the other hand, it is fairly straightforward to check that $(\mathbb{CP}^{2n}\uparrow G)^H$ is **nonequivariantly** diffeomorphic to $(\mathbb{CP}^{2n})^{|G/H|}$, and therefore we see that the map

$$j : V \to V \times (\mathbb{CP}^{2n}\uparrow G)$$

taking x to (x, y_0) for some fixed point y_0 induces a surjection

$$\check{j} : \pi(V) \to \pi(V \times (\mathbb{CP}^{2n}\uparrow G)).$$

This map has the property that $\check{p}\check{j} =$ identity; since $\check{j}$ is surjective, it follows that $\check{p}$ and $\check{j}$ are both isomorphisms.■

We can now state the periodicity theorems in both abstract and specific forms. In conformity with Section 1, the symbol I^a will refer to one of the following equivariant surgery theories:

(1) A theory $I^{c,CAT}$, where $c = ht$ or s and $CAT = DIFF$, PL, or TOP (but no theory is defined for $c = s$ and $CAT = TOP$).
(2) A theory $\mathcal{L}^c$ of Lück-Madsen type, where $c = h = ht$ or s.
(3) A theory $L^{BQ,c}$ of Browder-Quinn type, where $c = h = ht$ or s.

Furthermore, as in Section 1 we shall denote the indexing data or geometric reference of X by Λ_X or, when no danger of confusion seems likely, simply by Λ. Since the a geometric reference always has associated indexing data obtained by forgetting some of the underlying structure, in all cases it is meaningful to discuss indexing data. When we wish to stress that only properties of the indexing data are relevant we shall indicate this by referring to λ_X or λ. With these conventions, here is the abstract version of the Periodicity Theorem:

THEOREM 2.7. *Let X satisfy the conditions for defining $I^a(G; \Lambda_X)$ and constructing an abelian group operation, and also assume that the dimensions of all closed substrata are sufficiently large (at least 7 for the sake of simplicity). Let Y be a periodicity manifold such that the following hold:*

(i) $\Lambda \times \mathrm{data}\,(Y)$ *satisfies the Gap Hypothesis if* $I^a \neq L^{BQ,c}$.
(ii) Λ *is* Y*-preperiodic.*
(iii) *The set* $I^a(G; \mathbf{S}^{-2}\Lambda_X)$ *is nonempty.*
Then the product map

$$\mathbf{P}_Y : I^a(G;\Lambda) \to I^a(G;\Lambda \times \mathrm{data}\,(Y))$$

induces an isomorphism of abelian groups.

We shall prove this in Section 4. At this point we would like to formulate two special cases and explain how they follow from 2.7.

THEOREM 2.8. *Suppose that* X *satisfies the conditions in 2.7, and also assume*
(i) $\mathbf{S}^{4n}\lambda$ *satisfies the Gap Hypothesis if* $I^a \neq L^{BQ,c}$,
(ii) $I^a(G; \mathbf{S}^{-2}\Lambda_X) \neq \varnothing$.
Then $\mathbf{P}_Y$ *is an isomorphism for* $Y = \mathbf{CP}^{2n}$ *(with trivial* G*-action). If (ii) is not true but* $\mathbf{S}^{8n}\lambda$ *satisfies the Gap Hypothesis or* $I^a = L^{BQ,c}$*, then*

$$\mathbf{P}_Y : I^a(G;\Lambda_X \times \mathrm{data}\,(\mathbf{CP}^{2n})) \to I^a(G;\Lambda_X \times \mathrm{data}\,(\mathbf{CP}^{2n})^2)$$

induces an isomorphism of abelian groups.

If $I^a = L^{BQ,c}$ then this result reduces to a special case of [**BQ**, Cor. 3.4, p. 32].

PROOF OF 2.8: (*Assuming Theorem* 2.7). By 2.7 it is necessary to check that $\mathbf{S}^{4n}\lambda$ satisfies the Gap Hypothesis, λ is $\mathbf{CP}^{2n}$-preperiodic, and $\mathbf{CP}^{2n}$ is a periodicity manifold. The first is given, the second is noted after the definition of preperiodic, and the third is noted in (2.2). This proves the first assertion. To prove the second assertion, it suffices to show that $I^a(G; \mathbf{S}^{4n-2}\Lambda_X) \neq \varnothing$ and then apply the first assertion. But one can construct a representative for the latter bordism set by crossing the given surgery problem

$$(f : M \to X, \text{ other data })$$

with S^{4n-2}.■

Theorem 2.8 gives some fourfold periodicity properties for the equivariant surgery groups $I^a(G;\Lambda_X)$, but as noted earlier the sequence of isomorphisms

$$I^a(G; \mathbf{S}^{4k}\Lambda_X) \cong I^a(G; \mathbf{S}^{4k+4}\Lambda_X)$$

must terminate after finitely many k if $I^a = I^{c,CAT}$ or $\mathcal{L}^c$. The following result provides an infinite periodicity sequence for all the theories under consideration:

THEOREM 2.9. *Suppose that* X *satisfies the conditions in 2.7. Let* $Y = (\mathbf{CP}^{2n}\uparrow G)$. *Then there is an* $r_0 \geq 0$ *such that the product maps*

$$\mathbf{P}_Y : I^a(G;\Lambda \times \mathrm{data}\,(Y^r)) \to I^a(G;\Lambda \times \mathrm{data}\,(Y^{r+1}))$$

are isomorphisms for all $r \geq r_0$. If $\mathbf{S}^2\Lambda$ satisfies the Gap Hypothesis or $I^a = L^{BQ,c}$, then one can take $r_0 = 0$.

PROOF: (*Assuming Theorem 2.7*). Once again the idea is to apply 2.7. All the conditions on $I^a(\Lambda_X \times \mathrm{data}(Y^r))$ are immediate except for the Gap Hypothesis, and by 2.1 this will also be true for all r greater than or equal to some positive integer r_0. Furthermore, $\lambda \times \mathrm{data}(Y^r)$ is isogeneric for all $r \geq 1$ by Proposition 2.4.

It now suffices to show that $\lambda \times \mathrm{data}(Y^r)$ is Y-preperiodic for all $r \geq 1$ and that

$$I^a(G; \mathbf{S}^{-2}\Lambda_X \times \mathrm{data}(Y^r)) \neq \varnothing$$

for all $r \geq 1$. The first of these is true by Corollary 2.6. For the second statement, it suffices to consider the case $r = 1$ (cross the representative in this case with the identity on Y^{r-1} to prove the other cases). But one can construct an element of

$$I^a(G; \mathbf{S}^{-2}\Lambda_X \times \mathrm{data}(Y))$$

by crossing a Λ-indexed normal map

$$(f : M \to X; \text{other data})$$

with the unit sphere in the quotient representation

$$\mathbf{R}[G]^{4n}/\mathbf{R}^1$$

obtained from $4n$ copies of the regular representation by factoring out a 1-dimensional (trivial) submodule of the fixed point set.■

The proofs of Theorems 2.7–2.9 are essentially formal, and there are three main steps:

(1) Fix an equivariant surgery theory I^a, and view $I^b(G; \Lambda_X; \Sigma \subset \Sigma') := I^a(G; \Lambda_X \times \mathrm{data}(Y); \Sigma \subset \Sigma')$ as another equivariant surgery theory over X. Show that I^b satisfies the exactness condition (II.2.0), and view the product maps $\mathbf{P}_Y(\Sigma \subset \Sigma')$ as natural transformations of equivariant surgery theories $I^a \to I^b$ satisfying (II.3.1.A)–(II.3.1.B). The Y-preperiodicity assumption is needed here.
(2) Show that $\mathbf{P}_Y$ is an isomorphism for $\Sigma \subset \Sigma \cup \{\alpha\}$, where α is minimal in $\pi(X) - \Sigma$.
(3) Use (i) and (ii) together with the Comparison Theorem II.3.4 to show that $\mathbf{P}_Y(\Sigma)$ is an isomorphism for all Σ.

This formal approach also yields periodicity theorems like 2.7–2.9 for many other versions of equivariant surgery. In Chapter IV we shall extend 2.7–2.9 to the theories for equivariant surgery with coefficients. Our results cover

(1) the theories $I^{ht(R),CAT}$ discussed in (II.2.3) and the analogous Browder-Quinn theories $L^{BQ,h(R)}$ introduced in Subsection II.3F,
(2) certain equivariant homology surgery theories that correspond to the Γ-group homology surgery of [CS].

In Chapter V we shall consider the theories for equivariant surgery up to pseudoequivalence that are described in (II.2.5) and some variants of these theories. Many new technical difficulties arise, but in some cases it is still possible to prove analogs of the periodicity theorems for a reasonable class of finite groups.

Recent developments. As noted in the Summary, recent results of M. Yan [Ya] yield an analog of Theorem 2.9 for the isovariant stratified surgery theory developed by S. Weinberger [Wb]. In fact, the results of [Ya] also yield a periodicity theorem for the associated surgery sequences that generalizes the fourfold periodicity of the ordinary topological surgery sequence (see [KiS] and [Ni]), and similar periodicity results are established if $\mathbb{CP}^2 \uparrow G$ is replaced by $\mathbb{CP}^2 \uparrow S$ where S is an arbitrary finite G-set (compare the discussion following the statement of Theorem 2.3). If one combines the appropriate observations from [Ya] with the methods developed here, it is possible to extend Theorem 2.9 as follows:

(2.10) Complement to Theorem 2.9. *The conclusion of Theorem* 2.9 *remains valid if one takes* $Y = \mathbb{CP}^2 \uparrow S$ *(where* S *is some finite* G*-set) and assumes that all relevant indexing data are* $\mathbb{CP}^2 \uparrow S$*-preperiodic and satisfy the Gap Hypothesis or* $I^a = L^{BQ,c}$.

Dynamic indexing data/geometric references

At the end of Subsection II.3F we mentioned that the indexing data λ_X and the geometric reference R_X could be viewed as *static* referencing objects and that one has corresponding *dynamic* referencing objects. In particular, one can consider triples $(\lambda_1, \lambda_2, \check{f})$ where λ_1 and λ_2 represent indexing data and $\check{f}$ is an isomorphism of the underlying G-posets; these triples are the G-Poset (*sic*) pairs of [DR1]. In the theory of [LM] one can consider triples (R_1, R_2, Φ) where R_1 and R_2 are static geometric references and Φ describes some weak compatibility properties. The periodicity theorems extend directly to equivariant surgery groups equipped with dynamic (as opposed to static) reference data:

Complement 2.11. *Periodicity theorems analogous to* 2.7 – 2.9 *hold for dynamic indexing data if* $I^a = I^{c,CAT}$ *or dynamic references if* $I^a = \mathcal{L}^c$.

In fact, the proofs for static data generalize directly to the dynamic case; we have chosen to deal mainly with static data for the sake of notational simplicity.

3. Permutation actions on product manifolds

In this section we shall verify the properties of the G-manifolds $X \uparrow G$ that were stated in Section 2.

The first order of business is to show that

$\mathbb{CP}^2 \uparrow G$ *=product of* $|G|$ *(= order of* G*) copies of* $\mathbb{CP}^2$ *with*
G acting by permuting the coordinates

is a periodicity manifold in the sense of Section 2.

PROPOSITION 3.1. *Let G be a finite group, and let $Y = \mathbf{CP}^2\uparrow G$ as above.*

(i) For each subgroup H, the fixed point set Y^H is a product of $|G/H|$ copies of $\mathbf{CP}^2$. Furthermore, the induced action of $W(H) = N(H)/H$ on Y^H is a product of $|G/N(H)|$ copies of the $W(H)$-action on $\mathbf{CP}^2\uparrow W(H)$ by permuting the coordinates.

(ii) The homology groups $H_*(Y;\mathbf{Z})$ are G-isomorphic to direct sums of permutation modules.

(iii) If G has odd order, then the equivariant Witt invariant

$$Witt(G; \mathbf{CP}^2\uparrow G) \in Witt_+(\mathbf{Z}[G])$$

(as defined in [**ACH**]*) is equal to the Witt invariant of the "unit form" $\mathbf{Z}\otimes\mathbf{Z}\to\mathbf{Z}$ induced by multiplication (= the Witt invariant of $\mathbf{CP}^2$ with standard multiplication).*

Notice that (i) implies that Y^H is a 1-connected oriented manifold whose dimension is divisible by 4. Furthermore, (i) and (ii) combine to imply that

$$Witt(W(H); Y^H) = 1 \in Witt_+(\mathbf{Z}[W(H)])$$

if G has odd order. Therefore Y is a periodicity manifold as defined in Section 2, and thus Theorem 2.3 follows from Proposition 3.1.

PROOF: (i) This is just a general fact about the action of G on $X\uparrow G$. Recall that under this group action the group element g sends the point x with coordinates $x_a (a \in G)$ to a point y with coordinates

$$y_a = x_{ag^{-1}}.$$

The conclusions in (i) are elementary consequences of this formula.

(ii) By the Künneth formula the homology and cohomology of $Y = \mathbf{CP}^2\uparrow G$ are torsion free and concentrated in even dimensions. Furthermore, $H^2(Y;\mathbf{Z})$ is a free $\mathbf{Z}[G]$-module on some generator e, and each $H^{2k}(Y;\mathbf{Z})$ has a $\mathbf{Z}$-free basis of the form

$$(a_1^*e)\dots(a_r^*e)(b_1^*e)^2\dots(b_q^*e)^2$$

for distinct elements $a_1,\dots,a_r,b_1,\dots b_q \in G$ such that $r+2q=k$. Clearly this basis is G-invariant. If we dualize this we find that $H_{2k}(Y;\mathbf{Z})$ also has a G-invariant basis; *i.e.*, $H_{2k}(Y;\mathbf{Z})$ is a direct sum of G-permutation modules.

(iii) The basis for $H^{2|G|}(Y;\mathbf{Z})$ constructed in (ii) contains an element of the form $y = \Pi g^*e$, the product running through all $g \in G$. The class y is G-invariant, its cup square generates the top dimensional cohomology of $H^*(Y;\mathbf{Z})$, and in fact y is orthogonal to all the other basis elements for $H^{2|G|}(Y;\mathbf{Z})$ described in (ii). If $\mathcal{M}$ denotes the submodule generated by these remaining basis elements, then the Witt class of Y is clearly equal to the unit element plus the class of $\mathcal{M}$. Therefore it will suffice to prove that $\mathcal{M}$ is metabolic in the sense of [**ACH**].

It is useful to write the basis elements of $\mathcal{M}$ in the form $X_{A,B}$, where $A, B \subseteq G$ are disjoint subsets of G such that $B \neq \emptyset$ and $|A|+2|B| = |G|$ and $X_{A,B} = (\Pi_A a^*e)(\Pi_B(b^*e)^2)$. If y is the class defined in the previous paragraph, then the cup products of these basis elements are are given by

$$(3.2) \qquad X_{A,B}\cdot X_{C,D} = \begin{cases} y^2 & \text{if } D = G-(A\cup B) \text{ and } C = A \\ 0 & \text{otherwise} \end{cases}$$

This relationship has the following consequence:

(3.3) *For all* A, B *such that* $B \neq \emptyset$ *and all* $g \in G$ *the product* $(g^* X_{A,B}) \cdot X_{A,B}$ *vanishes.*

To see this, note first that $g^* X_{A,B}$ may be written as $X_{Ag,Bg}$, where if $S \subseteq G$ then Sg is the right translation of S via g. According to (3.2) the product is nontrivial only if $Bg = G - (A \cup B)$ and $Ag = A$. If this were the case then by a counting argument we must also have $(Bg)g = B$, or $Bg^2 = B$. It follows that $Bg^{2k} = B$ for all k.

But G has odd order, and therefore there is some k_0 such that $g^{2k_0} = g$. Therefore we have $B = Bg^{2k_0} = Bg$, which yields a contradiction. This means that the product $(g^* X_{A,B}) \cdot X_{A,B}$ must vanish. ∎

We now return to the proof of 3.1(iii). Let δ be the set of all ordered pairs of disjoint subsets (A, B) such that $B \neq \emptyset$ and $|A| + 2|B| = |C|$. There is a free involution on δ sending (A, B) to $(A, G-(A \cup B))$. This involution commutes with the action of G on δ by right translation. It follows that there is a G-invariant subset $\delta_0 \subset \delta$ which contains exactly one element of each orbit of the involution. Let $\mathcal{M}_0 \subseteq \mathcal{M}$ be the submodule generated by δ_0. Then $\mathcal{M}_0$ is a $\mathbf{Z}|G|$-submodule of $\mathcal{M}$ with

$$rank \mathcal{M}_0 = |\delta_0| = |\delta|/2 = rank \mathcal{M}/2.$$

and by (3.2) and (3.3) this submodule is self-annihilating under cup product. Therefore $\mathcal{M}$ is a metabolic with subkernel $\mathcal{M}_0$, and it follows that the Witt class of $H^{2|G|}(Y; \mathbf{Z})$ is precisely the unit element of $Witt_+(\mathbf{Z}[G])$. ∎

REMARKS 1. As indicated previously, results from [Ya] show that 3.1(*iii*) remains valid if $\mathbf{CP}^2 \uparrow G$ is replaced by $\mathbf{CP}^2 \uparrow S$ where S is an arbitrary finite G-set (provided that G has odd order).

Assertion 3.1(*iii*) fails for even order groups; in fact, if $G = \mathbf{Z}_2$ a formula for the G-signature is given in Proposition 3.7.

Similar reasoning yields the following generalizations:

COMPLEMENT 3.1.A. *Both Proposition 3.1 and Theorem 2.3 remain valid if* $\mathbf{CP}^2$ *is replaced by* $\mathbf{CP}^{2n}, \mathbf{KP}^{2n}$ *(= quaternionic projective 2n-space), or* $\mathbf{CayP}^2$ *(= the Cayley projective plane).* ∎

We shall also need the following properties of $\mathbf{CP}^{2n} \uparrow G$:

THEOREM 3.4. *Let* λ *denote the indexing data for a normal map* $f : M \to X$, *and let* $Y = \mathbf{CP}^{2n} \uparrow G$.

(i) If λ *satisfies the Gap Hypothesis and* M *and* X *have strongly saturated orbit structure, then* $\lambda \times \text{data}(Y^k)$ *also satisfies the Gap Hypothesis for all* $k \geq 1$.

(ii) If G *has odd order, then there is an* $r > 0$ *such that* $\lambda \times \text{data}(Y^{r+k})$ *satisfies the Gap Hypothesis for all* $k \geq 0$.

Theorem 2.1 is the second half of 3.4.

PROOF: (i) Notice that the fixed point sets Y^H are all connected and that every subgroup of G is an isotropy subgroup for some point of Y. Therefore if $f : M \to X$ has indexing data λ and $Q = M$ or X, then every closed substratum of $Q \times Y$ has the form $Q_\beta \times Y^H$, where Q_β is a closed substratum satisfying $G^\beta = H$ (notation as in (1.0iii)); in general one can only say that G^β contains H, but strong saturation implies that equality holds. The generic isotropy subgroup of this substratum is (not surprisingly) simply H. The singular points of $Q_\beta \times Y^H$ have the form (u, v) where $G_u \cap G_v$ properly contains H. Since H is the generic isotropy subgroup for $Q_\beta \times Y^H$ this means that u is a singular point of Q_β and v is a singular point of Y^H. If we let $\operatorname{Sing} Q_\beta \subseteq Q_\beta$ be the set of points u with $G_u \supsetneqq H$, then the Gap Hypothesis implies that

$$\dim \operatorname{Sing} Q_\beta < (\dim Q_\beta)/2.$$

We need a comparable result for Y.

(3.5) *Let* $\operatorname{Sing}(Y^H)$ *be the set of all points* $v \in Y^H$ *such that the isotropy subgroup* G_v *strictly contains* H. *Then*

$$\dim \operatorname{Sing}(Y)^H = \dim(Y^H)/C(G,H),$$

where $C(G,H) \geq 2$ *is the minimum of* $|K|/|H|$ *for all subgroups* K *of* G *properly containing* H.

PROOF OF 3.5: Suppose K properly contains H. Then $\dim Y^H = 2n|G/H|$ and $\dim Y^K = 2n|G/K|$. This shows that

$$\dim(Y^K) = \dim(Y^H)/(|K|/|H|).$$

Since $\operatorname{Sing} Y^H$ is the union of the Y^K's for K properly containing H, the equation above follows directly. Notice that $|K|/|H| \geq 2$ always holds, and this is why $C(G,H) \geq 2$.■

PROOF OF 3.4(*i*): (*Conclusion*). We now have

$$\operatorname{Sing}(Q_\beta \times Y^H) \subseteq \operatorname{Sing} Q_\beta \times \operatorname{Sing} Y^H,$$

so that

$$\begin{aligned}\dim \operatorname{Sing}(Q_\beta \times Y^H) &\leq \dim \operatorname{Sing} Q_\beta + \dim \operatorname{Sing}(Y^H)\\ &< (\dim Q_\beta + \dim(Y^H))/2,\end{aligned}$$

and hence the Gap Hypothesis holds for $\lambda \times \operatorname{data}(Y)$.■

PROOF OF 3.4(*ii*): The important point is that $C(G,H) \geq 3$ in this case; recall that $|G|, |K|$, and $|H|$ must be odd, and $|H|$ must divide $|K|$ properly.

Without loss of generality we may assume λ has strongly saturated domain and codomain because the latter holds for $\lambda \times \operatorname{data}(Y)$ even if it fails for λ. The argument of (i) then shows that $\lambda \times \operatorname{data}(Y)$ has closed substrata of the form $Q_\beta \times Y^H$, where $G(\beta) = H$. Furthermore, one also has

$$\operatorname{Sing}(Q_\beta \times Y^H) \subseteq \operatorname{Sing}(Q_\beta) \times \operatorname{Sing}(Y^H).$$

For an arbitrary closed substratum Z_γ define a *Gap Hypothesis balance*

$$\Delta(Z_\gamma) = \dim(Z_\gamma) - 2\dim(\mathrm{Sing}\,(Z_\gamma)). \tag{3.6}$$

This balance is positive if and only if the standard Gap Hypothesis holds for Z_γ. It follows that

$$\Delta(Q_\beta \times Y^H) \leq \Delta(Q_\beta) + \Delta(Y^H).$$

On the other hand, by (3.5) and $G(G,H) \geq 3$ we know that

$$\Delta(Y^H) \geq 2n,$$

and hence

$$\Delta(Q_\beta \times Y^H) \leq \Delta(Q_\beta) + 2n.$$

In other words,

($\star$) *crossing with Y raises all Gap Hypothesis balances by at least $2n$.*

By induction, if we cross with Y successively q times, we raise all Gap Hypothesis balances by at least $2qn$. Since there are only finitely many closed substrata, for all sufficiently large values of q the Gap Hypothesis balances will all be positive. But this means the Gap Hypothesis will hold for $\lambda \times \mathrm{data}\,(Y^q)$ if q is sufficiently large.■

Groups of even order

Theorems 2.1 and 2.3 fail badly if the group G has even order. In particular, repeated crossing with $\mathbb{C}P^{2n}\uparrow G$ will not create a problem in which the Gap Hypothesis holds from a problem where it does not. Furthermore, the following result shows that periodicity manifolds are more difficult to find and use if $G = \mathbb{Z}_2$:

PROPOSITION 3.7. *(i) Suppose that M^{4n} is a closed oriented smooth $\mathbb{Z}_2$-manifold such that the signature of M^{4n} is odd. Then (M, $\mathbb{Z}_2$-action) does not satisfy the Gap Hypothesis.*

(ii) If N^{2k} is a closed smooth manifold, then $\mathbb{Z}_2$ acts orientation-preservingly on $N\uparrow\mathbb{Z}_2$ and the equivariant signature of the latter is

$$(\mathrm{sgn}\ \mathrm{N})^2 + \chi(N)T \in \mathbb{Z}[\mathbb{Z}_2]$$

where sgn *denotes the ordinary signature, $\chi(N)$ is the Euler characteristic, and T represents the generator of $\mathbb{Z}_2$.*

PROOF: *(i)* Since the signature and Euler characteristic of a closed oriented $4m$-dimensional manifold are congruent mod 2, it follows that the Euler characteristic of M is also odd. Therefore an old result of Conner and Floyd [**CF**, Cor. 27.4, p. 71] implies that the fixed point set has a component of dimension at least $2n$.

(ii) The first assertion follows from standard theorems on the cohomology of cartesian products. Since $\{1, T\} \subset \mathbb{Z}_2$ forms a free basis for $\mathbb{Z}[\mathbb{Z}_2]$ we can express the equivariant signature of $N\uparrow\mathbb{Z}_2$ as a sum $a+bT$ where a and b are integers. By definition a equals the

ordinary signature of the ambient manifold, and therefore $a = (\mathrm{sgn}(N))^2$. Furthermore, it is well known that b is given by the Euler class of the fixed point set's equivariant normal bundle (compare [Hi], [JäO]), and this class is $\chi(N)$ times the orientation class of N.■

4. Orbit sequences and product operations

If I^a denotes one of the equivariant surgery theories considered in Section 2, then the stepwise obstructions of Chapters I and II provide a means for analyzing the groups $I^a(G; \Lambda_X)$ in terms of ordinary surgery obstruction groups. In this section we shall establish some basic relationships between these stepwise obstructions and the product operations described in Section 1. These relationships will play an important role in the proofs of the periodicity theorems.

Each equivariant surgery theory is formulated in a specific category of locally linear G-manifolds (smooth, piecewise linear, or topological), and for each equivariant surgery theory the zero objects are given by equivariant homotopy equivalences or some variation of this concept. We shall say that the isomorphisms in the underlying category are I^a-**isomorphisms** and the equivalences defining zero objects are I^a-**equivalences**.

Review of stepwise obstructions. Let I^a and Λ_X be as above, and let $\pi(X)$ be the geometric poset. If $\Sigma \subset \pi(M)$ is a closed G-invariant subset, we shall say that $(f : M \to X, \text{other data})$ is Σ-*adjusted* if for all $\alpha \in \Sigma$ the induced maps f_α satisfy the conditions for an I^a-equivalence in the definition of $I^a(G; \Lambda_X)$. The group of all Σ-adjusted classes will be denoted by $I^a(G; \Lambda; \Sigma)$; if $\Sigma \subset \Sigma'$ is a pair of closed G-invariant subsets one also has relative groups $I^a(G; \Lambda; \Sigma \subset \Sigma')$ and exact sequences as in (II.2.0). If $\Sigma' = \Sigma \cup \{\alpha\}$ where α is minimal in $\pi(X) - \Sigma$, then these sequences are particularly useful because the groups $I^a(G; \Lambda_X; \Sigma \subset \Sigma')$ are generally isomorphic to ordinary Wall groups or (in exceptional cases) quotients of such groups; explicit statements of these results appear in Sections I.5 and II.1–2. If $\Sigma' = \Sigma \cup \{\alpha\}$ the exact sequence is called an **Orbit Sequence** in the sections cited.

SIMPLIFYING HYPOTHESIS: We shall assume all fixed point sets Y^H are connected and that Y is Λ_X-preperiodic; *i.e.*, $\pi(X \times Y) \cong \pi(X)$. In this situation we shall identify the two geometric posets via projection onto the X coordinate. It is possible to work in greater generality, but the results become considerably more complicated to state.

If the simplifying hypothesis holds, it follows immediately that the product of a Σ-adjusted map $f : M \to X$ with the identity on Y is always Σ-adjusted. Furthermore, the results of Section 1 can be extended to yield homomorphisms

$$\mathbf{P}_Y(\Sigma \subset \Sigma') : I^a(G; \Lambda_X; \Sigma \subset \Sigma') \to I^a(G; \Lambda_X \times \mathrm{data}(Y); \Sigma \subset \Sigma').$$

Furthermore, these product constructions satisfy the expected compatibility conditions:

THEOREM 4.1. *In the setting and notation described above, let $\Omega \subset \Omega'$ be another pair of closed G-invariant subsets of $\pi(X)$ such that $\Sigma \subset \Omega$ and $\Sigma' \subset \Omega'$. Then the following diagrams are commutative:*

(i) The diagram arising from forgetful maps corresponding to the inclusion of the pair $(\Sigma \subset \Sigma')$ in $(\Omega \subset \Omega')$:

$$\begin{array}{ccc} I^a(G;\Lambda_X;\Omega\subset\Omega') & \longrightarrow & I^a(G;\Lambda_X;\Sigma\subset\Sigma') \\ {\scriptstyle \mathbf{P}_Y(\Omega\subset\Omega')}\downarrow & & {\scriptstyle \mathbf{P}_Y(\Sigma\subset\Sigma')}\downarrow \\ I^a(G;\Lambda_{X\times Y};\Omega\subset\Omega') & \longrightarrow & I^a(G;\Lambda_{X\times Y};\Sigma\subset\Sigma') \end{array}$$

(ii) The diagram arising from the boundary homomorphism for the pair $(\Sigma \subset \Sigma')$:

$$\begin{array}{ccc} I^a(G;\mathbf{S}\Lambda_X;\Sigma\subset\Sigma') & \longrightarrow & I^a(G;\Lambda_X;\Sigma') \\ {\scriptstyle \mathbf{P}_Y(\Sigma\subset\Sigma')}\downarrow & & {\scriptstyle \mathbf{P}_Y(\Sigma')}\downarrow \\ I^a(G;\mathbf{S}\Lambda_{X\times Y};\Sigma\subset\Sigma') & \longrightarrow & I^a(G;\Lambda_{X\times Y};\Sigma') \end{array}$$

IDEA OF PROOF: The results follow directly from the geometric definitions of the groups under consideration.■

Suppose now that $\Sigma' = \Sigma \cup \{\alpha\}$, where α is minimal in $\pi(X) - \Sigma$. By the results of Chapters I and II we have isomorphisms

$$\sigma_\alpha : I^a(G;\Lambda_X;\Sigma\subset\Sigma\cup G\{\alpha\}) \cong L^c_{\dim\alpha}(\mathbb{Z}[E_\alpha], w_\alpha)/B$$

where $c = h$ or s depending upon the type of equivalence considered and $B = 0$ if $I^a = L^{BQ,c}$ or if $\mathbf{S}^2\Lambda$ satisfies the Gap Hypothesis. Furthermore, there are maps

$$\gamma_\alpha : L^c_{\dim\alpha}(\mathbb{Z}[E_\alpha], w_\alpha) \to I^a(G;\Lambda_X;\Sigma\subset\Sigma\cup G\{\alpha\})$$

such that the composites $\sigma_\alpha\gamma_\alpha$ are the usual quotient projections.

THEOREM 4.2. *In the situation described above there is a commutative diagram as follows:*

$$\begin{array}{ccccc} L^c_{\dim\alpha}(\mathbb{Z}[E_\alpha];w_\alpha) & \xrightarrow{\gamma_\alpha} & I^a(G;\Lambda_X;\Sigma\subset\Sigma\cup\{\alpha\}) & \xrightarrow[\cong]{\sigma_\alpha} & L^c_{\dim\alpha}(\mathbb{Z}[E_\alpha];w_\alpha)/B \\ {\scriptstyle \mu(Y^\#)}\downarrow & & {\scriptstyle \mathbf{P}_Y(\Sigma\subset\Sigma\cup\{\alpha\})}\downarrow & & \\ L^c_{\dim\alpha+q}(\mathbb{Z}[E_\alpha];w_\alpha) & \xrightarrow{\gamma_\alpha} & I^a(G;\Lambda_{X\times Y};\Sigma\subset\Sigma\cup\{\alpha\}) & \xrightarrow[\cong]{\sigma_\alpha} & L^c_{\dim\alpha+q}(\mathbb{Z}[E_\alpha];w_\alpha)/B' \end{array}$$

In this diagram $\mu(-)$ refers to the Yoshida twisted product, $Y^{\#}$ denotes the fixed point set of the generic isotropy subgroup of X_α, and q is the dimension of $Y^{\#}$.

REMARK: Yoshida's twisted product maps of [Yo] are homomorphisms

$$\mu^s : Witt_m(G) \otimes L_n^s(\pi, w) \to L_{m+n}^s(\pi, w')$$

that can be described geometrically as follows: First take a normal map $(f,-)$ such that the codomain has fundamental group π, its first Stiefel-Whitney class is w, and its surgery obstruction is v; let $\widetilde{f}$ be the associated map to the universal covering of the codomain. Next, take a smooth G-manifold U with equivariant Witt invariant u; part of the data for the twisted product is a surjection from π to G, and use this to view the G-action as a π action. The class $\mu^s(u \otimes v)$ is then represented by the balanced product $(\widetilde{f} \times_\pi \mathrm{id}_U)$.

Some cases of Theorem 4.2 require analogous twisted product maps μ^h involving the L^h-groups rather than the L^s-groups. In fact, everything in [Yo] goes through if one replaces L^s by L^h. The quickest way to see this is to use the Shaneson splitting [Sh]

$$L_k^s(\pi \times \mathbf{Z}, w \circ \mathrm{proj}_\pi) \cong L_k^s(\pi, w) \oplus L_{k-1}^h(\pi, w)$$

to embed $L_{k-1}^h(\pi, w)$ into $L_k^s(\pi \times \mathbf{Z}, w \circ \mathrm{proj}_\pi)$ and to notice that the twisted product commutes with this embedding. In Chapter IV we shall show that Yoshida's twisted product formulas also hold in many other situations.

PROOF OF THEOREM 4.2: We begin by noting an elementary excision property of stepwise obstructions. Let Σ and α be as in the statement of the theorem. Suppose that we are given an I^a normal map $(h : A \to B, -)$ that is Σ-adjusted, and suppose that the associated map $h_\alpha : A_\alpha \to B_\alpha$ splits W_α-equivariantly as a map of triads $(A_\alpha; A'_\alpha, A''_\alpha) \to (B_\alpha; B'_\alpha, B''_\alpha)$ such that A''_α and B''_α contain the singular sets of A_α and B_α, the associated maps $h''_\alpha : A''_\alpha \to B''_\alpha$ and $\partial h''_\alpha : \partial A''_\alpha \to \partial B''_\alpha$ satisfy the conditions for a $\Sigma \cup G\{\alpha\}$-adjusted map, and $B'_\alpha \subset B_\alpha$ is 2-connected. Then $h^*_\alpha : A'_\alpha / W_\alpha \to B'_\alpha / W_\alpha$ defines an ordinary normal map that is a homotopy equivalence on the boundary, and the stepwise obstruction $\sigma_\alpha(h, -)$ is equal to the ordinary surgery obstruction $\sigma(h^*_\alpha, -)$.

Here is one way of seeing this: First make h^*_α connected up to the middle dimension by ordinary surgery away from the boundary; call the resulting map h^{**}_α. Next, lift the surgeries to equivariant surgeries on h'_α, replacing the latter by a map $\widehat{h}_\alpha$ that is connected up to the middle dimension. Finally, use the bundle data to extend $\widehat{h}_\alpha$ to a map $\widehat{h} : \widehat{A} \to B$. It follows by excision that the form or formation determined by $\mathrm{Cone}\,(\widehat{h}_\alpha)$ is isomorphic to the corresponding form or formation associated to $\mathrm{Cone}\,(h^{**}_\alpha)$. Therefore $\sigma_\alpha(\widehat{h})$ and $\sigma(h^{**}_\alpha)$ are represented by the same form or formation, and it follows that their classes in the Wall group are also equal. But the ordinary surgery obstructions determined by h^*_α and h^{**}_α are equal by construction, and the stepwise obstructions of h and $\widehat{h}$ are the same because these maps are normally bordant; therefore it follows that $\sigma_\alpha(h, -) = \sigma(h^*_\alpha, -)$ as claimed.

Consider now the composite $\sigma_\alpha^{X \times Y} \mathbf{P}_Y \gamma_\alpha^X$. By construction the map γ_α^X is given as follows: Take an element ω in the appropriate Wall group and a G-manifold V such that $\Lambda_V = \mathbf{S}^{-1}\Lambda_X$. Next, construct a cobordism $(W_\alpha; V_\alpha, V'_\alpha)$ such that RelSing $W_\alpha \cong$ RelSing $V_\alpha \times I$ and there is a suitable map of triads F_α from W_α into $V_\alpha \times I$ that is an I^a isomorphism on RelSing W_α, maps V_α to itself by the identity, and maps V'_α to itself by an I^a equivalence. Having done this, use the standard "top hat" extension process for normal cobordisms (compare [**DS**, pp. 284–285]) to construct an extension of F_α to a Σ-adjusted map of triads, say F, such that $F|V$ is the identity and $F|V'$ is $(\Sigma \cup \{\alpha\})$-adjusted. It follows that $\sigma_\alpha^{X \times Y} \mathbf{P}_Y \gamma_\alpha^X$ is the equivariant surgery obstruction for the restriction of $F \times 1_Y$ to the α-stratum of $W \times Y$. But this restriction is just $F_\alpha \times \mathbf{id}(Y^\#)$, and therefore it suffices to show that the equivariant surgery obstruction of $F_\alpha \times \mathbf{id}(Y^\#)$ is the twisted product $\mu(Y^\#)v$.

By construction $F_\alpha \times \mathbf{id}(Y^\#)$ defines a map of triads that is an I^a-isomorphism near RelSing $(W_\alpha \times Y^\#)$; in fact, $F_\alpha \times \mathbf{id}(Y^\#)$ is an isomorphism near the slightly larger set $(\text{RelSing}\,(W_\alpha)) \times Y^\#$. Therefore the previously stated excision property shows that the equivariant surgery obstruction for $F_\alpha \times \mathbf{id}(Y^\#)$ is equal to $\mu(Y^\#)v$ as required.■

Proof of the abstract periodicity theorem (= 2.7)

We begin by recalling the statement of this result. Let X be a smooth G-manifold to which I^a equivariant surgery theory applies; in the terminology of Section II.2, we want X to belong to the appropriate class of manifolds $\mathcal{F}^a$. If Y is a smooth oriented G-manifold such that the Witt invariants of all fixed point sets are units in the appropriate Witt rings $Witt_+(\mathbb{Z}[N(H)/H])$, then the product map $\mathbf{P}_Y$ defines an isomorphism

$$I^a(G;\Lambda) \to I^a(G;\Lambda \times \text{data}\,(Y)).$$

Assume that G, X, Y satisfy the conditions described above; as elsewhere in this section we also assume that $\pi(X)$ is Y-preperiodic. The results of [**Yo**] and the remark following Theorem 4.2 now show that the twisted product maps $\mu^c(Y^\#)$ are isomorphisms. Therefore by Theorem 4.2 the equivariant product maps are isomorphisms if $\mathbf{S}^2\Lambda_X$ and $\mathbf{S}^2\Lambda_X \times \text{data}\,(Y)$ satisfy the Gap Hypothesis and epimorphisms if we only know that the first suspensions of Λ_X and $\Lambda_X \times \text{data}\,(Y)$ satisfy the Gap Hypothesis.

As in Section 2, we define $I^b(G;\Lambda)$ to be $I^a(G;\Lambda \times \text{data}\,(Y))$. Then the product construction $\mathbf{P}_Y$ defines maps from $I^a(G;\Lambda;\Sigma \subset \Sigma')$ to $I^b(G;\Lambda;\Sigma \subset \Sigma')$, and by Theorem 4.1 these maps satisfy the commutativity properties described in (II.3.1.A) and (II.3.1.B). The discussion in the preceding paragraph shows that the maps $\mathbf{P}_Y(\Sigma \subset \Sigma')$ are isomorphisms except for a few borderline cases, and in these cases the maps are epimorphisms. Therefore by Comparison Theorem II.3.4 the maps $\mathbf{P}_Y(\Lambda)$ are isomorphisms and the maps $\mathbf{P}_Y(\mathbf{S}^{-1}\Lambda)$ are monomorphisms.■

Problems involving equivariant product formulas

In ordinary surgery theory one knows that the product homomorphisms

$$\mathbf{P}_*(\mathbf{CP}^{2m} \times \mathbf{CP}^{2n}) : L^c_*(\mathbb{Z}[\pi], w) \to L^c_{*+4(n+m)}(\mathbb{Z}[\pi], w)$$

induced by crossing with $\mathbf{CP}^{2m} \times \mathbf{CP}^{2n}$ are equal to the maps $\mathbf{P}_*(\mathbf{CP}^{2(m+n)}$ induced by crossing with $\mathbf{CP}^{2(m+n)}$ (*e.g.*, see [**Mrg**]). It is natural to ask if this result extends to equivariant surgery theory. Similarly, it is natural to ask if $\mathbf{P}_*((\mathbf{CP}^2\uparrow G)^n)$ is equal to $\mathbf{P}_*(\mathbf{CP}^{2n}\uparrow G)$. However, currently existing techniques do not seem powerful enough to establish such relationships. Since the algebraic methods developed by Ranicki have proved to be particularly effective for studying products (compare [**Ra2**, Section 8]), one approach to these questions could involve an equivariant version of Ranicki's techniques.

5. Periodic stabilization

The Gap Hypothesis is technically convenient because it leads to a well-behaved surgery theory and there are many situations in transformation groups where it either holds or can be realized quite easily. Since equivariant surgery with the Gap Hypothesis yields information about many basic questions about group actions (compare [**DPS**], [**MP**], and [**Sc2**]), the introduction of the Gap Hypothesis can also be justified by its applications. On the other hand, the Gap Hypothesis is unquestionably a restrictive assumption, and there are many important questions for which the usefulness of the Gap Hypothesis is at best uncertain. Thus one would like an approximation to an equivariant surgery theory that has the desirable features implied by the Gap Hypothesis, but without assuming the Gap Hypothesis itself. As Browder noted in his 1976 lecture at Stanford, Theorem 2.3 suggests the following idea:

(5.0) *Suppose G has odd order. Given a G-surgery problem $f : M \to X$, choose r so large that $f \times \mathbf{id}(\mathbf{CP}^2\uparrow G)^r$ satisfies the Gap Hypothesis, and study the surgery-theoretic invariants of this new problem.*

One objection to this idea is that the invariants of $f \times \mathbf{id}(\mathbf{CP}^2\uparrow G)^r$ may depend upon r in some unpredictable fashion. The periodicity theorems—especially Theorem 2.9—imply that the surgery-theoretic invariants of $f \times \mathbf{id}(\mathbf{CP}^2\uparrow G)^r$ are indeed independent of r provided r is sufficiently large.

We shall limit ourselves to three illustrations based upon (5.0). First, we shall show that stepwise surgery obstructions are frequently definable even if the Gap Hypothesis does not hold. Second, we shall consider surgery obstruction groups in some cases where the Gap Hypothesis does not hold, define a periodic stabilization map by crossing with $\mathbf{id}(\mathbf{CP}^2\uparrow G)^r$, and show that these maps are often split surjective. Finally, we shall use periodic stabilization and results of [**LM**] to show that if $c = h$ or s then the Browder-Quinn groups $L^{BQ,c}(G;\Lambda_X)$ are generally isomorphic to direct sums of stepwise surgery obstruction groups; it is known that similar splittings do not exist for groups of even order (compare [**LM**, Part II, Example 1.18]). In a subsequent paper the second author will use periodic stabilization to construct an infinite stable G-surgery exact sequence (compare [**Sc3**] and the remarks in the introduction) if G has odd order.

5A. Stable stepwise surgery obstructions

Although the Gap Hypothesis plays an important role in analyzing the sets $I^a(G,\Lambda)$, the definitions of these sets do not require that the Gap Hypothesis hold (compare [**DP1**,

(3.16), page 27] and Part I of [**LM**]). In some theories one needs the Gap Hypothesis to show that the abelian group structure in [**Dov1**] is well-defined, but in this subsection we shall not attempt to use this extra structure on $I^a(G;\Lambda)$.

In fact, the entire geometric discussion of the product map

$$\mathbf{P}_Y : I^a(G;\Lambda) \to I^a(G;\Lambda \times \mathrm{data}(Y))$$

(except possibly the discussion of addition) makes no use of the Gap Hypothesis. Therefore if we let r_0 be so large that $\lambda \times (data(\mathbf{CP}^2\!\uparrow\! G))^r$ satisfies the Gap Hypothesis if $r \geq r_0$, we obtain an r-fold stabilization map

$$\mathbf{P}_r : I^a(G;\Lambda) \to I^a(G;\Lambda \times (\mathrm{data}(\mathbf{CP}^2\!\uparrow\! G))^r)$$

by crossing with $(\mathbf{CP}^2\!\uparrow\! G)^r$. The periodicity Theorems (especially 2.9) say that this map is essentially independent of r, and thus it is meaningful to define the *stable set* $I^{a,\infty}(G;\Lambda)$ to be the codomain of $\mathbf{P}_r$ for all sufficiently large values of r. Of course, one also has a stabilization map

$$\mathbf{P}_\infty : I^a(G;\Lambda) \to I^{a,\infty}(G;\Lambda)$$

equal to $\mathbf{P}_r$ for r sufficiently large.

In order to use these stabilized objects effectively, it is necessary to know that they have many formal properties in common with ordinary surgery obstruction groups. We begin with the most basic observation.

PROPOSITION 5.1. *Let G be an odd order finite group. Then the set $I^{a,\infty}(G;\Lambda)$ has a natural abelian group structure given by the construction of* [**Dov1–2**] *and periodicity.*

PROOF: It suffices to observe that for r sufficiently large the data $\lambda \times (\mathrm{data}(\mathbf{CP}^2\!\uparrow\! G))^r$ satisfies the conditions for a group structure. Roughly speaking, these are the Gap Hypothesis and lower bounds on the dimension of closed substrata. But crossing with $(\mathbf{CP}^2\!\uparrow\! G)$ raises the dimensions of all closed substrata by at least 4, and therefore it is fairly simple to achieve any conceivable dimension bounds by taking products repeatedly (this also uses 2.6 to show that the G-posets in the indexing data $\lambda \times \mathrm{data}(\mathbf{CP}^2\!\uparrow\! G)^r$) stabilize for $r \geq 2$).■

The stable groups $I^{a,\infty}(G;\Lambda)$ have another familiar property.

PROPOSITION 5.2. *Crossing with $\mathbf{CP}^2$ induces an isomorphism from $I^{a,\infty}(G;\Lambda)$ to $I^{a,\infty}(G;\mathbf{S}^4\Lambda)$. Thus the groups $I^{a,\infty}(G;\mathbf{S}^k\Lambda)$ $(k \in \mathbb{Z})$ depend only upon k modulo 4.*

PROOF: For r sufficiently large one has that $\mathbf{S}^4\Lambda \times \mathrm{data}(\mathbf{CP}^2\!\uparrow\! G)^r$ satisfies the Gap Hypothesis, and therefore by Theorem 2.8 the product map

$$\mathbf{P}_Y : I^a(G;\Lambda \times \mathrm{data}(\mathbf{CP}^2\!\uparrow\! G)^r) \to I^a(G;\mathbf{S}^4\Lambda \times \mathrm{data}(\mathbf{CP}^2\!\uparrow\! G)^r)$$

is an isomorphism for $Y = \mathbf{CP}^2$. Furthermore, the associativity and commutativity of Cartesian product imply a compatibility relation

$$\mathbf{P}(\mathbf{CP}^2)\mathbf{P}(\mathbf{CP}^2\!\uparrow\! G) = \mathbf{P}(\mathbf{CP}^2\!\uparrow\! G)\mathbf{P}(\mathbf{CP}^2),$$

and therefore $\mathbf{P}(\mathbf{CP}^2)$ passes to a map $\mathbf{P}^* : I^{a,\infty}(\Lambda) \to I^{a,\infty}(\mathbf{S}^4\Lambda)$ in the limit. Since the individual maps $\mathbf{P}(\mathbf{CP}^2)$ are bijective on

$$I^a(\Lambda \times \text{data}(\mathbf{CP}^2\uparrow G)^r)$$

for r large, it follows that the limit map $\mathbf{P}^*$ is also bijective.■

Most of our discussion of products has dealt with G-manifolds X such that Λ_X is $(\mathbf{CP}^2\uparrow G)$-preperiodic. However, we want to state the main result on stepwise surgery obstructions in greater generality. If $\beta \in \pi(X)$ corresponds to the closed substratum X_β, then the associated closed substratum $\mathbf{P}(\beta) \in \pi(X \times \mathbf{CP}^2\uparrow G)$ is $X_\beta \times \text{Fix}(G^\beta, \mathbf{CP}^2\uparrow G)$. Suppose now that $\Sigma \subset \pi(X)$ is closed and G-invariant, and let α be minimal in $\pi(X) - \Sigma$. If $T_\alpha \subset \pi(X \times \mathbf{CP}^2\uparrow G)$ is the set

$$T_\alpha := \{\gamma \in \pi(X \times \mathbf{CP}^2\uparrow G) |\ \gamma \leq \mathbf{P}(\beta), \text{ some } \beta \in \Sigma \cup G\{\alpha\}\}$$

then it is a routine exercise to verify that $T_0 := T_\alpha - G\{\mathbf{P}(\alpha)\}$ is a closed and G-invariant subset of $\pi(X \times \mathbf{CP}^2\uparrow G)$ and $\mathbf{P}(\alpha)$ is minimal in $\pi(X \times \mathbf{CP}^2\uparrow G) - G\{\mathbf{P}(\alpha)\}$. Furthermore, it follows that the assignment $f \to f \times \mathbf{id}(\mathbf{CP}^2\uparrow G)$ sends Σ- and $(\Sigma \cup \{\alpha\})$-adjusted maps to T_0- and T_α-adjusted maps respectively.

THEOREM 5.3. *Assume the setting of the preceding paragraph, and let* (f, other data) *be Σ-adjusted and represent a class in $I^a(G; \Lambda)$. Then the following hold:*

(i) The maps $(f \times (\mathbf{CP}^2\uparrow G)^r, -)$ are all T_0-adjusted, where T_0 is as above.

(ii) There is a $(T_0 \subset T_\alpha)$ stepwise obstruction $\sigma_\alpha(f \times (\mathbf{CP}^2\uparrow G)^r$, other data) that depends only on the class of $(f, -)$ and vanishes if f is $\Sigma \cup G(\alpha)$-adjustable. This obstruction lies in $L^h_*(\mathbf{Z}[W(\alpha)])$ and is independent of r (i.e., it is compatible with the periodicity isomorphisms).

PROOF: (*Sketch*) As noted above, the closed substrata of $X \times (\mathbf{CP}^2\uparrow G)^r$ all have the form

$$X_\beta \times \text{Fix}(G^\beta, (\mathbf{CP}^2\uparrow G)).$$

¿From this description it is immediate that $f \times (\mathbf{CP}^2\uparrow G)^r$ is T_0-adjusted if f is Σ-adjusted. Therefore the methods of Sections I.4–5 yield a stepwise surgery obstruction

$$\sigma_\alpha(f \times \mathbf{id}, -) \in L(\alpha),$$

where $L(\alpha) = L^h_{\dim \alpha}(\mathbf{Z}[W_\alpha])$. Since taking products preserves geometric bordism, it follows that this obstruction only depends upon the bordism class of $(f, -)$ in $I^a(G; \Lambda)$. Furthermore, $(f \times \mathbf{id}, -)$ is T_α-adjusted if $(f, -)$ is $(\Sigma \cup G\{\alpha\})$-adjusted, and this proves everything except independence of r. For this one applies the results on stepwise obstructions (*e.g.*, Theorem 4.2), the result of 3.1(iii) on the Witt class of $(\mathbf{CP}^2\uparrow G)$, and the Witt invariance of the twisted product in [Yo] (plus the extension to L^h given in Section 4 of this chapter).

The stepwise obstructions in 5.3 look exactly like the obstructions one encounters if the Gap Hypothesis holds or the maps are transverse linear and isovariant. On the other hand, in general the stable stepwise obstruction in 5.3 is unlikely to be the *entire* obstruction to converting f into a $\Sigma \mathrm{U}G(\alpha)$-adjusted map; often there will be still further obstructions to completing equivariant surgery on X_α. We shall explain what happens in some special cases at the end of the next subsection (see 5.7).

If G has even order such stepwise stable obstructions do not necessarily exist (compare [**Dov3**, Example 4.13, p. 281] and the discussion in [**DS**, pp. 281–282]; also see [**Mmto**, Corollary C, page 468]).

5B. Unstable surgery obstruction groups

In Chapter II we noted that the Browder-Quinn groups $L^{BQ,c}$ and the Lück-Madsen groups $\mathcal{L}^c$ are defined even if the Gap Hypothesis does not hold. Furthermore, even though the basic construction of the Type I groups $I^{c,CAT}$ requires the Gap Hypothesis, there are some natural generalizations of these groups to cases where the Gap Hypothesis fails. Specifically, if X is a smooth manifold satisfying the Codimension ≥ 3 Gap Hypothesis and all the closed substrata of X are simply connected of dimension ≥ 5, we shall describe a sequence of abelian groups $\mathcal{U}I_k^{c,CAT}(G;X)$ such that $\mathcal{U}I_k^{c,CAT}(G;X) \cong I^{c,CAT}(G;\mathbf{S}^k\lambda_X)$ if $\mathbf{S}^{k+3}\lambda_X$ satisfies the Gap Hypothesis. Similar constructions can be performed using the Lück-Madsen groups $\mathcal{L}^c$ instead of the groups $I^{c,CAT}$. A few examples of this type were considered by P.Löffler in an early version of [**Lö**] and from a somewhat different viewpoint by the second author (unpublished). Nonequivariant versions of these constructions have also been known for some time (compare Rothenberg [**Ro**]).

REMARK: If the Gap Hypothesis holds, then the existence of an isomorphism from $\mathcal{U}I_k^{c,CAT}(G;X)$ to $I^{c,CAT}(G;\mathbf{S}^k\lambda_X)$ shows that the groups $\mathcal{U}I_k^{c,CAT}(G;X)$ depend only on the indexing data. This isomorphism is essentially a formal consequence of the equivariant $\pi-\pi$ Theorem (*i.e.*, Theorem I.5.1; compare [**DR**, Lemma 8.2, p. 68]). The examples of Section I.6 show that a similar argument will not work if the Gap Hypothesis fails. In these cases it seems quite likely that the unstable groups $\mathcal{U}I_k^{c,CAT}(G;X)$ depend upon more than just the indexing data.

DEFINITION: Let X be a closed smooth G-manifold with simply connected closed substrata and codimension ≥ 3 gaps between strata. Define the set $\mathcal{J}_k(G;X)$ to be the set of all G-normal maps of triads (in the sense of Dovermann-Rothenberg [**DR**])

$$(f;\partial_+f,\partial_-f):(W;\partial_+W,\partial_-W)\to(D^k;D_+^{k-1},D_-^{k-1})\times X$$

such that ∂_+f is a G-homotopy equivalence and ∂_-f is a diffeomorphism. Define a bordism relation on maps of triads via normal maps of manifolds with faces indexed by the faces of the square $[0,1]\times[0,1]$ (this terminology is borrowed from Jänich [**J**], M. Davis [**Dav1**], and Verona [**V**]). The induced maps of faces satisfy

(1) the triad maps $(f_{0I};f_{01},f_{00})$ and $(f_{1I};f_{11},f_{10})$ are in $\mathcal{J}_k(G;X)$,
(2) the map $(f_{I0};f_{10},f_{00})$ is a G-diffeomorphism,
(3) the map $(f_{I1};f_{11},f_{01})$ is a G-homotopy equivalence.

The picture below might be helpful:

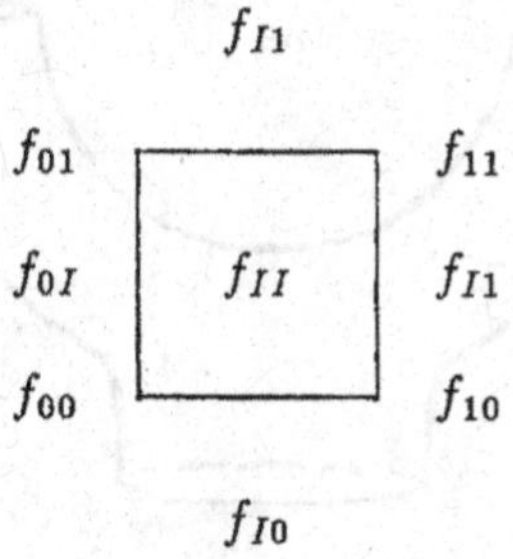

Figure I

Set $\mathcal{UI}_k^{c,CAT}(G;X)$ equal to the set of bordism classes of triad maps in $\mathcal{J}_k(G;X)$.

For the sake of definiteness and technical simplicity we shall assume $CAT = DIFF$ henceforth.

THEOREM 5.4. *The set $\mathcal{UI}_k^{c,DIFF}(G;X)$ has a group structure if $k \geq 2$, and this group structure is abelian if $k \geq 3$. If $\mathbf{S}^{k+2}\lambda_X$ satisfies the Gap Hypothesis, then $\mathcal{UI}_k^{c,DIFF}(G;X)$ is isomorphic to $I^{c,DIFF}(G;\mathbf{S}^k\lambda_X)$ with the addition operation defined as in* **[Dov1]** *and* **[Dov2]**. *Finally, there is a canonical homomorphism from the Browder-Quinn group $L^{BQ,c}(\mathbf{S}^k\Lambda_X)$ into $\mathcal{UI}_k^{c,DIFF}(G;X)$, and this map is an isomorphism if $\mathbf{S}^k\lambda_X$ satisfies the Gap Hypothesis and G has odd order.*

PROOF: (*Sketch*) Suppose that $(f_i; \partial_+ f_i, \partial_- f_i)$ are in $\mathcal{J}_k(G;X)$ for $i = 0,1$. Up to bordism we may assume that the maps $\partial_+ f_i$ are G-diffeomorphisms on neighborhoods of the boundary (for the maps $\partial_- f_i$ are G-diffeomorphisms and $\partial\partial_+ = \partial\partial_-$). Choose a smoothly embedded disk $E \subseteq S^{k-1}$ such that $E \times X$ is contained in these neighborhoods, ∂E meets a great sphere S^{k-2} transversely and $(E; D_+^{k-1}, D_-^{k-1})$ is diffeomorphic to

$$(D^{k-1}; \{x_{k-1} \geq 0\}, \{x_{k-1} \leq 0\});$$

if $k = 3$ and one considers the $(k-2)$-sphere corresponding to the $0°$ and $180°$ meridians, then one might choose E to the region determined by the North Pole and the Arctic Circle. Attach the maps $(f_0; f_{0+}, f_{0-})$ and $(f_1; f_{1+}, f_{1-})$ along E and smooth out the corners; this is possible because the restrictions $F_i | f_i^{-1}(E \times X)$ are diffeomorphisms. Of course, this is a standard method for constructing geometric sums, and the usual arguments imply that this yields a group structure if $k \geq 2$ and the group is abelian if $k \geq 3$.

Suppose now that the Gap Hypothesis holds and $I^{c,DIFF}(\mathbf{S}^k\lambda_X)$ has a group structure by the construction of **[Dov1]**. There is a standard fork (or "pair of pants") cobordism P from $D^k \amalg D^k$ to $D^k \simeq D^k \cup_E D^k$ (see the figure below), and one can construct a map F into $P \times X$

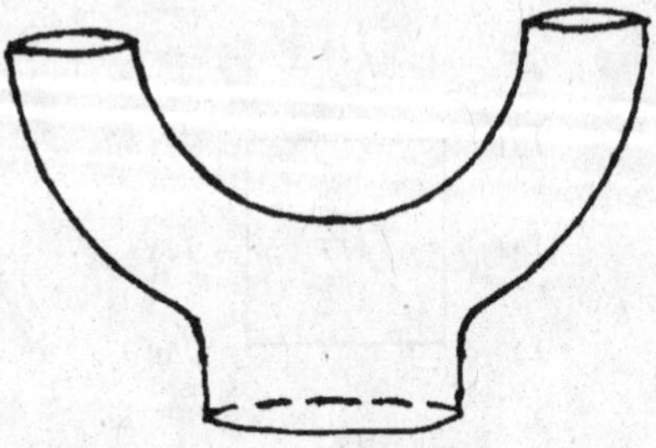

Figure II

such that F over $(D^k \amalg D^k)\times X$ is $f_0 \amalg f_1$ and F over $(D^k \cup_E D^k)\times X$ is a representative for the previously defined sum of $[f_0]+[f_1]$ in the group $\mathcal{U}I_k^{c,DIFF}(G;X)$. But by 1.2 we also know that $F|F^{-1}((D^k \cup_E D^k)\times X)$ is a representative for the sum of $[f_0]$ and $[f_1]$ in $I^{c,DIFF}(\mathbf{S}^k\lambda_X)$. Therefore we have that the natural map

$$j : \mathcal{U}I_k^{c,DIFF}(G;X) \to I^{c,DIFF}(G;\mathbf{S}^k\lambda_X)$$

is a homomorphism. On the other hand, in the remark preceding the theorem we observed that j is bijective, and therefore j is a group isomorphism.

To construct the map from $L^{BQ,c}(\mathbf{S}^k\lambda(\mathrm{id}_X))$ into $\mathcal{U}I_k^{c,CAT}(G;X)$, use the Browder-Quinn $\pi-\pi$ theorem (compare [**BQ**, (3.1), p. 31]) to show that the group $L^{BQ,c}(\mathbf{S}^k\Lambda_X)$ is definable by Browder-Quinn maps of triads into $D^k\times X$ such that one has a Browder-Quinn equivalence over $D^k_+\times X$ and a G-diffeomorphism over $D^k_-\times X$. The results of Subsection II.3F imply that Browder-Quinn maps yield elements of $\mathcal{J}_k(G;X)$; in general the passage from the Browder-Quinn maps to $\mathcal{J}_k(G;X)$ depends upon choices and is not well-defined, but any two choices yield cobordant representatives, and accordingly there is a well-defined map $h : L^{BQ,c}(G;\mathbf{S}^k\Lambda_X) \to \mathcal{U}I_k^{c,DIFF}(G;X)$. The group operation $[f_0]+[f_1]$ in $L^{BQ,c}(G;\mathbf{S}^k;\Lambda_X)$ is generally given by disjoint union, but in this case one can use the cobordism P of Figure II to construct a cobordism from the disjoint union to a class in $\mathcal{J}_k(G;X)$ that represents the sum $h[f_0]+h[f_1]$ in $\mathcal{U}I_k^{c,DIFF}(G;X)$. Thus h is a homomorphism.

Finally, we must show that h is an isomorphism if $\mathbf{S}^{k+2}\lambda_X$ satisfies the Gap Hypothesis. But jh is just the canonical map from $L^{BQ,c}$ to $I^{c,DIFF}$, and thus the results of Subsection II.3F show that jh is an isomorphism; since j is an isomorphism by the preceding discussion, it follows that h must also be an isomorphism.■

5C. Splitting theorems

If G has odd order there is an interesting relationship between the unstable groups $\mathcal{U}I_k^{c,DIFF}(G;X)$ and the stable groups $I^{c,DIFF,\infty}(\mathbf{S}^k\lambda_X)$ introduced in Subsection 5A above. We begin with some elementary observations.

Proposition 5.5. *(i) If G is a finite group and Y is a closed smooth G-manifold, then there is a product homomorphism*

$$\mathbf{P}_Y : \mathcal{U}I_k^{c,DIFF}(G;X) \to \mathcal{U}I_k^{c,DIFF}(G;X\times Y)$$

defined on representatives by taking $(f;\partial_+ f,\partial_- f)$ to $(f\times \mathbf{id}_Y;\partial_+ f\times \mathbf{id}_Y,\partial_- f\times \mathbf{id}_Y)$.
(ii) If G has odd order, then periodic stabilization defines a homomorphism

$$\mathbf{P}_\infty : \mathcal{U}I_k^{c,DIFF}(G;X) \to I^{c,DIFF,\infty}(G;\mathbf{S}^k\lambda_X).\blacksquare$$

The following result states that $\mathbf{P}_\infty$ carries a significant amount of information about $\mathcal{U}I_k^{c,DIFF}(Y)$ in many cases.

Proposition 5.6. *If X has strongly saturated orbit structure and the dimensions of all closed substrata are ≥ 7, then $\mathbf{P}_\infty$ is split surjective.*

Proof: We have a commutative diagram

$$\begin{array}{ccc}
L^{BQ,c}(G;\mathbf{S}^k\Lambda_X) & \xrightarrow{h(X)} & \mathcal{U}I_k^{c,DIFF}(G;X) \\
\mathbf{P}_Y^{BQ,c}\big\downarrow & & \mathbf{P}_Y^{ht}\big\downarrow \\
L^{BQ,c}(G;\mathbf{S}^k\Lambda_{X\times Y}) & \xrightarrow{h(X\times Y)} & I^{c,DIFF}(G;\mathbf{S}^k\lambda_{X\times Y})
\end{array}$$

in which $Y = (\mathbf{C}P^2\uparrow G)^r$ for r sufficiently large, the map $\mathbf{P}_Y^{ht}$ is essentially periodic stabilization and $h(X\times Y)$ is an isomorphism by 5.4. On the other hand, X is Y-preperiodic by 2.5 since X has strongly saturated orbit structure, and therefore the Browder-Quinn product map $\mathbf{P}_Y^{BQ,c}$ is an isomorphism by 2.9. It follows that the composite $\mathbf{P}_Y^{ht}h(X)$ is bijective and therefore $\mathbf{P}_Y^{ht}$ must be split surjective.$\blacksquare$

Remark: The proof of 5.5 also shows that the map

$$h(X) : L^{BQ,c}(\mathbf{S}^k\Lambda_X)) \to \mathcal{U}I_k^{c,DIFF}(G;X)$$

is split injective *if G has odd order*. In contrast, results of the first named author of this paper show that $h(X)$ need not be injective if $G = \mathbf{Z}_2$ and $S^k\times X$ lies just outside the Gap Hypothesis range; *i.e.*,

$$2(\dim X^{\mathbf{Z}_2} + k) = \dim X + k,$$

where $\dim X + k$ is odd (see **[Dov3]**, **[DS]**, and **[Mmto]** for further information).

Comments on the unstable summand

Of course the kernel of $\mathbf{P}_\infty$ measures the additional difficulties in equivariant surgery when the Gap Hypothesis fails. In some cases it should be possible to obtain considerable information on the kernel of $\mathbf{P}_\infty$. One test case is $D(V)$, where V is a free representation of $G = \mathbb{Z}_p$, and $D(V)$ and $S(V)$ denote the unit disk and sphere. It is possible to construct variants of the previous groups

$$\mathcal{U}I_k^{t,sp}(D(V), S(V))$$

in which all restricted maps into $D^k \times S(V)$ are G-diffeomorphisms and all normal maps are *equivariantly stably tangential* (compare Madsen-Taylor-Williams [**MTW**]) and *specially framed* in the sense of Löffler [**Lö**]. In this case one again has a split surjection from $\mathcal{U}I_k^{t,sp}(D(V), S(V))$ to $L^{BQ,c}(\mathbf{S}^k \Lambda_{D(V)})$, and the kernel is isomorphic to

$$\pi_k(F_G, F_G(V)),$$

where F_G and $F_G(V)$ are defined in [**BeS**]. Thus there is an isomorphism

$$\mathcal{U}I_k^{t,sp}(D(V), S(V)) \simeq L^{BQ,c}(\mathbf{S}^k \Lambda_{D(V)}) \oplus \pi_k(F_G, F_G(V)). \tag{5.7}$$

This will be proved and studied further in a detailed version of [**Sc3**].

A splitting theorem for Browder-Quinn groups

In [**LM**] Lück and Madsen show that the groups $\mathcal{L}^s(G; R_X)$ split as a direct sum of stepwise obstruction groups if G has odd order and X satisfies a form of the Gap Hypothesis and a *slice extension property* described below. We shall conclude this section with an extension of this result to Browder-Quinn groups.

DEFINITION: Let X be a compact locally linear G-manifold, let x be a point in X with isotropy subgroup H, and let $NH(x)$ be the component of the normalizer of H that sends the component of Fix(H, X) containing x into itself. The manifold X satisfies the **slice extension property** at x if the slice representation at x extends to a representation of $NH(x)$.

EXAMPLE: If G is an abelian group of odd order, then X satisfies the slice extension property at every point.

THEOREM 5.8. (*cf.* [**LM**, Thm. II.2.11]). *Suppose that the group G has odd order, and let X be a compact locally linear G-manifold such that all closed substrata have dimension at least 5, $\mathbf{S}^3\Lambda_X$ satisfies the Gap Hypothesis, and X has the slice extension property at every point. Then $\mathcal{L}^s(G; R_X)$ is isomorphic to a direct sum of stepwise groups*

$$\bigoplus_{\beta \in \pi'} L^s_{\dim \beta}(\mathbb{Z}[E_\beta], w_\beta)$$

where $\pi' \subset \pi(X)$ contains one element from each orbit in the G-poset $\pi(X)$. ■

Since the Lück-Madsen and Browder-Quinn groups are isomorphic if $\mathbf{S}^3\Lambda_X$ satisfies the Gap Hypothesis, it follows that the corresponding Browder-Quinn groups also split in this range. In fact, one also has such splittings outside the Gap Hypothesis range.

THEOREM 5.9. *Suppose that G is an abelian group of odd order, and let X be a compact smooth G-manifold such that all closed substrata have dimension at least 5, Λ_X satisfies the Codimension ≥ 3 Gap Hypothesis, and X has the slice extension property at every point. Then $L^{BQ,s}_{\dim X}(G; R_X)$ is isomorphic to a direct sum of stepwise groups*

$$\bigoplus_{\beta\in\pi'} L^{s}_{\dim \beta}(\mathbf{Z}[E_\beta], w_\beta)$$

where $\pi' \subset \pi(X)$ contains one element from each orbit in the G-poset $\pi(X)$.

It seems quite likely that one could prove this by systematically modifying the argument in [**LM**, Part II, Section 2]. However, the periodicity theorems of this chapter yield a much simpler proof.

PROOF OF 5.9: Let $Y(r) = (\mathbf{CP}^2\uparrow G)^r$ where r is so large that $\mathbf{S}^3\Lambda_{X\times Y(r)}$ satisfies the Gap Hypothesis. We then have the following commutative diagram, in which the horizontal arrows are given by taking products with $Y(r)$, the vertical arrows are the canonical maps from Browder-Quinn to Lück-Madsen groups, and the arrows marked by $\cong$ are isomorphisms by Periodicity Theorem 2.9 and the results of Section II.3:

$$\begin{array}{ccc}
L^{BQ,s}(G;\Lambda_X) & \xrightarrow[\cong]{\text{stabilize}} & L^{BQ,s}(G;\Lambda_{X\times Y(r)}) \\
\downarrow & & \cong\downarrow \\
\mathcal{L}^{s}(G;\Lambda_X) & \xrightarrow{\text{stabilize}} & \mathcal{L}^{s}(G;\Lambda_{X\times Y(r)})
\end{array}$$

Theorem 5.8 states that the group in the lower right hand corner splits as a sum of stepwise obstruction groups. Since the groups in the upper left and lower right corners are isomorphic, it follows that the group in the upper left hand corner is also a direct sum of ordinary Wall groups. Furthermore, the naturality properties of the canonical maps $L^{BQ,s} \to \mathcal{L}^{s}$ imply that the summands of the induced splitting of $L^{BQ,s}(G;\Lambda_X)$ correspond to stepwise obstruction groups in the Browder-Quinn theory.■

Splittings of unstable Lück-Madsen groups

An argument similar to the proof of Theorem 5.9 yields the following splitting theorems for the unstable Lück-Madsen groups if X satisfies the Codimension ≥ 3 Gap Hypothesis, all closed substrata have dimension at least five, and as usual the group G has odd order.

THEOREM 5.10. *If G has odd order and the closed smooth G-manifold X satisfies the conditions in the preceding paragraph, then periodic stabilization defines homomorphisms*

$$\mathbf{P}_\infty : \mathcal{L}^{c}(G;\Lambda_X;\Sigma\subset\Sigma') \to \mathcal{L}^{c,\infty}(G;\Lambda_X;\Sigma\subset\Sigma')$$

that are **split surjective**. *Furthermore, a one sided inverse is given by the canonical map from the corresponding Browder-Quinn theory $L^{BQ,c}(G;\Lambda_X;\Sigma\subset\Sigma')$.*

Note that the Browder-Quinn group $L^{BQ,c}(G;\Lambda_X;\Sigma\subset\Sigma')$ is isomorphic to the periodically stabilized Lück-Madsen group because both are isomorphic to the periodically stabilized Browder-Quinn groups $L^{BQ,c,\infty}(G;\Lambda_X;\Sigma\subset\Sigma')$. Specifically, the equivalence between ordinary and stabilized Browder-Quinn groups follows from the equivariant periodicity theorems of this chapter and the equivalence between stabilized Browder-Quinn and Lück-Madsen groups follows from Theorem II.3.15 (i).

PROOF: *(Sketch)* We shall use the notation of the previous proof except that the superscript c will denote either h or s. Choose r so large that the group $L^{BQ,c}(G;\Lambda_X\times Y(r);\Sigma\subset\Sigma')$ and $\mathcal{L}^c(G;\Lambda_X\times Y(r);\Sigma\subset\Sigma')$ maps isomorphically to its periodic stabilization, and consider the commutative square constructed in the proof of Theorem 5.9. Once again the upper and right hand arrows are isomorphisms. Therefore the lower arrow is split surjective and the left hand arrow is split injective.■

References for Chapter III

[ACH] J. P. Alexander, P. E. Conner, and G. C. Hamrick, "Odd Order Group Actions and Witt Classification of Inner Products," Lecture Notes in Mathematics Vol. 625, Springer, Berlin-Heidelberg-New York, 1977.

[AS] M. F. Atiyah and I. M. Singer, *The index of elliptic operators. III*, Ann. of Math. **87** (1968), 546–604.

[BeS] J. C. Becker and R. E. Schultz, *Equivariant function spaces and stable homotopy theory I*, Comment. Math. Helv. **49** (1974), 1-34.

[Bre] G. Bredon, "Introduction to Compact Transformation Groups," Pure and Applied Mathematics Vol. 46, Academic Press, New York, 1972.

[Bro1] W. Browder, "Surgery on Simply Connected Manifolds," Ergeb. der Math. (2) 65, Springer, New York, 1972.

[Bro2] ———, "Surgery and group actions," lectures at A.M.S. Symposium on Algebraic and Geometric Topology, Stanford University, 1976.

[BQ] W. Browder and F. Quinn, *A surgery theory for G-manifolds and stratified sets*, in "Manifolds–Tokyo, 1973," (Conf. Proc. Univ. of Tokyo, 1973), University of Tokyo Press, Tokyo, 1975, pp. 27–36.

[CS] S. Cappell and J. Shaneson, *The codimension two placement problem and homology equivalent manifolds*, Ann. of Math. **99** (1974), 277–348.

[CF] P. E. Conner and E. E. Floyd, "Differentiable Periodic Maps," Ergeb. der Math. Bd. 33, Springer, Berlin-Göttingen-Heidelberg, 1963.

[Dav] M. Davis, *Smooth G-manifolds as collections of fiber bundles*, Pac. J. Math. **77** (1978), 315–363.

[DH] M. Davis and W. C. Hsiang, *Concordance classes of regular $U(n)$ and $Sp(n)$ actions on homotopy spheres*, Ann. of Math. **105** (1977), 325–341.

[DHM] M. Davis, W. C. Hsiang, and J. Morgan, *Concordance classes of regular $O(n)$ actions on homotopy spheres*, Acta Math. **144** (1980), 153–221.

[tD] T. tom Dieck, *Orbittypen und äquivariante Homologie II*, Arch. Math. (Basel) **26** (1975), 650–662.

[Dov1] K. H. Dovermann, "Addition of equivariant surgery obstructions," Ph.D. Thesis, Rutgers University, 1978 *(Available from University Microfilms, Ann Arbor, Mich.: Order Number DEL79-10380.)*—Summarized in Dissertation Abstracts International **39** (1978/1979), 5406..

[Dov2] ———, *Addition of equivariant surgery obstructions*, in "Algebraic Topology, Waterloo 1978 (Conference Proceedings)," Lecture Notes in Mathematics Vol. 741, Springer, Berlin-Heidelberg-New York, 1979, pp. 244-271.

[Dov3] ———, *Z_2 surgery theory*, Michigan Math. J. **28** (1981), 267–287.

[**DP1**] K. H. Dovermann and T. Petrie, *G-Surgery II*, Mem. Amer. Math. Soc. **37** (1982), No. 260.

[**DP2**] ———————, *An induction theorem for equivariant surgery (G-Surgery III)*, Amer. J. Math. **105** (1983), 1369–1403.

[**DPS**] K. H. Dovermann, T. Petrie, and R. Schultz, *Transformation groups and fixed point data*, Proceedings of the A.M.S. Summer Research Conference on Group Actions (Boulder, Colorado, 1983), Contemp. Math. 36 (1985), 161-191..

[**DR**] K. H. Dovermann and M. Rothenberg, *An Equivariant Surgery Sequence and Equivariant Diffeomorphism and Homeomorphism Classification*, Mem. Amer. Math. Soc. **71** (1988), No. 379.

[**DS**] K. H. Dovermann and R. Schultz, *Surgery on involutions with middle dimensional fixed point set*, Pac. J. Math. **130** (1988), 275–297.

[**Hi**] F. Hirzebruch, *Involutionen auf Mannigfaltigkeiten*, in "Proceedings of the Conference on Transformation Groups (Tulane, 1967)," Springer, Berlin-Heidelberg-New York, 1968, pp. 148–166.

[**J**] K. Jänich, *On the classification of O_n-manifolds*, Math. Ann. **176** (1978), 53–76.

[**JäO**] K. Jänich and E. Ossa, *On the signature of an involution*, Topology **8** (1969), 27–30.

[**KiS**] R. C. Kirby and L. C. Siebenmann, "Foundational Essays on Topological Manifolds, Smoothings, and Triangulations," Annals of Mathematics Studies Vol. 88, Princeton University Press, Princeton, 1977.

[**Le**] W. Lellmann, *Orbiträume von G-Mannigfaltigkeiten und stratifizierte Mengen*, Diplomarbeit, Universität Bonn, 1975.

[**Lö**] P. Löffler, *Homotopielineare $\mathbb{Z}_p$ Operationen auf Sphären*, Topology **20** (1981), 291–312.

[**LM**] W. Lück and I. Madsen, *Equivariant L-theory I*, Aarhus Univ. Preprint Series (1987/1988), No. 8; [same title] *II*, Aarhus Univ. Preprint Series (1987/1988), No. 16 (to appear in Math. Zeitschrift).

[**MTW**] I. Madsen, L. Taylor, and B. Williams, *Tangential homotopy equivalence*, Comment. Math. Helv. **55** (1980), p. 445–484.

[**MP**] M. Masuda and T. Petrie, *Lectures on transformation groups and Smith equivalence*, Proceedings of the A.M.S. Summer Research Conference on Group Actions (Boulder, Colorado, 1983), Contemp. Math. 36 (1985), 193-244.

[**MS**] J. Milnor and J. Stasheff, "Characteristic Classes," Annals of Mathematics Studies Vol. 76, Princeton University Press, Princeton, 1974.

[**Mrg**] J. Morgan, *A Product Formula for Surgery Obstructions*, Mem. Amer. Math. Soc. **14** (1978), No. 201.

[**Mmto**] M. Morimoto, *Bak groups and equivariant surgery*, *K*-Theory **2** (1989), 456–483.

[Ni] A. Nicas, *Induction Theorems for Groups of Homotopy Manifold Structures*, Memoirs Amer. Math. Soc. **39** (1982). No. 267.

[P] T. Petrie, *G-Surgery I–A survey*, in "Algebraic and Geometric Topology (Conference Proceedings, Santa Barbara, 1977)," Lecture Notes in Mathematics Vol. 644, Springer, Berlin-Heidelberg-New York, 1978, pp. 197–223.

[Ra1] A. Ranicki, *The total surgery obstruction*, in "Algebraic Topology," (Sympos. Proc., Aarhus, 1978) Lecture Notes in Mathematics, Springer, Berlin-Heidelberg-New York, 1979, pp. 271–316.

[Ra2] ——————, *The algebraic theory of surgery I: Foundations*, Proc. London Math. Soc. (3) **40** (1980), 87–192.

[Ra3] ——————, *The algebraic theory of surgery II: Applications to Topology*, Proc. London Math. Soc. (3) **40** (1980), 193–283.

[Ra4] ——————, "Exact sequences in the algebraic theory of surgery," Princeton Mathematical Notes No. 26, Princeton University Press, Princeton, N. J., 1981.

[Ro] M. Rothenberg, *Differentiable group actions on spheres,*, in "Proc. Adv. Study Inst. on Algebraic Topology (Aarhus, 1970)," Various Publications Series Vol. 13, Matematisk Institut, Aarhus Universitet, 1970, pp. 455–475.

[Sc1] R. Schultz, *Homotopy sphere pairs admitting semifree differentiable actions*, Amer. Math. J. **96** (1974), 308–323.

[Sc2] ——————, *Homotopy invariants and G-manifolds: A look at the past 15 years*, Proceedings of the A.M.S. Summer Research Conference on Group Actions (Boulder, Colorado, 1983), Contemp. Math. 36 (1985), 17-83.

[Sc3] ——————, *An infinite exact sequence in equivariant surgery*, Mathematisches Forschungsinstitut Oberwolfach Tagungsbericht 14/1985 (Surgery and L-theory), 4–5.

[Sh] J. Shaneson, *Wall's surgery obstruction groups for* $G \times \mathbb{Z}$, Ann. of Math. **90** (1969), 296–334.

[Sh2] ——————, *Product formulas for* $L_n(\pi)$, Bull. Amer. Math. Soc. **76** (1970), 787–791.

[Su] D. Sullivan, *On the Hauptvermutung for manifolds*, Bull. Amer. Math. Soc. **73** (1967), 598–600.

[Th] R. Thom, *Ensembles et morphismes stratifiés*, Bull. Amer. Math. Soc. **75** (1969), 240–284.

[V] A. Verona, "Stratified Mappings–Structure and Triangulability," Lecture Notes in Mathematics Vol. 1102, Springer, Berlin-Heidelberg-New York, 1984.

[WL] C. T. C. Wall, "Surgery on Compact Manifolds," London Math. Soc. Monographs Vol. 1, Academic Press, London and New York, 1970.

[Wb] S. Weinberger, *The topological classification of stratified spaces*, preprint, University of Chicago, 1989.

[Wi] R. E. Williamson, *Surgery in $M \times N$ with $\pi_1(M) \neq 1$*, Bull. Amer. Math. Soc. **75** (1969), 582–585.

[Ya] M. Yan, *Periodicity in equivariant surgery and applications*, Ph. D. Thesis, University of Chicago, in preparation.

[Yo] T. Yoshida, *Surgery obstructions of twisted products*, J. Math. Okayama Univ. **24** (1982), 73–97.

CHAPTER IV

TWISTED PRODUCT FORMULAS FOR SURGERY WITH COEFFICIENTS

The periodicity theorems of Chapter III depend substantially upon Yoshida's work on twisted product formulas [**Yo**] and Ranicki's algebraic approach to product formulas in ordinary surgery (see [**Ra3**]–[**Ra5**]). In this chapter we shall extend Ranicki's and Yoshida's results to a larger class of surgery obstruction groups, and we shall use these extensions to prove periodicity theorems for various equivariant surgery theories with coefficients. The results of this chapter will also be used in Chapter V to prove a periodicity theorems in the theories for equivariant surgery up to pseudoequivalence described in [**DP1**] and [**DP2**].

Most of the results of this chapter deal with nonequivariant surgery theory. Since it is possible that the results on twisted products could prove useful for questions of independent interest, we shall not work with equivariant surgery in this chapter until we reach the applications in the final section.

For our purposes it will be essential to distinguish between various definitions of surgery groups. We shall denote the usual simple and homotopy surgery obstruction groups of [**Wa1**] by $L_*^X(\mathbb{Z}[\pi], w)$, where $X = s$ or h, and we shall denote the algebraic groups of Ranicki [**Ra1–6**] by ${}^R L_*^X(\mathbb{Z}[\pi], w)$. The results of [**Ra4**] and [**Ra5**] yield an isomorphism from ${}^R L_*^X$ to L_*^X. This isomorphism is a fundamental link between algebraic and geometric surgery; in particular, the isomorphism is crucial to the proof of Yoshida's formula in [**Yo**]. A key technical step in this paper is to establish a similar isomorphism when $\mathbb{Z}[\pi]$ is replaced by $R[\pi]$ for some subring R of the rationals. Although the conclusion is predictable, the proof contains some unanticipated complications. We shall also generalize Yoshida's formula to certain projective and subprojective L groups (such as the groups L_*^p of [**Mau1,Ra4,PR**], the groups $L_*^{D(G)}$ discussed in Section 4, and the so-called negative Wall groups L_*^{-i} [**AP,Q2,Ra2**]), and also to the Cappell–Shaneson Γ–groups [**CS**].

The first section of this paper develops our notation and contains the precise statements of our results (see Proposition 1.3, Theorem 1.5, Propositions 1.8–9, and Theorem 1.11). In Section 2 we verify that the appropriate algebraic and geometric surgery obstruction groups are isomorphic. Proofs of the basic product formulas appear in Section 3; from a conceptual viewpoint, the main problem is to formulate suitable generalizations of Ranicki's stable homotopy theoretic constructions in [**Ra5**]. Section 4 discusses the extensions to projective L–groups, intermediate groups L_*^X where X is a "reasonable" subgroup of $K_1(\mathbb{Z}[\pi])$ or $\widetilde{K}_0(\mathbb{Z}[\pi])$, and several other families of a similar nature.

In Section 5 we shall describe extensions to the homology surgery obstruction groups Γ_*^X of Cappell and Shaneson [**CS**] using the algebraic description of the latter groups due to Ranicki [**Ra6**]. Finally, in Section 6 we shall apply the formulas of Sections 1-5 to obtain periodicity theorems for several different types of equivariant surgery theories not covered in Chapter III.

Acknowledgments. We are grateful to Andrew Ranicki for helpful comments on an early version of this chapter and to Bruce Williams for kindly providing accounts of unpublished work on supersimple L–theories due to himself, his collaborators, and his student G. T. Kennedy [**BW,K**].

1. Basic definitions and results

We shall begin by describing the appropriate analogs of Wall's geometric L–groups in [**Wa1**], Section 9. In order to streamline our notation we shall adopt some conventions introduced by G. Anderson [**An**] (see Section 5.3, pages 86–87). The geometric L–groups depend upon five pieces of data:

(i) The group π.
(ii) A homomorphism $w : \pi \to \mathbb{Z}_2$.
(iii) A subring $R \subset \mathbb{Q}$ (with unit).
(iv) If $R = \mathbb{Z}$, an appropriate self–dual subgroup A of $K_1(\mathbb{Z}[\pi])$, in our setting either $\langle -1, \pi\rangle$ or all of K_1; if $R \neq \mathbb{Z}$, we shall only consider the case where the relevant subgroup is all of $K_1(R[\pi])$.
(v) A unit d in R.

As in Anderson [**An**, page 87], we shall write

$$\varPi = (K(\pi,1), w_*; R; A).$$

The relevant groups will be denoted by

$$\Omega_n(\varPi, d).$$

In defining these groups we shall follow Wall's approach involving restricted objects. Elements in the group $\Omega_n(\varPi, d)$ are represented by normal maps

$$\mathcal{W} = (X, f, b)$$

and reference maps $k : Y \to K(\pi,1)$ where

(1.1a) $f : (X, \partial X) \to (Y, \partial Y)$ is a degree d map of compact, connected CAT-manifolds, where $CAT = DIFF$, PL or TOP,
(1.1b) $f_* : \pi_1(X) \to \pi_1(Y)$ is bijective,
(1.1c) k_* induces an isomorphism from $\pi_1(Y)$ to π such that w corresponds to the Stiefel–Whitney class $w_1(X)$ and $w = w_1(X) = f_* w_1(Y)$,
(1.1d) b is a bundle map from the stable normal (equivalently, tangent) bundle of X to some bundle ξ over Y, and b covers f,
(1.1e) the map $\partial f : \partial X \to \partial Y$ is an R-homology equivalence with twisted coefficients in $R[\pi]$. If $R = \mathbb{Z}$ and A is the subgroup of $K_1(\mathbb{Z}[\pi])$ specified by $\varPi$ in (iv) then we require the torsion $\tau(\partial f)$ to lie in A. The case $A = \langle -1, \pi\rangle$ corresponds to the setting in Section 9 of Wall's book.

There is an obvious definition of normal cobordism for such objects. If $c = h$, a normal map (X, f, b) represents zero if it is cobordant rel boundary to an R homology equivalence (with arbitrary twisted coefficients). If $c = s$ we also require that the torsion $\tau(f)$ vanishes in $Wh(\mathbb{Z}[\pi])$. Therefore two objects $\mathcal{W}_0$ and $\mathcal{W}_1$ represent the same class in $\Omega_n(\Pi, d)$ if there exists a normal map $\mathcal{U} = (V, F'' : V \to Z, B)$ such that

(i) the boundary $\partial\mathcal{U}$ is the union of $\mathcal{W}_0 \amalg -\mathcal{W}_1$ (minus means reverse orientation), and a zero object $\mathcal{W}_2$,

(ii) the inclusions induce isomorphism from $\pi_1(Y_i)$ to $\pi_1(Z)$ for $i = 0, 1$.

We must strengthen our notion of equivalence to include bundle data. Specifically, if (ξ, b) and (ξ', b') are bundle data for $f : X \to Y$, then we identify (X, f, b) and (X, f, b') if there is a bundle isomorphism $a : \xi \to \xi'$ such that $f^*(a)b = b'$. The equivalence classes of this equivalence relation are denoted by $\Omega_n(\Pi, d)$; if $R = \mathbb{Z}$ and $d = 1$ these are the groups $L_n^c(K(\pi, 1), w)$ as defined in Section 9 of Wall's book [**Wa1**].

PROPOSITION 1.2. *If $n \geq 5$, then the set*

$$\Omega_n(\Pi, d)$$

is an abelian group.

This type of result is quite well understood, but we shall prove it in Section 2 for the sake of completeness. In fact, the result is implicit in [**Do**], but in our case the ideas really go back to Section 9 of Wall's book [**Wa1**].

The next step is to compare the bordism groups of Proposition 1.2 to the algebraic surgery obstruction groups $L_n^c(R[\pi], w)$ defined by Wall ([**Wa1**], Sections 5, 6, and 17) and Ranicki [**Ra1–6**] (also see [**An**] and [**T**]).

Suppose that $\mathcal{W} = (X, f, b)$ is a normal map which represents a class in $\Omega_n(\Pi, d)$. Apply surgery below the middle dimension to make f connected up to the middle dimension, and let $\widetilde{f} : \widetilde{X} \to \widetilde{Y}$ denote the induced map of universal covering spaces. If $n = 2k$ we obtain a $(-1)^k$-symmetric quadratic form on a free $R[\pi]$-module as follows: By general principles we have that

$$H^i(\widetilde{X}; R) = H^i(\widetilde{Y}; R) \oplus \widetilde{H}^{i+1}(\mathrm{Cone}(\widetilde{f}); R)$$

and in fact the second summand is trivial unless $i = k$. Furthermore, for all practical purposes the only possibly nontrivial second summand $H^{k+1}(\mathrm{Cone}(\widetilde{f}); R)$ may be assumed to be a free $R[\pi]$-module (compare [**Wa1**], Section 2), and the cup product defines a $(-1)^k$-symmetric bilinear form on the group $\widetilde{H}^{k+1}(\mathrm{Cone}(\widetilde{f}); R)$. A more careful analysis show that one in fact obtains the entire quadratic form data as in [**Wa1**], Section 5. Such forms define elements of $L_{2k}^c(R[\pi], w)$, and one can proceed to show that this defines a homomorphism

$$\sigma : \Omega_{2k}(\Pi, d) \to L_{2k}^c(R[\pi], w).$$

If $n = 2k + 1$ then one can try to kill

$$H_{k+1}(\mathrm{Cone}(\widetilde{f}); R) \cong H^k(\mathrm{Cone}(\widetilde{f}, \partial\widetilde{f}); R)$$

by finding a finite set of pairwise disjointly embedded k-spheres to generate the first summand of

$$H_k(\widetilde{X}) \simeq \widetilde{H}_{k+1}(\mathrm{Cone}(\widetilde{f})) \oplus H_k(\widetilde{Y})$$

as an $R[\pi]$-module. Usually this attempt will not succeed, for new classes may well be created in killing the summand; on the other hand, one does obtain a splitting of $f : X \to Y$ into a map of triples

$$f = f_0 \cup_\partial f_1,$$

where

(i) the map f_1 sends neighborhoods of the embedded k-spheres into a coordinate disk by a map of pairs $\{X_1 = \amalg S^k \times (D^{k+1}, S^k)\} \to \{(D^{2k+1}, S^{2k}) = Y_1\}$,

(ii) the map f_0 sends

$$X_0 = X - \mathrm{Int}(\amalg S^k \times (D^{k+1}, S^k))$$

to $Y_0 = Y - \mathrm{Int}\ D^{2k+1}$ such that $f_0|\partial X_0 = f_1|\partial X_1$.

By construction, the intersection $f_{01} = f_0|\partial X_0 = f_1|\partial X_1$ defines a $2k$-dimensional surgery problem, and this problem turns out to be trivial *for two independent reasons.* Namely, f_{01} bounds both f_0 and f_1. Since elements of $L^c_{2k+1}(R[\pi], w)$ may be defined to be $(-1)^k$-symmetric Hermitian forms with two trivializations (*i.e.*, **formations**), it follows that in the odd-dimensional case one also obtains a map

$$\sigma : \Omega_{2k+1}(\Pi, d) \to L^c_{2k+1}(R[\pi], w)$$

as before (compare Ranicki [**Ra1,Ra4**]). Once again it is possible to verify that σ defines a homomorphism. For further details see Sections 5, 6, and 9 of Wall [**Wa1**] in the case $R = \mathbb{Z}$; also see the papers of Cappell–Shaneson [**CS**], G. Anderson [**An**], Dovermann–Petrie [**DP2**], and Taylor–Williams [**TW**] for analogous discussions in the case of surgery with coefficients.

PROPOSITION 1.3. *If $n \geq 5$, then σ is an isomorphism.*

Of course, this is the crucial link between geometry and algebra. An appropriate version of the $\pi - \pi$ theorem is an important ingredient in the proof of 1.3. Here is the basic setting:

DEFINITION 1.4: Let $\mathcal{W}_0 = (X, f, b)$ be a normal map with $f : X \to Y$. We say $\mathcal{W}$ is an equivalence if f is an R homology equivalence with twisted $R[\pi]$-module coefficients. If $R = \mathbb{Z}$ and A is the subgroup of $K_1(\mathbb{Z}[\pi])$ specified by Π in (iv) then we require the torsion $\tau(f)$ to lie in A.

Now suppose $\mathcal{W} = (V, F, B)$ is a normal map of triads and the boundary $\partial\mathcal{W}$ of $\mathcal{W}$ is the union of two normal maps $\mathcal{W}_0$ and $\mathcal{W}_1$. Let $F : V \to Z$. Suppose that $\mathcal{W}_1$ is an equivalence.

THEOREM 1.5 ($\pi - \pi$ THEOREM). *Given the above assumptions, suppose also that* $\dim V \geq 6$ *and* $\pi_1(Y_1) \to \pi_1(Z)$ *is an isomorphism. Then $\mathcal{W}$ is normally cobordant to an equivalence* rel $\mathcal{W}_1$, *and $\mathcal{W}_0$ is normally cobordant to an equivalence* rel *the boundary.*■

The proof is a standard generalization of Wall's proof in [**Wa1**]; to be more explicit, the proof in [**DP1**, Section 5], goes through word for word in the setting of 1.5.

Finally we have the algebraic L-groups of Ranicki based upon quadratic and symmetric algebraic Poincaré complexes. These are defined for any ring A with a conjugation $a \mapsto \overline{a}$ and an element $e \in A$ such that $\overline{e} = e^{-1}$. We denote these groups by ${}^R L_n^c(A)$ and ${}^R L_c^n(A)$, where $c = h$ or s (compare [**Ra1**] and [**Ra4**]). For our purposes the main examples of rings with involution are given by group rings $R[\pi]$ and homomorphisms $w : \pi \to \mathbf{Z}_2$; as usual, the involution takes g to $(-1)^{w(g)} g^{-1}$.

There is a slight differencc between the normal maps (X, f, b) as described above the normal maps (f, b) as defined by Ranicki. We shall recall the latter because it will be needed later in this chapter. Following Quinn [**Q1**], one defines an n-dimensional *normal space* (X, ν_X, ρ_X) to be an n-dimensional finitely dominated CW complex X with $(k-1)$-spherical fibration $\nu_X : X \to BG_k$ and an element

$$\rho_X \in \pi_{n+k}(\mathrm{Th}(\nu_X)) \qquad \text{("Th" denotes Thom complex).}$$

The normal space triple determines a fundamental class

$$[X] \in H_n(\widetilde{X}; {}^w\mathbf{Z}),$$

where $\widetilde{X}$ is a (π, w) covering of X, the map $w : \pi \to \mathbf{Z}_2$ describes the orientation behavior, and ${}^w\mathbf{Z}$ denotes twisted coefficients.

In Ranicki's paper [**Ra5**] a normal map of n-dimensional normal spaces (f, b) consists of a map $f : M \to Y$ of the underlying spaces and a stable fiber homotopy equivalence $b : \nu_M^k \to \nu_Y^k$ over f such that

$$\text{(1.6)} \qquad \mathrm{Th}(b)\rho_M = \rho_Y \in \pi_{n+k}(\mathrm{Th}(\nu_Y))$$

for sufficiently large k. The last equation implies that f has degree 1. For a map f of degree $d > 1$ we must replace (1.6) by the following condition:

$$\text{(1.7)} \qquad d\rho_Y = \mathrm{Th}(b)\rho_M + y, \text{ where } d^\ell y = 0 \text{ for some } \ell.$$

With the notation above we can state and prove analogs of many basic results from [**Ra5**]. The most fundamental of these is the identification of the Ranicki and Wall surgery obstructions.

PROPOSITION 1.8. *Suppose $n \geq 5$. Then the quadratic signature of an n-dimensional degree d normal map (f, b) is the Wall surgery obstruction*

$$\sigma_*(f, b) = \theta(f, b) \in {}^R L_n^c(R[\pi_1(Y)]) = L_n^c(R[\pi_1(Y)], w_1(Y)).$$

(Compare [**Ra5**, 7.1] for the case $R = \mathbf{Z}$).

It is also possible to state a generalization of Ranicki's abstract product formula. Let $\sigma^*(X) \in {}^R L^n(\mathbf{Z}[\pi_1(X)])$ be the symmetric signature of X in the sense of Ranicki.

PROPOSITION 1.9. (Compare [**Ra5**, page 256]). *Let $f : M \to X$ and $g : N \to Y$ determine continuous (resp., normal) maps of degrees d and d' respectively, where d*

and d' are units in R. Then the symmetric (resp. quadratic) signature of the product is given as follows:

$$\begin{aligned}\sigma^*(f\times g) &= \sigma^*(f)\otimes\sigma^*(g)+\sigma^*(X)\otimes\sigma^*(g)\\ &\quad+\sigma^*(f)\otimes\sigma^*(Y)\in {}^RL^{m+n}(R[\pi_1(X\times Y)]).\\ \sigma_*(f\times g, b\times c) &= \sigma_*(f,b)\otimes\sigma_*(g,c)+\sigma^*(X)\otimes\sigma_*(g,c)\\ &\quad+\sigma_*(f,b)\otimes\sigma^*(Y)\in {}^RL^c_{m+n}(R[\pi_1(X\times Y)],w,d\cdot d').\end{aligned}$$

The proofs of 1.8 and 1.9 are presented in Section 3.

The desired generalizations of Yoshida's results [**Yo**] follow directly from the notation and results described above. As in [**Yo**], let $\chi : G \to \mathbb{Z}_2$ be a homomorphism and let $\Omega^\chi_m(G)$ be an appropriate bordism group of oriented m–dimensional G–manifolds for which χ describes the effect on orientation, and let $W^\chi_m(G;\mathbb{Z})$ be the appropriate Witt group. Specifically, if m is even the latter is basically a standard Witt group as in [**ACH**] with orientation taken into account, while if m is odd then one considers analogous groups of linking forms. There is a natural map

$$\rho : \Omega^\chi_*(G;\mathbb{Z}) \to W^\chi_*(G,\rho)$$

defined via the middle–dimensional homology with bilinear forms induced by cup product or linking numbers. One of the main results in [**Yo**] relates the codomain of ρ to Ranicki's symmetric L–groups; following [**Yo**] we shall denote the symmetric (simple) L-group associated to the pair $(\mathbb{Z}[G],\chi : G\to\mathbb{Z}_2)$ by $L^*_{G,\chi}(\mathbb{Z})$:

THEOREM 1.10. ([**Yo**, Theorem 1, Section 7]). *The groups $L^*_{G,\chi}(\mathbb{Z})$ and $W^\chi_*(G;\mathbb{Z})$ are isomorphic.*■

Note: The orientation map χ in Ranicki's symmetric L–theory is often suppressed in [**Ra5**].

Let $\varphi : \pi \to G$ be a surjection and let $\chi : G \to \mathbb{Z}_2$ satisfy $w = \chi\varphi$. Then a twisted product construction is defined on page 73 of [**Yo**]:

$$\begin{array}{ccc} \Omega^\chi_m(G)\otimes\Omega_n(\Pi,d) & \xrightarrow{\;\psi\;} & \Omega_{m+n}(\Pi',d) \\ {\scriptstyle\text{external}}\big\downarrow{\scriptstyle\text{product}} & & \big\| \\ \Omega_{m+n}(\Pi'',d) & \xrightarrow{\text{transfer}} & \Omega_{m+n}(\Pi',d)\end{array}$$

(The data for Π'' consists of the group $\pi\times G$ and the map $w\chi : \pi\times G\to\mathbb{Z}_2$, with R and A as before.) The transfer is taken with respect to the embedding $\gamma : \pi\to\pi\times G$ given by $\gamma(x) = (x,\varphi(x))$. The product ψ is defined by first taking M in $\Omega^\chi_m(G)$ and a normal map (X,f,b), next lifting to the universal covering $\widetilde{f} : \widetilde{X}\to\widetilde{Y}$, and finally taking the product $\widetilde{f}\times 1_M$. Then π acts freely on $\widetilde{X}\times\widetilde{M}$ and $\widetilde{Y}\times M$ by the action $\alpha(u,m) = (\alpha(u), \varphi(\alpha)(m))$. Factor out this π-action to obtain the normal map representing a class in

$$\Omega_{n+m}(\Pi^{\chi}).$$

(*Note:* $\Pi^{\chi} = \Pi$ except that $w : \pi \to \mathbf{Z}_2$ is replaced by $w\chi$). In analogy with [**Yo**, Theorem 1] we have the following relationship:

THEOREM 1.11. *The diagram below commutes if $n \geq 5$, $\mathbf{Z} \subseteq R \subseteq \mathbf{Q}$ and d is a positive unit in R:*

$$\begin{array}{ccc}
\Omega^{\chi}_m(G) \otimes \Omega_n(\Pi, d) & \longrightarrow & \Omega_{m+n}(\Pi^{\chi}, d) \\
\big\downarrow & & \big\downarrow \sigma \\
W_m(G;\mathbf{Z}) \times L^c_n(R[\pi], w) & & L^c_{m+n}(R[\pi], w_{\chi}) \\
\big\uparrow \cong & & \big\| \\
L^m_{G,\chi}(\mathbf{Z}) \otimes L^c_n(R[\pi], w) & \longrightarrow & L^c_{m+n}(R[\pi], w_{\chi}).
\end{array}$$

The crucial step in [**Yo**] is the proof of 1.10; similarly, Theorem 1.11 will be an immediate formal consequence of 1.10 and the abstract product formula of 1.9 (take $(f,b) =$ identity so that $\sigma(f,b) = 0$).

2. Algebraic description of geometric *L*-groups

The idea behind the proofs of 1.2 and 1.3 is as follows. First we show that the sets $\Omega_m(\Pi, d)$ are nonempty for $m \geq 5$. Then we describe the addition process in detail; with this description the proof of 1.2 becomes a routine exercise. The description of addition is explicit enough to show that the middle dimensional data will be unaffected, and therefore σ is additive for suitably chosen representatives of the geometric bordism classes. The $\pi - \pi$ theorem then implies 1.2 and the injectivity assertion of 1.3.

THEOREM 2.1. *If $m \geq 5$, then the set $\Omega_m(\Pi, d)$ is nonempty for arbitrary π (finitely presented), $w : \pi \to \mathbf{Z}_2$, and d.*

Note. A similar result is true if $m = 4$ but the proof is longer and the conclusion is not needed in this paper. A nearly identical construction appears in [**Lö**, p. 144].

PROOF: If π is any finitely presented group and $w : \pi \to \mathbf{Z}_2$ is a homomorphism it is well known that one can construct a closed smooth $(m-1)$-manifold $X(\pi, w)$ with fundamental group π and first Stiefel-Whitney class w. Define $X_0(\pi, w)$ to be $X(\pi, w) - \operatorname{Int} D^{m-1}$, so that $\partial X_0(\pi, w) = S^{m-2}$. Let $\psi_d : D^2 \to D^2$ be the map $\psi_d(z) = z^d$ (complex multiplication), and let

$$f_d : X_0(\pi, w) \times D^2 \to X_0(\pi, w) \times D^2$$

be the map $f_d(x,z) = (x, z^d)$. Notice that f_d maps $\partial(X_0 \times D^2)$ into itself by a map of degree d. An application of the Seifert-van Kampen theorem shows that $\pi_1(\partial(X_0 \times D^2)) = \pi$. Furthermore, f_d and ∂f_d are twisted $R[\pi]$-homology equivalences, and in fact one can easily construct a bundle map $b : X_0 \times D^2 \to X_0 \times D^2$ covering f. The object

$$(\partial(X_0 \times D^2), \partial f_d, \partial b)$$

then represents an element in $L^c_m(R[\pi], w, d)$.■

REMARK: In the smooth category we must straighten the corners of $\partial(X_0 \times D^2)$ in the standard fashion.

We now describe the addition in L. Suppose $\mathcal{W}_0 = (X_0, f_0, b_0)$ and $\mathcal{W}_1 = (X_1, f_1, b_1)$ are representatives of classes in $\Omega_n(\Pi, d)$, where $f_i : X_i \to Y_i$.

STEP 1: Choose points $y_i \in Y_i$, and make f_i transverse to y_i so that

$$f_i^{-1}(y_i) = \{x_i^1, \ldots, x_i^d\}.$$

Now form the triad map with bundle data $\mathcal{U}_1 = (V_1, F_1 : V_1 \to Z_1, B_1)$ as follows:

(2.2a) $Z_1 = (Y_0 \times I \amalg Y_1 \times I) \cup_\beta D^1 \times D^{n+1}$, where β is the embedding of a tubular neighborhood $\nu(\{y_0\} \amalg \{y_1\}, (Y_0 \amalg -Y_1) \times \{1\})$ and the latter is identified with $S^0 \times D^n$.

(2.2b) $V_1 = (X_0 \times I \amalg X_1 \times I) \cup_\alpha d(S^0 \times D^n)$, where α is the embedding of a tubular neighborhood $\nu(\amalg x_j^i, (X_0 \amalg X_1) \times \{1\})$ and the latter is identified with $d(S^0 \times D^n)$.

(2.2c) The bundles ξ_0 and ξ_1 (which are suppressed in our notation) extend to a bundle ξ over Z_1, and $f_0 \amalg f_1$ and $b_0 \amalg b_1$ extend in a natural way to F_1 and B_1.

The top end of $\mathcal{U}$ defines a map of degree d with bundle data. We shall call this $\mathcal{W}_2 = (X_2, f_2, b_2)$ and write $f_2 : X_2 \to Y_2$. It follows that $\pi_1(Y_2) = \pi_1(Y_0) * \pi_1(Y_1)$ and

$$\pi_1(X_2) = \pi_1(X_0) * \pi_1(X_1) * (*^{d-1}(\mathbb{Z})),$$

where $*$ denotes the group-theoretic free product.

STEP 2: Now kill off the excess $\mathbb{Z}$'s in $\pi_1(X_2)$. This is easily done by surgery. Call the resulting map $\mathcal{W}_3 = (X_3, f_3, b_3)$ with $f_3 : X_3 \to Y_3$, and let $\mathcal{U}_2$ denote the associated cobordism between $\mathcal{W}_2$ and $\mathcal{W}_3$. Observe that f_3 induces an isomorphism in fundamental groups.

STEP 3: In this step we adjust the fundamental groups. Recall that the groups $\pi_1(X_i)$ and $\pi_1(Y_i)$ have canonical identifications with π. Then f_3 induces an isomorphism of fundamental groups, and $\pi_1(X_3)$ and $\pi_1(Y_3)$ are identified with the free product $\pi * \pi$. Consider the sequence

$$1 \to T_\pi \to \pi * \pi \xrightarrow{\gamma} \pi \to 1$$
$$(\gamma = \text{ identity on each summand}).$$

We want to kill T_π. Let $a \in T_\pi$, and choose disjoint embeddings $a_1, \ldots, a_d : S^1 \to X_3$ each representing a. Represent $f_{3*}(a)$ by an embedding $b : S^1 \to Y_3$. After a suitable homotopy we can suppose that

$$f_3^{-1}(b(S^1)) = \coprod a_i(S^1)$$

and f_3 induces vector bundle isomorphisms from $\nu(a_i(S^1), X_3)$ to $\nu(b(S^1), Y_3)$. Extend a_i to $\alpha_i : S^1 \times D^{n-1} \to X_3$ and b to $\beta : S^1 \times D^{n-1} \to Y_3$; these extensions can be made compatible with the derivative $D(f_3)$. Form

$$V_3 = X_3 \times I \cup d(D^2 \times D^{n-1}),$$

where $d(S^1 \times D^{n-1})$ is attached along α_i and form

$$Z_3 = Y_3 \times I \cup D^2 \times D^{n-1}$$

where $S^1 \times D^{n-1}$ is attached along β. The assumptions on the α_i and β imply that $f_3 \times I$ extends to a map $F_3 : V_3 \to Z_3$. A choice of bundle isomorphism $T(D^2 \times D^{n-1}) \cong D^2 \times D^{n-1} \times \mathbf{R}^{n+1}$ and the given bundle data b_3 provide the instructions for extending the bundle $\xi_3 \times I$ to Z_3.

A finite number of such steps will kill T_π. The result of all these steps is a bordism $\mathcal{U}_3$ with $\partial \mathcal{U}_3 = \mathcal{W}_3 \amalg \mathcal{W}_4$.

STEP 4: By Step 3 we have

$$[\text{kernel}\,(H_2(X_4) \to H_2(Y_4))] = [\text{kernel}\,(H_2(X_3) \to H_2(Y_3))] \oplus B_2$$

for some free $\mathbf{Z}$–module B_2. Specifically, B_2 is a direct summand of $\pi_3(f_4)$. We may kill B_2 by surgery without creating higher dimensional classes. Call the resulting map

$$\mathcal{W}_5 = (X_5, f_5, b_5),$$

and let $\mathcal{U}_4$ be the associated cobordism from $\mathcal{W}_4$ to $\mathcal{W}_5$. It follows that $\mathcal{W}_5$ is a normal map which represents a class in $\Omega_n(\Pi, d)$. Furthermore, if $\mathcal{U}$ is given by taking the cobordisms $\mathcal{U}_i$ obtained in Step i $(1 \le i \le 4)$, and joining them along their common boundaries, then $\mathcal{U}$ defines a normal cobordism from $\mathcal{W}_0 \amalg \mathcal{W}_1$ to $\mathcal{W}_5$. Thus $\mathcal{W}_1$ represents the sum of $\mathcal{W}_0$ and $\mathcal{W}_1$.■

PROOF OF 1.2: (*Group structure on* $\Omega_n(\Pi, d)$.) The proof is a straightforward exercise which has been carried out more generally in **[Do]** for the case of finite groups. We shall show that the addition is well-defined and leave the other parts to the reader. Suppose that the normal maps $\mathcal{W}_i$ and $\mathcal{W}_i'$ $(i = 0, 1)$ represent the same classes in $L_n^c(R[\pi], w)$. Let the equivalences between them be realized by the normal cobordisms

$$\mathcal{U}_i = (W_i, F_i, B_i),$$

let $\mathcal{U} = (W, F, B)$ be the cobordism between $\mathcal{W}_0 \amalg \mathcal{W}_1$ and $\mathcal{W}_0 + \mathcal{W}_1$, and let $\mathcal{U}' = (W', F', B')$ be the cobordism between $\mathcal{W}_0' \amalg \mathcal{W}_1'$ and $\mathcal{W}_0' + \mathcal{W}_1'$. Join the cobordisms $\mathcal{U} \amalg -\mathcal{U}'$ to $\mathcal{U}_1 \amalg \mathcal{U}_2$ along

$$\mathcal{W}_0 \amalg \mathcal{W}_1 \amalg -\mathcal{W}_0' \amalg -\mathcal{W}_1'$$

to obtain a new cobordism

$$\mathcal{U}_T = (W_T, F_T : W_T \to Z_T, B_T)$$

where the manifold W_T is given by

(1) $W_T = W' \cup_{X_0' \amalg X_1'} (W_0 \amalg W_1) \cup_{X_0' \amalg X_1'} W$,
(2) $\partial W' \cap (\partial W_0 \amalg \partial W_1) = X_0' \amalg X_1'$,
(3) $\partial W \cap (\partial W_0 \amalg \partial W_1) = X_0 \amalg X_1$,

and the mapping $F_T : W_T \to Z_T$ and the bundle data B_T have analogous descriptions. The decomposition of W_T in terms of (1) is illustrated by the accompanying figure:

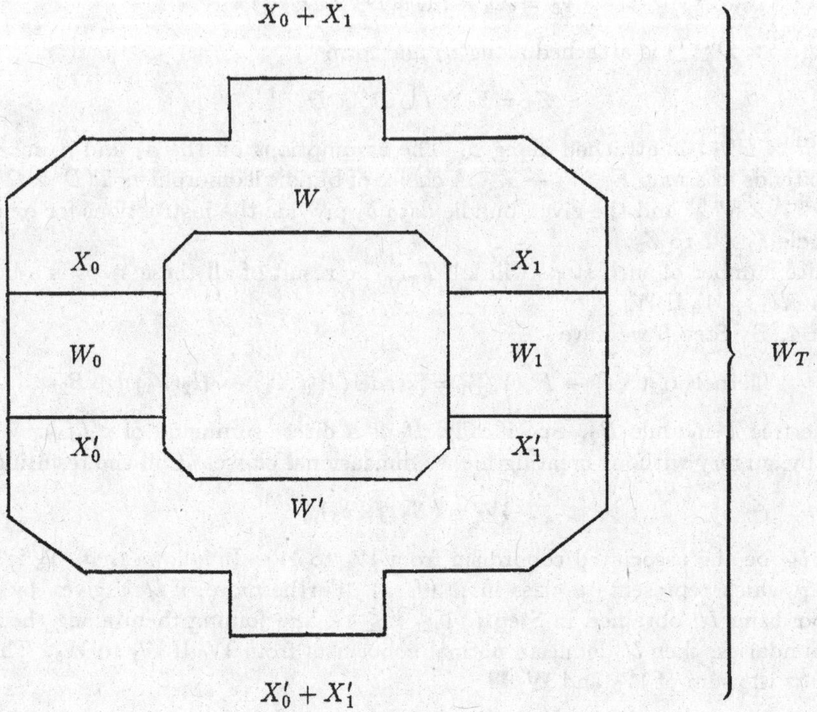

Figure I

As in Steps 3 and 4 of the addition we apply surgery to kill homotopy groups, changing $\pi_1(Z_T)$ and $\pi_1(W_T)$ to π and ensuring that F_T induces an isomorphism on fundamental groups. An additional application of steps 3 and 4 is needed to kill a single extra copy of $\mathbf{Z}$ that corresponds to a "hole" in the middle of W_T (see Figure I). From this we obtain a new cobordism between $\mathcal{W}_0 + \mathcal{W}_1$ and $\mathcal{W}_0' + \mathcal{W}_1'$ which represents an equivalence in $L^c_m(R[\pi], w, d)$ between these two maps.

PROOF OF 1.3: There are essentially three steps. The first is to justify our definition of σ from the geometric groups to the algebraic groups; *i.e.*, a normal map which is connected up to the middle dimension actually defines an algebraic object which represents a class in the appropriate group. The details are essentially a special case of [**DP2**, 5.1′], and the argument generalizes Wall's approach from [**Wa1**, Sections 5-6].

The second step is to show that σ is additive on the level of representatives. The additivity of σ follows from the construction of addition on the level of representatives. If $\mathcal{W}_0$ and $\mathcal{W}_1$ are connected up to the middle dimension, then by construction the same is true for $\mathcal{W}_0 + \mathcal{W}_1$. Since none of the constructions used to form $\mathcal{W}_0 + \mathcal{W}_1$ from

$\mathcal{W}_0 \amalg \mathcal{W}_1$ affect the middle-dimensional homology, it is immediate that $\sigma(\mathcal{W}_0 + \mathcal{W}_1) = \sigma(\mathcal{W}_0) + \sigma(\mathcal{W}_1)$.

The third step is to show that σ is well-defined. One way to show this is to generalize the proofs in Wall's book [**Wa1**, Sections 5-6]. It is also possible to do this more formally using the proof of the $\pi - \pi$ theorem. The argument may be summarized as follows: Let $\mathcal{W}$ and $\mathcal{W}'$ be normal maps which are connected up to the middle dimension and which represent the same class in $\Omega_n(\Pi, d)$. Let $\mathcal{U}$ be the cobordism between $\mathcal{W}$ and $\mathcal{W}'$, and $\mathcal{U}^+$ the cobordism between $\mathcal{W} \amalg -\mathcal{W}'$ and the sum $\mathcal{W}^+$ of $\mathcal{W}$ and $-\mathcal{W}'$. Let $\mathcal{U}_1$ be the union of $\mathcal{U}$ and $\mathcal{U}^+$ along $\mathcal{W} \amalg -\mathcal{W}'$. Some surgeries in the interior as in the addition process will modify $\mathcal{U}_1$ such that it almost becomes a normal map that we shall call $\mathcal{U}^*$; we say almost because $\mathcal{U}^*$ restricted to the boundary is not an equivalence. Observe that $\mathcal{W}^+$ is the only boundary piece of $\mathcal{U}^*$ on which the maps are not assumed to be equivalences. It follows from Step 2 that $\sigma(\mathcal{W}^+) = \sigma(\mathcal{W}) - \sigma(\mathcal{W}')$. We then apply the $\pi - \pi$ theorem to the pair $(\mathcal{W}^*, \mathcal{W}^+)$, and from this thus conclude that $\mathcal{W}^+$ can be converted to surgery into an equivalence. The analysis of each step shows that $\sigma(\mathcal{W}^+)$ is not changed in this process. Hence $\sigma(\mathcal{W}^+) = 0$.

To see that σ is injective, let $\mathcal{W}$ be a highly connected normal map with $\sigma(\mathcal{W}) = 0$. The explicit algebraic representative of $\sigma(\mathcal{W})$ contains precise instructions on the homotopy classes to be killed by surgery, and it is straightforward to see that these surgeries are possible.

To see that σ is surjective, modify the argument in [**Wa1**] as follows: Begin with an equivalence in $\Omega_{m-1}(\Pi, d)$, say $f_0 : X_0 \to Y_0$. By the proof of Theorem 2.1 such an equivalence exists. Suppose that m is even, so that a class in $L^c_m(R[\pi], w)$ is represented by a $\pm$-symmetric form (K, λ, μ). Then λ and μ are given by matrices if one chooses a basis for K, and one can clear these matrices of fractions to obtain representative matrices in $\mathbb{Z}[\pi]$. Then one can realize λ and μ by attaching handles to $X_0 \times [0, 1]$ along the top component of the boundary; i.e., $X_0 \times \{1\}$. As in [**Wa1**], this will yield a cobordism $(W; X_0, X_1)$ and a normal map $g : W \to Y_0 \times I$ extending f_0. If m is odd one modifies the argument of [**Wa1**, Theorem 6.5, page 66] similarly.

3. Technical remarks

In this section we shall describe the proofs of (1.8) and (1.9). Specifically, we shall give a setting for generalizing the arguments of Ranicki [**Ra4-5**] to maps of degree greater than 1. We shall refer the reader to Ranicki's papers [**Ra4-5**] for details.

We have already described Ranicki's notion of a normal map in (1.6), and in (1.7) we gave an extension of (1.6) to maps of degree greater than 1. Unfortunately, for some purposes this definition is too restrictive. However, the following elementary result implies that (1.7) is adequate for our purposes:

PROPOSITION 3.1. *Let M and N be CAT(= DIFF, TOP, or PL) manifolds, and suppose that $f : M \to N$ is a map such that*

(i) *degree* $(f) = d$,

(ii) *f is stably tangential; i.e., one has $\tau_M \simeq f^*\tau_N$ (stably in CAT).*

Then f and a stable isomorphism in (ii) define a normal map in the sense of (1.7).

PROOF: The stable isomorphism from (ii) defines a stable fiber homotopy equivalence $b_0 : \nu_M \to \nu_N$. Let ρ_N and ρ_M be the canonical classes in $\pi_{n+k}(\mathrm{Th}(\nu_N))$ and $\pi_{n+k}(\mathrm{Th}(\nu_M))$ respectively, and $\mathrm{Th}(b_0)$ be the associated map of Thom spaces.
Consider the difference

$$\Delta = d\rho_N - \mathrm{Th}(b_0)\rho_M \in \pi_{n+k}(\mathrm{Th}(\nu_N)).$$

Let $\Gamma : \mathrm{Th}(\nu_N) \to S^{n+k}$ be a collapsing map which identifies an appropriate $(n+k-1)$–skeleton to a point; denote this skeleton by X_{n+k-1}. Then the composite $\Gamma\Delta : S^{n+k} \to S^{n+k}$ has degree zero by construction, and therefore Δ lies in the image of

$$\pi_{n+k}(X_{n+k-1}),$$

which by S–duality is isomorphic to the stable cohomotopy group

$$\{N, S^0\}. \tag{3.2}$$

The latter is finite, and therefore one can write the preimage Δ' of Δ in (3.2) as a sum

$$\Delta' = dy + z$$

where $d^E z = 0$ for some E.

The group in (3.2) also has a natural interpretation as the classes of fiber homotopy self–equivalences of the product bundle

$$X \times S^k \to X.$$

Given $y_0 \in \{X, S^0\}$ we may construct a fiber homotopy self equivalence c of the stable normal sphere bundle $S(\nu_N)$ by taking joins with the identity on $S(\nu_N)$. A purely formal calculation then shows that

$$\mathrm{Th}(cb_0) - \mathrm{Th}(b_0)\rho_M = dj_*(\mathrm{Dual}\;\; y_0),$$

where j is the inclusion of X_{n+k-1} in $\mathrm{Th}(\nu_N)$. Take y_0 to be the class y defined in the preceding paragraph. Then one has

$$\begin{aligned}\mathrm{Th}(cb_0)\rho_M - d\rho_N &= d(\mathrm{Dual}(y)) - \Delta \\ &= j_*\,\mathrm{Dual}(-z).\end{aligned}$$

But if $d^E z = 0$, then $d^E\,\mathrm{Dual}(-j_* z) = 0$ also holds.■

COMPLEMENT TO 3.1: Notice that y and hence c (and therefore $b = cb_0$ as well!) may be uniquely defined up to homotopy if one insists that y has order prime to d.■

The next step is a little awkward. We require yet another geometric definition of Wall groups using *stably tangential normal maps* (compare Madsen, Taylor, and Williams [MTW]). We shall denote the relevant groups by

$$\Omega^t_n(\Pi, d).$$

The representative objects are normal map triples (X, f, b) as before, BUT we now assume that

(i) the bundle ξ in (2.1.d) must be the stable normal bundle of Y,
(ii) in the definition of equivalence for bundle data, the CAT bundle isomorphism

$$a : \{\nu_Y = \xi\} \to \{\xi' = \nu_Y\}$$

must be *stably CAT isotopic to the identity.*

The whole discussion involving Proposition 1.1–1.3 carrier over to the groups Ω^t; for example, the proof of Theorem 2.1 actually shows that the sets Ω^t are nonempty. Of course there is a natural map from Ω^t to Ω such that the following diagram commutes:

$$\begin{array}{ccc} \Omega_n(\Pi, d) & \xrightarrow{\sigma} & L_n(R[\pi], w) \\ \gamma\uparrow & & =\uparrow \\ \Omega^t_n(\Pi, d) & \xrightarrow{\sigma'} & L_n(R[\pi], w) \end{array}$$

(We have deleted some of the notational data for the sake of simplicity.) Therefore the methods of (1.1)–(1.3) imply that σ and σ' and hence γ are all isomorphisms.

The groups Ω^t_n are valuable because one can easily define a well-behaved map from the tangential geometric surgery groups to the Ranicki L-groups. Proposition 3.1 contains a basic first step in this direction. The next step is to show that a normal map as in (1.7) defines a geometric Umkehr map as in [**Ra5**, Theorem 4.2, pp. 228-229]. The proof of the following result is a straightforward generalization of the argument in [**Ra5**]; for the sake of completeness we mention that normalized Poincaré complexes are defined on page 228 of [**Ra5**].

THEOREM 3.3. *Let (f, b) be a degree d normal map of normalized n-dimensional geometric Poincaré complexes*

$$(M, \nu_M, \rho_M) \to (X, \nu_X, \rho_X),$$

and let $\widetilde{X}$ be a cover of X with group of covering transformations π. Then there is induced a π-map of Thom π-spaces

$$\mathrm{Th}\pi(b) : \mathrm{Th}\pi(\nu_M) \to \mathrm{Th}\pi(\nu_X)$$

such that the $S\pi$-dual of $\mathrm{Th}\pi(b)$ with respect to the fundamental $S\pi$-duality maps

$$\alpha_M : S^N \to \widetilde{M}_+ \wedge_\pi \mathrm{Th}\pi(\nu_M),$$
$$\alpha_X : S^N \to \widetilde{X}_+ \wedge_\pi \mathrm{Th}\pi(\nu_X)$$

is an $S\pi$-homotopy class of geometric Umkehr maps $F : \Sigma^p \widetilde{X}_+ \to \Sigma^p \widetilde{M}_+$ such that $(\Sigma^p f_+)F = d \cdot (\mathrm{Identity}) + z' : \Sigma^p \widetilde{X}_+ \to \Sigma^p \widetilde{X}_+$, where $d^K z' = 0$, up to stable π-homotopy.

Note. The maps α_M, α_X are defined on page 227 of [**Ra5**].

REMARK ON THE PROOF: The proof is based upon a large commutative diagram on page 229 of [**Ra5**]. In our situation the corresponding diagram only commutes up to elements of order a power of d. Furthermore, the maps $\Sigma^p \rho_X$ and $\Sigma^p \alpha_X$ must be replaced by $d\Sigma^p \rho_X$ and $d\Sigma^p \alpha_X$ respectively. Once these changes are made the argument on page 229 of [**Ra5**] generalizes directly.■

This result gives us a geometric realization of the chain level Umkehr homomorphism

$$f^! : C(\widetilde{X}) \otimes R \to C(\widetilde{M}) \otimes R$$

exactly as in [**Ra5**]. At several points in Ranicki's work one needs to know that $f^!$ and F induce chain homotopic chain maps. In [**Ra5**] this follows because $(\Sigma^p f_+)F$ and 1 are π-stably homotopic. Although in our situation we only know that $(\Sigma^p f_+)F$ and $d(\text{identity}) + z'$ are π-stably homotopic, this condition suffices to show that the induced chain maps *with R coefficients* are chain homotopic (recall that d is a unit in R and $d^E z' = 0$).

Application to the proofs of (1.8) *and* (1.9)

The observations above allow one to define a quadratic signature for a tangential degree d normal map (f, b) exactly as in [**Ra5**, pages 229–231]. The remainder of the proof of (1.8) proceeds in close analogy to [**Ra5**, Proposition 7.1]. Inspection of the argument shows that the proof uses Proposition 4.2, 5.4, and 6.6. We have described the necessary modifications of 4.2. But these modifications allow one to carry through the arguments of [**Ra5**, Sections 5 and 6] without significant further changes. Strictly speaking, at this point we have only established (1.8) for degree d normal maps that are stably tangential. However, using the isomorphism $\Omega^t(\Pi, d) = \Omega(\Pi, d)$ and a diagram chase it is elementary to establish (1.8) in general.

The product formula (1.9) may now be proved as in Section 8 of [**Ra5**] if one restricts to tangential degree d normal maps. Since $\Omega^t(\Pi, d) = \Omega(\Pi, d)$, this immediately implies the general case.

4. Projective and subprojective Wall groups

In Chapter V we shall need Yoshida-type product formulas for the projective surgery obstruction groups

$$L_n^p(\mathbb{Z}[\pi], w)$$

(*e.g.*, see Ranicki [**Ra4**], Maumary [**Mau1-2**], Taylor [**T**], or Pedersen-Ranicki [**PR**]) and for variants of these groups of the form $L_n^A(\mathbb{Z}[\pi], w)$, where $A \subseteq \widetilde{K}_0(\mathbb{Z}[\pi])$ is a suitable subgroup. Therefore we shall discuss such formulas in this section. The methods also yield an analogous result for the *subprojective surgery obstruction groups*

$$L_n^{-i}(\mathbb{Z}[\pi], w)$$

which appear in the work of Ranicki [**Ra2**] (also see Anderson and Pedersen [**AP**] and Quinn [**Q2**]).

For the sake of notational convenience we shall abbreviate $L_n^c(\mathbf{Z}[\pi], w)$ to $L_n^c(\pi, w)$.

The first step is to consider twisted product formulas for $L_n^A(\pi, w)$ where A is a suitable subgroup of the Whitehead group $Wh(\pi)$. As in Section 1 let $\varphi : \pi \to G$ be a fixed surjection onto a fixed finite group G. Then we shall insist that A have the following two properties:

(4.1a) *A is closed under the usual involution on $Wh(\pi)$ given by w.*

(4.1b) *If $P \in GL(n, \mathbf{Z}[\pi])$ represents an element of A and $H \subseteq G$ is a subgroup, then $P \otimes \mathbf{Z}[G/H]$ also represents an element of A.*

Property (4.1b) implies that A is a module over the Burnside ring of G in a natural way. We may then generalize the entire discussion of Sections 1–3 to the case of $L_n^A(\pi, w)$.

THEOREM 4.2. *If $A \subseteq Wh(\pi)$ satisfies* (4.1a)–(4.1b), *then the twisted product construction defines a map*

$$\Omega_m^\chi(G) \otimes L_n^A(\pi, w) \to L_{n+m}^A(\pi, w\chi).$$

This construction is natural for inclusions $A \subseteq A'$ and suitable homomorphisms $\pi \to \pi'$ over G.■

We now generalize to projective Wall groups. Presumably one could attack this question from first principles, but it is simpler to use the natural splitting

$$L_n^h(\pi \times \mathbf{Z}, w) \simeq L_n^h(\pi, w) \oplus L_{n-1}^p(\pi, w) \tag{4.3}$$

(see [**Ra4**, Section 10]) as a guide.

More generally, suppose that $B \subseteq \widetilde{K}_0(\mathbf{Z}[\pi])$ is closed under conjugation and satisfies

(4.4) *If M represents a class in B, then so does $M \otimes \mathbf{Z}[G/H]$.*

There are two basic examples that are of interest. The first is $B = \widetilde{K}_0(\mathbf{Z}[\pi])$; in this case $L_n^B = L_n^p$. In Chapter V it will be necessary to take B to be the defect group

$$D(\pi) = \text{Kernel } \widetilde{K}_0(\mathbf{Z}[\pi]) \to \widetilde{K}_0(M_\pi)$$
$$(\text{where } M_\pi \text{ is a maximal order in } \mathbf{Q}[\pi])$$

that arises in many geometric contexts (compare Oliver [**OL**] and Dovermann–Petrie [**DP1-2**]).

It is well–known that $\widetilde{K}_0(\mathbf{Z}[\pi])$ is a natural direct summand of $Wh(\pi \times \mathbf{Z})$, and it is elementary to check that if B satisfies (4.4), then its image B' in $Wh(\pi)$ satisfies (4.1.b). We then have the following generalization of (4.3) and Shaneson's earlier splitting

$$L_n^s(\pi \times \mathbf{Z}, w) = L_n^s(\pi) \oplus L_{n-1}^h(\pi)$$

(see [**Sh**]).

PROPOSITION 4.5. *There is a natural split injection*

$$C : L^B_{n-1}(\pi, w) \to L^{B'}_n(\pi \times \mathbb{Z}, w)$$

given geometrically by crossing with S^1 or algebraically by tensoring with the symmetric Poincaré complex over $\mathbb{Z}[t, t^{-1}]$ defined by S^1.

This follows from an examination of the Rothenberg sequences for $L^h(\pi) \to L^B(\pi)$ and $L^s(\pi \times \mathbb{Z}) \to L^{B'}(\pi \times \mathbb{Z})$.

To prove a twisted product formula for $L^B_*(\pi, w)$ we must do two things. First we must show that one can define twisted product with P for any P representing an element of $\Omega^X_m(G)$. Next we must show that if P and Q represent the same class, then they induce the same map. Finally, we should verify that crossing with P is a homomorphism.

The first and last steps may be done as follows. Geometrically one can define a map

$$\mu_P : L^B_n(\pi, w) \to L^B_{n+m}(\pi, w)$$

by taking a proper surgery problem (as in [**Mau1**], [**PR**], or [**T**]) and crossing it with P. Relationship (4.3) ensures that μ_P sends elements of L^B to elements of L^B. One can also do this algebraically by taking a form or formation in the sense of Ranicki and tensoring with a good cellular G–chain complex for P. Let us call this map μ'_P. Then for $\alpha = \mu$ or μ' and c as in 4.5 one has a commutative diagram as follows:

(4.6)
$$\begin{array}{ccc} L^B_n(\pi, w) & \xrightarrow{C} & L^{B'}_{n+1}(\pi \times \mathbb{Z}, w) \\ \downarrow{\scriptstyle \alpha_{P(B)}} & & \downarrow{\scriptstyle \alpha_{P(B')}} \\ L^B_{n+m}(\pi, w_X) & \xrightarrow{C} & L^{B'}_{n+m+1}(\pi \times \mathbb{Z}, w) \end{array}$$

Since C is injective and $\mu_P(B') = \mu'_P(B')$, it follows that μ_P is also a homomorphism.

Finally, if P and Q represent the same class in $\Omega^X_m(G)$, then by (4.2) we know that

$$\mu_P(B') = \mu_Q(B'),$$

and therefore $\mu_P(B) = \mu_Q(B)$ by another diagram chase. To summarize, we have proved the following result:

THEOREM 4.7. *If $B \subseteq \widetilde{K}_0(\mathbb{Z}[\pi])$ satisfies (4.4) and is closed under conjugation, then the algebraic and geometric product constructions define a pairing*

$$\Omega^X_m(G) \otimes L^B_n(\pi, w) \to L^B_{n+m}(\pi, w_X)$$

that is natural in B and in homomorphisms $\pi \to \pi'$ over G.■

Since the definition of the subprojective groups $L^{-i}_n(\pi, w)$ was motivated by (4.3), it is not surprising that a similar argument applies to these groups. Following Ranicki [**Ra2**], we define $L^{-k}_n(\pi, w)$ for $k \leq 0$ inductively by setting $L^0_n = L^p_n$ and

(4.8)
$$L^{-k}_n(\pi, w) = \operatorname{Coker}\{i_* : L^{-k+1}_{n+1}(\pi, w) \to L^{-k+1}_{n+1}(\pi \times \mathbb{Z}, w)\}.$$

By naturality the homomorphism i_* splits, and therefore one has splittings

$$L_n^{-k+1}(\pi \times \mathbf{Z}, w) = L_n^{-k+1}(\pi, w) \oplus L_n^{-k}(\pi, w)$$

analogous to Shaneson's and Ranicki's; thus we shall write $L_n^1 = L_n^h$ and $L_n^2 = L_n^s$ when it is convenient to do so. Using (4.8), the standard forgetful map $L_n^p \to L_n^h$, Ranicki's formula (4.3), and finite induction it is fairly straightforward to construct natural homomorphisms

$$\theta_{-k} : L_n^{-k}(\pi, w) \to L_n^{-k-1}(\pi, w).$$

Similarly, given a smooth G–manifold P representing an element of $\Omega_m^\chi(G)$ we can inductively define twisted product maps μ_P^{-k} on L^{-k} groups as before; it is fairly elementary to verify that the θ–homomorphisms and the μ_P–homomorphisms commute.

One can also define $L_n^{-\infty}$ to be the direct limit of the L_n^{-k}'s (compare [**Q**2]). It follows that one has a twisted product map $\mu_P^{-\infty}$ and the diagrams

$$\begin{array}{ccc} L_n^{-k}(\pi, w) & \xrightarrow{\theta} & L_n^{-\infty}(\pi, w) \\ \mu_P \downarrow & & \downarrow \mu_P \\ L_{n+m}^{-k}(\pi, w) & \xrightarrow{\theta} & L_{n+m}^{-\infty}(\pi, w) \end{array}$$

commute for all k.

A purely formal argument based upon Theorem 4.7 and the discussion above shows that the homomorphisms $\mu_P^{-k}(-\infty \le k < 0)$ depend only on the class of P in $\Omega_m^\chi(G)$. This fact and similar formal considerations imply the following result:

THEOREM 4.9. *For $-2 \le k \le \infty$ there are twisted product pairings*

$$\Omega_m^\chi(G) \otimes L_n^{-k}(\pi, w) \to L_{n+m}^{-k}(\pi, w)$$

as in Theorem 4.7. *These are natural with respect to forgetful maps $\theta : L^{-k} \to L^{-\ell}(-2 \le k < \ell)$ and homomorphisms $\pi \to \pi'$ over G.*■

FINAL REMARK: For every $i > 2$ one can define *supersimple* surgery groups $L_*^{s(i)}(\pi, w)$ with $K_i(\mathbf{Z}[\pi])$–type invariants similar to the Whitehead torsion invariants on $L_*^h(\pi, w)$; related groups in the case $i = 2$ are considered in the thesis of G. T. Kennedy [**K**]. Forthcoming work of A. Bak and B. Williams will define an explicit construction for such groups [**BW**]. It seems natural to conjecture that one has good twisted product pairings in these cases also.

5. Cappell-Shaneson Γ-groups

In this section we shall explain how the methods of the preceding sections can be modified to deal with the homology surgery obstruction groups $\Gamma_n(\varphi)$ of [**CS**]. Since our arguments in this paper depend heavily upon Ranicki's algebraic methods for analyzing such groups, it is not surprising that the Γ-group version of Ranicki's machinery [**Ra6**, Sections 2.4 and 7.7] plays a crucial role in our discussion.

We begin with the same data Π and d introduced at the beginning of Section 1, and in addition we take an involution-preserving epimorphism (or local epimorphism) $\varphi : \mathbb{Z}[\pi] \to \Lambda$. Wall's bordism-theoretic definitions of L-groups can be extended directly to this setting. The objects of interest will be degree d maps of manifolds as in Section 1 such that induced the map of boundaries is a zero object, and zero objects are homology equivalences with twisted coefficients in Λ. An abelian group structure can be described as in Section 2, replacing the results of [**Wa1**] for L-groups by the results of [**CS**] for Γ-groups as needed. The groups that arise in this fashion will be denoted by $\Omega^\Gamma_n(\Pi, d, \varphi)$. Very few changes are needed to apply the discussion of Section 1 to this setting. Of course, the appropriate notion of equivalence is a homology equivalence with twisted coefficients in Λ (often abbreviated to Λ-equivalence). At various points one must appeal to [**CS**] for extensions of Wall's methods to the Cappell-Shaneson setting; for example, the appropriate version of the $\pi - \pi$ theorem is Theorem 3.1 of [**CS**]. Ultimately, these observations and the results of [**Ra6**] yield a quadratic signature homomorphism

$$\sigma^\Gamma_* : \Omega^\Gamma_n(\Pi, d, \varphi) \to \Gamma_n(\varphi),$$

and the proof of Proposition 1.8 can be combined with the results of [**Ra6**, Section 7.7] to show that σ^Γ_* is an isomorphism of abelian groups.

Although the product maps for Γ-groups are not discussed explicitly in [**Ra6**], it is not difficult to retrieve the information we need. From a geometrical viewpoint it is clear that product maps exist, for if $f : (M, \partial M) \to (X, \partial X)$ is a suitable degree d map that is a Λ-equivalence on the boundary and N is an arbitrary compact unbounded manifold, then $f \times 1_N$ is a degree d map, and the map on the boundary is a homology equivalence with twisted coefficients in $\Lambda \otimes \mathbb{Z}[\pi_1(N)]$.

PROPOSITION 5.1. (compare [**Ra5**, Prop. 8.1]) *The Γ-quadratic signature of a product map $(f \times 1_N, b \times 1) : M \times N \to X \times N$ is given by*

$$\sigma^\Gamma_*(f, b) \otimes \sigma^*(N) \in \Gamma_{m+n}(h \otimes \text{Identity on } \mathbb{Z}[\pi_1(N)]).$$

The proof of this result is a straightforward combination of Proposition 1.9 and the techniques of [**Ra5**, Section 8].■

If $\Omega^X_m(G)$, $L^*_{G,X}(\mathbb{Z})$, and $W^X_*(G; \mathbb{Z})$ are defined as in Section 1, then it is immediate that the twisted product constructions of [**Yo**] have analogs for Γ-groups. Furthermore, the observations of this section lead directly to the following analog of Theorem 1.11:

THEOREM 5.2. *The diagram below commutes if $n \geq 5$, $\mathbb{Z} \subseteq R \subseteq \mathbb{Q}$ and d is a positive unit in R:*

$$\begin{array}{ccc}
\Omega^X_m(G)\otimes\Omega_n(\Pi,d,\varphi) & \longrightarrow & \Omega_{m+n}(\Pi^X,d,\varphi)\\
\big\downarrow & & \big\downarrow{\scriptstyle \sigma^\Gamma_*}\\
W_m(G;\mathbf{Z})\times\Gamma_n(\varphi) & & \Gamma_{m+n}(\varphi\otimes 1)\\
\big\uparrow{\scriptstyle \simeq} & & \big\|\\
L^m_{G,X}(\mathbf{Z})\otimes\Gamma_n(\varphi) & \longrightarrow & \Gamma_{m+n}(\varphi\otimes 1).
\end{array}$$

As noted at the end of Section 1, Theorem 1.11 follows formally from 1.10 (Yoshida's main result) and Ranicki's product formula; similarly, Theorem 5.2 is an immediate formal consequence of 1.10 and the product formula of 5.1.■

6. Applications to periodicity theorems

Let R be a subring of the rationals containing the integers, and let X be a compact CAT locally linear G-manifold such that X satisfies the Gap Hypothesis and all closed substrata are simply connected with dimension ≥ 5. As noted in Section II.2, the methods of [**DR**] yield *equivariant R-homology surgery obstruction groups*

$$I^{ht(R),CAT}(G;\lambda_X)$$

that carry the obstructions to converting an equivariant surgery problem ($f : M \to X$, other data) into a cobordant problem ($h : N \to X$, other data) such that the closed substrata of N are simply connected and h induces an isomorphism in homology with coefficients in R over each closed substratum. By the methods and results of [**MMT**], such a map h induces a homotopy equivalence from the equivariant R-localization of N to the equivariant R-localization of X, and therefore we shall call h an equivariant R-local homotopy equivalence. As suggested in (II.2.3), the stepwise surgery obstruction groups are the Wall groups $L^h_{\dim\alpha}(R[W_\alpha],w_\alpha)$ if $\mathbf{S}^2\lambda$ satisfies the Gap Hypothesis, or quotients of these groups if $\mathbf{S}\lambda$ satisfies the Gap Hypothesis but $\mathbf{S}^2\lambda$ does not. Further information in the case $R=\mathbf{Q}$ can be found in [**DR**, Section 10]. The following result shows that the theories $I^{ht(R),CAT}$ have the same periodicity properties as the theories considered in Chapter III.

THEOREM 6.1. *Assume that X satisfies the conditions of Theorems* III.2.7–9 *and the conditions needed to construct $I^{ht(R),CAT}(G;\lambda_X)$.*

(*i*) *If Y is a periodicity manifold in the sense of Section* III.2 *such that λ_X is Y-preperiodic and $\lambda_{X\times Y}$ satisfies the Gap Hypothesis, then the product map*

$$\mathbf{P}_Y : I^{ht(R),CAT}(G;\lambda_X)\to I^{ht(R),CAT}(G;\lambda_{X\times Y})$$

induces an isomorphism of abelian groups.

(*ii*) *If $\mathbf{S}^{4n+2}\lambda_X$ satisfies the Gap Hypothesis then $\mathbf{P}_Y$ is an isomorphism if $Y=\mathbf{C}\mathrm{P}^{2n}$.*

(ii) If $\mathbf{S}^2\lambda_X$ *satisfies the Gap Hypothesis then* $\mathbf{P}_Y$ *is an isomorphism if* $Y = \mathbb{C}P^2\!\uparrow\! G$.

PROOF: *(Sketch)* The proofs of Theorems III.2.7–9 in Sections III.2 and III.4 are based upon (*a*) the exactness properties of equivariant surgery obstruction groups, (*b*) the compatibility of products with these exactness properties, (*c*) the correspondence between equivariant products and twisted products given by Theorem III.4.2, and (*d*) Yoshida's results stating that the twisted products $\mu^c(Y) : L^c_*(\mathbb{Z}[\pi], w) \to L^c_{*+\dim Y}(\mathbb{Z}[\pi], w')$ only depend upon the Witt invariant of Y. The first three of these generalize immediately to the theories $I^{ht(R),CAT}$, and the invariance property in (*d*) is contained in Theorem 1.11 of this chapter. Therefore assertions (*i*) – (*iii*) follow by straightforward modifications of the proofs of the periodicity theorems in Chapter III.■

COMPLEMENT TO 6.1: In analogy with Complement III.2.10 the results of [Ya] yield an extension of 6.1(*ii*) to the case $Y = \mathbb{C}P^2\!\uparrow\! S$ where S is a finite G-set and all relevant indexing data are $\mathbb{C}P^2\!\uparrow\! S$-preperiodic and satisfy the Gap Hypothesis.

Near the end of Subsection II.3F we mentioned analogous Browder-Quinn theories $L^{BQ,h(R)}$ (see subheading titled *Localized equivariant surgery*). An argument similar to the proof of Theorem 6.1 yields a periodicity theorem in this case too.

THEOREM 6.2. *Assume that* X *satisfies the conditions of Theorems* III.2.7–9 *and the conditions needed to construct* $L^{BQ,h(R)}(G;\Lambda_X)$.

(*i*) *If* Y *is a periodicity manifold in the sense of Section* III.2 *such that* λ_X *is* Y-*preperiodic, then the product map*

$$\mathbf{P}_Y : L^{BQ,h(R)}(G;\Lambda_X) \to L^{BQ,h(R)}(G;\Lambda_{X\times Y})$$

induces an isomorphism of abelian groups.

(*ii*) *If* G *has odd order, then the map* $\mathbf{P}_Y$ *is an isomorphism if* $Y = \mathbb{C}P^{2n}$ *or* $\mathbb{C}P^2\!\uparrow\! G$.■

VARIANTS OF THEOREM 6.2: As noted in Subsection II.3F, there are variants of the Browder-Quinn groups $L^{BQ,h(R)}$ such that the fixed coefficient ring $R \subset \mathbb{Q}$ is replaced by a family of rings $R_\alpha \subset \mathbb{Q}$ indexed by the closed substrata (see [**DHM**] for a particularly significant example). The statement and proof of Theorem 6.2 generalize in a straightforward manner to such groups. Furthermore, in analogy with Complement III.2.10 the results of [Ya] yield an extension of 6.2(*ii*) to the case $Y = \mathbb{C}P^2\!\uparrow\! S$ where S is a finite G-set and Λ_X is $\mathbb{C}P^2\!\uparrow\! S$-preperiodic.■

Our next objective is to extend the Periodicity Theorems III.2.7–9 to equivariant analogs of the Cappell-Shaneson homology surgery groups $\Gamma^c_n(\varphi : \mathbb{Z}[\pi] \to A, w)$ for homology surgery with local coefficients in A; by definition A is a ring with involution and $\varphi : \mathbb{Z}[\pi] \to A$ is an epimorphism (or almost epimorphism) of rings with involution. In order to state the results we need a logically sound concept of equivariant local coefficients. Our definition is perhaps old-fashioned (compare [St] in the nonequivariant case), but it requires a minimum of extra categorical terminology. More algebraic definitions of equivariant local coefficients are contained in recent work of I. Moerdijk and J.-A. Svensson on equivariant Serre spectral sequences [**MSv**] and a recent book by W. Lück [**Lü**].

We shall need the notion of *equivariant fundamental groupoid system* of a G-space X as described in [**tD**, Section I.10]; to simplify the discussion we assume that G is finite. The objects of this category $\mathrm{FGPD}^G(X)$ are pairs (x, H) where $x \in X$ and H is a subgroup of the isotropy subgroup G_x. A morphism $\gamma : (x, H) \to (y, K)$ is represented by a pair $(h : G/H \times I \to X, \alpha)$ where h is an equivariant map satisfying $h(\{H\}, 0) = x$ and $\alpha : G/K \to G/H$ is an equivariant map such that $h(\alpha(\{K\}), 1) = y$. Morphisms are defined by G-homotopy classes of such objects leaving the endpoints fixed, and composition is defined by the usual concatenation (or addition) of curves. This category is not really a groupoid, but for all H the full subcategory generated by all objects (x, H) is a groupoid. In particular, the set of morphisms from an object to itself is always a group. There is a canonical functor from $\mathrm{FGPD}^G(X)$ to the category **Rng** of associative rings with unit, taking each object (x, H) to the integral group ring $\mathbb{Z}[\mathrm{Aut}(x, H)]$, where $\mathrm{Aut}(x, H)$ is the set of all self maps of (x, H) in $\mathrm{FGPD}^G(X)$. If X is a locally linear G-manifold and $\beta(x)$ is the unique closed substratum so that $x \in X_{\beta(x)} - \mathrm{RelSing}\, X_{\beta(x)}$, then the group $\mathrm{Aut}(x, H)$ is in fact isomorphic to the group $E_{\beta(x)}$ considered in Proposition II.1.4, and thus the associated ring is merely $\mathbb{Z}[E_{\beta(x)}]$. We shall call this functor the **equivariant fundamental groupoid ring** and denote it by $\mathbb{Z}[\mathrm{FGPD}^G(X)]$.

Definition: **A system of equivariant local coefficients** for X is a covariant functor

$$A : \mathrm{FGPD}^G(X) \to \mathbf{Rng}$$

and a natural transformation

$$\psi : \mathbb{Z}[\mathrm{FGPD}^G(X)] \to A$$

such that $\psi_{(x,H)}$ is an epimorphism (or almost epimorphism in the sense of [**CS**]) for all objects (x, H).

If $x \in X_\beta - \mathrm{RelSing}\, X_\beta$, then the isomorphism types of $\mathbb{Z}[\mathrm{Aut}(x, G^\beta)]$ and $A(x, G^\beta)$ do not depend upon x; furthermore, the naturality properties of ψ show that the maps $\psi_{(x,H)}$ are conjugate to each other in a canonical fashion. Thus if we set $A_\beta := A_{(x,H)}$ and use the isomorphism $\mathrm{Aut}\,(x, \beta) \cong E_\beta$, it follows that we have a canonical map $\psi_\beta : \mathbb{Z}[E_\beta] \to A_\beta$ that is well-defined up to canonical conjugacy.

If $\psi : \mathbb{Z}[\mathrm{FGPD}^G(X)] \to A$ describes an equivariant system of local coefficients, then one can define the *Browder-Quinn* Γ*-groups*

$$\Gamma_n^{BQ,c}(G; \Lambda_X; \psi : \mathbb{Z}[\mathrm{FGPD}^G(X)] \to A)$$

(where $c = h$ or s) by taking objects in [**BQ**] but taking equivalences to be transverse linear isovariant maps $f : M \to X$ such that the map of geometric posets $\check{f} : \pi(M) \to \pi(X)$ is bijective and the induced maps of open substrata

$$f_\beta^\# : M_\beta - \mathrm{RelSing}\, M_\beta \longrightarrow X_\beta - \mathrm{RelSing}\, X_\beta$$

determines homology isomorphisms with local coefficients in the $\mathbb{Z}[E_\beta]$-module A_β determined by ψ_β. One can also define Σ- and $(\Sigma \subset \Sigma')$-adjusted versions of these groups

$$\Gamma_n^{BQ,c}(G; \Lambda_X; \psi : \mathbb{Z}[\mathrm{FG{\small PD}}^G(X)] \to A; \Sigma)$$

$$\Gamma_n^{BQ,c}(G; \Lambda_X; \psi : \mathbb{Z}[\mathrm{FG{\small PD}}^G(X)] \to A; \Sigma \subset \Sigma')$$

and these groups determine the usual sorts of long exact sequences as in (II.1.3) and (II.2.0). Furthermore, there is a $\pi - \pi$ theorem for the Browder-Quinn Γ-groups that is analogous to [**BQ**, 3.1, p. 31]; the proof of the latter extends directly modulo using the $\pi - \pi$ theorem for ordinary Γ-groups [**CS**, also 3.1] instead of the $\pi - \pi$ theorem for ordinary Wall groups. Reasoning by analogy with Section II.1, one would expect that the stepwise surgery obstructions for $\Gamma_n^{BQ,c}$ lie in ordinary Γ-groups. In order to state a result of this type we need to relate the concepts of this section to those of Section II.1. If X_α is a closed substratum of X, let NonSing X_α^* denote the image of X_α − RelSing X_α in the orbit space $X^* = X/G$. If X satisfies the Codimension ≥ 2 Gap Hypothesis, it follows that NonSing X_α^* is a connected open substratum of X^*; conversely, if X satisfies the Codimension ≥ 2 Gap Hypothesis then every open connected substratum of X^* arises in this way.

PROPOSITION 6.3. *In the preceding notation, assume that $\Sigma' - \Sigma$ only contains $G\{\alpha\}$, and assume* $\dim \mathrm{NonSing}\, X_\alpha^* + n - \dim X \geq 5$. *Then there is an isomorphism σ_α from* $\Gamma_n^{BQ,c}(G; X; \psi : \mathbb{Z}[\mathrm{FG{\small PD}}^G(X)] \to A; \Sigma \subsetneq \Sigma')$ *to* $\Gamma_{n(\alpha)}^c(\mathbb{Z}[\pi_1(\mathrm{NonSing}\, X_\alpha^*)] \to A_\alpha, w_\alpha)$, *where* $n(\alpha) = \dim X_\alpha + n - \dim X$ *and w_α is given by the first Stiefel-Whitney class.*

The proof is a straightforward modification of the argument establishing Proposition II.1.4 with Cappell-Shaneson Γ-groups replacing Wall L-groups at the appropriate points.■

We can now use the usual sorts of methods to show that the Browder-Quinn Γ-theory has the standard periodicity properties:

THEOREM 6.4. *Assume that X satisfies the conditions of Theorems* III.2.7–9 *and the conditions needed to construct* $\Gamma_n^{BQ,c}(G; \Lambda_X)$.

(*i*) *If Y is a periodicity manifold in the sense of Section* III.2 *such that λ_X is Y-preperiodic, then the product map*

$$\mathbf{P}_Y : \Gamma_n^{BQ,c}(G; \Lambda_X) \to \Gamma_n^{BQ,c}(G; \Lambda_{X \times Y})$$

induces an isomorphism of abelian groups.

(*ii*) *If G has odd order, then the map $\mathbf{P}_Y$ is an isomorphism if $Y = \mathbb{C}P^{2n}$ or $\mathbb{C}P^2 \uparrow G$.*

PROOF: (*Sketch*) The argument is very similar to the proofs of 6.1 and Theorems III.2.7–9; it suffices to establish analogs of the formal properties (*a*)–(*d*) listed in the proof of 6.1. The first two properties deal with exactness properties and the compatibility of products with respect to exact sequences; these generalize immediately to the theories $\Gamma^{BQ,c}$. The third and fourth properties for $\Gamma^{BQ,c}$ deal with the compatibility between stepwise obstructions in Γ-groups and Yoshida's twisted products and a result stating that the twisted products $\mu^c(Y) : \Gamma_*^c(\mathbb{Z}[\pi] \to A, w) \to \Gamma_{*+\dim Y}^c(\mathbb{Z}[\pi] \to A, w')$

only depend upon the Witt invariant of Y. The third property is a straightforward generalization of the results in Section III.4, and the invariance property for twisted products on Γ-groups is given by Theorem 5.2 of this chapter.■

In analogy with Complement III.2.10 and the discussion following the statement of Theorem 6.2, the results of [Ya] yield an extension of 6.4(*ii*) to the case $Y = \mathbf{CP}^2 \uparrow S$ where S is a finite G-set and Λ_X is $\mathbf{CP}^2\uparrow S$-preperiodic.■

Lück-Madsen Γ-groups

If ψ as above is a G-equivariant local coefficient system, it is also possible to construct Lück-Madsen Γ-groups

$$\mathcal{L}\Gamma_n^c(G; X; \psi : \mathbf{Z}[\mathrm{FG}\textsc{pd}^G(X)] \to A; \Sigma \subset \Sigma')$$

that are analogous to the Browder-Quinn Γ-groups. Once again it is straightforward to show that such groups have the usual properties and that analogs of 6.3 and 6.4 hold if $\mathbf{S}^2\Lambda_X$ satisfies the Gap Hypothesis. Furthermore there are canonical maps $\varphi(n; G; X; \psi; \Sigma \subset \Sigma')$ from the Browder-Quinn Γ-groups

$$\Gamma_n^{BQ,c}(G; X; \psi : \mathbf{Z}[\mathrm{FG}\textsc{pd}^G(X)] \to A; \Sigma \subseteq \Sigma')$$

to the Lück-Madsen Γ-groups

$$\mathcal{L}\Gamma_n^c(G; X; \psi : \mathbf{Z}[\mathrm{FG}\textsc{pd}^G(X)] \to A; \Sigma \subset \Sigma')$$

and these maps are isomorphisms if $\mathbf{S}^2\Lambda_X$ satisfies the Gap Hypothesis and Λ_X desuspends in the usual fashion. Finally, the methods of Subsection III.5C show that the canonical maps $\varphi(n; G; X; \psi; \Sigma \subset \Sigma')$ are split injective if G has odd order.

References for Chapter IV

[**ACH**] J. P. Alexander, P. E. Conner, and G. C. Hamrick, "Odd Order Group Actions an Witt Classification of Inner Products," Lecture Notes in Mathematics Vol. 625, Springer, Berlin-Heidelberg-New York, 1977.

[**AP**] D. R. Anderson and E. K. Pedersen, *Semifree topological actions of finite groups on spheres*, Math. Ann. **265** (1983), 23–44.

[**An**] G. A. Anderson, "Surgery with Coefficients," Lecture Notes in Mathematics Vol. 591, Springer, Berlin-Heidelberg-New York, 1977.

[**BW**] A. Bak and B. Williams, *Surgery and higher K–theory*, in preparation.

[**Br**] W. Browder, "Surgery on Simply Connected Manifolds," Ergeb. der Math. (2) 65, Springer, Berlin-Heidelberg-New York, 1972.

[**CS**] S. Cappell and J. Shaneson, *The codimension two placement problem and homology equivalent manifolds*, Ann. of Math. **99** (1974), 277–348.

[**DHM**] M. Davis, W. C. Hsiang, and J. Morgan, *Concordance of regular $O(n)$-actions on homotopy spheres*, Acta Math. **144** (1980), 153–221.

[**tD**] T. tom Dieck, "Transformation Groups," de Gruyter Studies in Mathematics Vol. 8, W. de Gruyter, Berlin and New York, 1987.

[**Do**] K. H. Dovermann, *Addition of equivariant surgery obstructions*, Algebraic Topology, Waterloo 1978 (Conference Proceedings), Lecture Notes in Mathematics Vol. 741, Springer, Berlin-Heidelberg-New York; *same title*, Ph.D. Thesis. Rutgers University, 1978 *(Available from University Microfilms, Ann Arbor, Mich.: Order Number DEL79-10380.)* Summarized in Dissertation Abstracts International **39** (1978/1979), 5406.

[**DP1**] K. H. Dovermann and T. Petrie, *G–Surgery II*, Memoirs Amer. Math. Soc. **37** (1982), No. 260.

[**DP2**] ______________, *An induction theorem for equivariant surgery (G-Surgery III)*, Amer. J. Math. **105** (1983), 1369–1403.

[**DR**] K. H. Dovermann and M. Rothenberg, *Equivariant Surgery and Classification of Finite Group Actions on Manifolds*, Memoirs Amer. Math. Soc. **71** (1988), No. 379.

[**Ke**] G. T. Kennedy, "Foundations of supersimple surgery theory," Ph. D. Thesis, University of Notre Dame, 1986 *(Available from University Microfilms, Ann Arbor, Mich.: Order Number DA8612997.)*—Summarized in Dissertation Abstracts International **47** (1986/1987), 1095B.

[**Lö**] P. Löffler, *Über rationale Homologiesphären*, Math. Ann. **249** (1980), 141–151.

[**Lü**] W. Lück, "Transformation Groups and Algebraic K-Theory," Lecture Notes in Mathematics No. 1408, Springer, Berlin-Heidelberg-New York, 1989.

[**LM**] W. Lück and I. Madsen, *Equivariant L-theory I*, Aarhus Univ. Preprint Series (1987/1988), No. 8; [*same title*] *II*, Aarhus Univ. Preprint Series (1987/1988), No. 16.

[**MTW**] I. Madsen, L. Taylor, and B. Williams, *Tangential homotopy equivalence*, Comment. Math. Helv. **55** (1989).

[**Mau1**] S. Maumary, *Proper surgery groups and Wall–Novikov groups*, in "Algebraic K–theory III (Battelle Inst. Conf., 1972)," Lecture Notes in Mathematics Vol. 343, Springer, Berlin-Heidelberg-New York, 1973, pp. 526–539.

[**Mau2**] ———, *Proper surgery groups for noncompact manifolds of finite dimension*, preprint, University of California, Berkeley, 1972.

[**MMT**] J. P. May, J. McClure, and G. Triantafillou, *Equivariant localization*, Bull. London Math. Soc. **14** (1982), 223–230.

[**MSv**] I. Moerdijk and J.-A. Svensson, *The equivariant Serre spectral sequence*, (to appear).

[**Mrg**] J. Morgan, *A Product Formula for Surgery Obstructions*, Memoirs Amer. Math. Soc. **14** (1978), 201.

[**OL**] R. Oliver, *G–actions on disks and permutation representations I*, J. Algebra **50** (1978), 44–62.

[**PR**] E. K. Pedersen and A. A. Ranicki, *Projective surgery theory*, Topology **19** (1980), 239–254.

[**Q1**] F. Quinn, *Surgery on Poincaré and normal spaces*, Bull. Amer. Math. Soc. **78** (1972), 262–267.

[**Q2**] ———, in "Current Problems in Homotopy Theory and Related Topics," Proceedings of the Northwestern Homotopy Theory Conference (Evanston, Ill., 1982), Contemporary Mathematics, 1983, pp. 453–454.

[**Ra1**] A. A. Ranicki, *Algebraic L–theory I: Foundations*, Proc. London Math. Soc. **3:27** (1973), 101–125.

[**Ra2**] ———, *Algebraic L–theory II: Laurent extensions*, Proc. London Math. Soc. **3:27** (1973), 126–158.

[**Ra3**] ———, *The total surgery obstruction*, in "Algebraic Topology," (Sympos. Proc., Aarhus, 1978) Lecture Notes in Mathematics, Springer, Berlin-Heidelberg-New York, 1979, pp. 271–316.

[**Ra4**] ———, *The algebraic theory of surgery I: Foundations*, Proc. London Math. Soc. **3:40** (1980), 87–192.

[**Ra5**] ———, *The algebraic theory of surgery II: Applications to topology*, Proc. London Math. Soc. **3:40** (1980), 193–283.

[**Ra6**] ———, "Exact sequences in the algebraic theory of surgery," Princeton Mathematical Notes No. 26, Princeton University Press, Princeton, 1981.

[**Sh1**] J. Shaneson, *Wall's surgery obstruction groups for* $G \times Z$, Ann. of Math. **98** (1969), 296–334.

[**Sh2**] ———, *Product formulas for* $L_n(\pi)$, Bull. Amer. Math. Soc. **76** (1970), 787–791.

[**Sp**] E. H. Spanier, "Algebraic Topology," McGraw-Hill, New York, 1967.

[**St**] N. Steenrod, "The Topology of Fibre Bundles," Princeton Mathematical Series Vol. 14, Princeton University Press, Princeton, 1951.

[**Su**] D. Sullivan, *On the Hauptvermutung for manifolds*, Bull. Amer. Math. Soc. **73** (1967), 598–600.

[**T**] L.R. Taylor, "Surgery on paracompact manifolds," Ph.D. Thesis, University of California, Berkeley, 1972.

[**TW**] L. Taylor and B. Williams, *Local surgery: Foundations and applications*, Algebraic Topology (Sympos. Proc., Aarhus (1978), in "Lecture Notes in Mathematics," Springer, Berlin-Heidelberg-New York, 1979, pp. 673–695.

[**Wa1**] C. T. C. Wall, "Surgery on Compact Manifolds," London Math. Soc. Monographs No. 1, Academic Press, London and New York, 1970.

[**Wa2**] ———, *Classification of Hermitian forms VI. Group rings*, Ann. of Math. **103** (1976), 1–80.

[**Wa3**] ———, *Formulæfor surgery obstructions*, Topology **15** (1976), 189–210. Correction, **16** (1977), 495–496

[**Wi**] R. E. Williamson, *Surgery in* $M \times N$ *with* $\pi_1(M) \neq 1$, Bull. Amer. Math. Soc. **75** (1969), 582–585.

[**Ya**] M. Yan, *Periodicity in equivariant surgery and applications*, Ph. D. Thesis, University of Chicago, in preparation.

[**Yo**] T. Yoshida, *Surgery obstructions of twisted products*, J. Math. Okayama Univ. **24** (1982), 73–97.

CHAPTER V

PRODUCTS AND PERIODICITY FOR SURGERY UP TO PSEUDOEQUIVALENCE

Most of the equivariant surgery theories in Chapters I–IV are designed to study group actions that are more or less similar to some given example. However, a few equivariant surgery theories are designed to yield very exotic group actions with few similarities to obvious models. The main examples of such theories are the theories I^h and $I^{h(R)}$ for equivariant surgery up to pseudoequivalence that are mentioned in (II.2.5) and developed in [**DP1–2**] and [**PR**]; the power of such theories is illustrated by their applications to group actions on spheres with one fixed point [**Pe2**] and numerous other questions (see [**DPS**] for further information). In order to study existence questions for very exotic actions it is necessary to study extremely general classes of equivariant surgery problems, and the required level of generality leads to many new technical complications. It is more or less predictable that some of these will cause difficulties in any attempt to extend the periodicity theorems of Chapter III to the theories I^h and $I^{h(R)}$. In this chapter we shall study the difficulties that arise and deal with them effectively enough to obtain periodicity theorems for the theories $I^h(G;-)$ and $I^{h(R)}(G;-)$ in many important cases; for example, there are periodicity theorems if G is an odd order abelian group. We shall also consider some variants of the basic theories I^h and $I^{h(R)}$ from [**DP1–2**].

Here is a more specific description of this chapter's contents. In Section 1 we shall describe the indexing data and normal maps for I^h and $I^{h(R)}$, and in Section 2 we shall state the main results on the existence, additivity, and periodicity properties of product constructions for these theories. The next four sections (3 through 6) deal with formal properties of the I^h and $I^{h(R)}$ obstruction groups such as the stepwise obstruction homomorphisms, the maps given by restriction to closed substrata, and the properties of certain projective class group obstructions that do not appear in the other equivariant surgery theories considered in this book. Products in I^h and $I^{h(R)}$ are discussed in Section 7, and in Section 8 we combine the results of Sections 3–7 with the methods of Chapter III to prove the main results from Section 2. Finally, in Section 9 we prove a vanishing theorem for Wall groups that underlies the entire approach of this chapter; namely, if G is an odd order group and R is a subring of the rationals, then the odd (homotopy) Wall groups of the group ring $R[G]$ are trivial. It seems likely that this result had been known to some researchers in the area, but the literature does not seem to contain a result stated at the required level of generality. A reader who is primarily interested in the main results should begin by reading Sections 1, 2, and 8; the remaining material is more of a technical nature.

1. The setting

As indicated in the introduction, this chapter deals mainly with the theories I^h and $I^{h(R)}$ for surgery up to pseudoequivalence described in [**DP1–2**] and [**PR**]. These theories are very similar to the theory $I^{ht,DIFF}$ of Chapter I and [**DR1**] in many respects, but there are numerous complications. As noted in (II.2.5) the first complications arise in formulating an appropriate notion of indexing data, and there are further difficulties in defining the sorts of groups, manifolds, and mappings to be considered. In this section we shall deal with the main technical complications as part of our summary of the theories I^h and $I^{h(R)}$ for actions of odd order groups. At the end of this section we shall describe the modifications needed to treat even order groups. Fortunately, a relatively limited discussion of technicalities will suffice to formulate the most basic definitions and prove what we need, and it will not be necessary to mention many subtle and unpleasant points. The complete references for this material are [**DP1**, Sections 1–4, 9, and Appendix A–1] and [**PR**].

Indexing data

We shall assume throughout that G is a finite group and $f : M \to X$ is an equivariant map of compact, smooth G-manifolds. Furthermore, we shall assume that all smooth G-manifolds satisfy the Codimension ≥ 2 Gap Hypothesis of Section I.2 unless indicated otherwise.

In Section I.2 we defined the *indexing data*

$$\lambda_M = (\pi(M), d_M, s_M, w_M)$$
$$\lambda_X = (\pi(X), d_X, s_X, w_X)$$

associated to M and X. Specifically, the *geometric posets* $\pi(M)$ and $\pi(X)$ list the components of the various fixed point sets and their interrelations, the *dimensions* d_M and d_X give the dimensions of these components, the *normal slice data* s_M and s_X describe the oriented normal slices on the sets in $\pi(M)$ and $\pi(X)$, and the *orientation data* w_M and w_X describe the group action's effect on orientations over each set in $\pi(M)$ and $\pi(X)$. For the sake of relative simplicity we shall assume that M and X satisfy the Codimension ≥ 3 Gap Hypothesis and all closed substrata of M and X (in the sense of Section I.2) are oriented. The **functional indexing data** of f will be the r-tuple

$$\lambda(f) = (\lambda_M, \lambda_X, \check{f}, \mu)$$

where λ_M and λ_X are as before, the map $\check{f} : \pi(M) \to \pi(X)$ is the induced map of G-posets described in Section I.2, and $\mu : \pi(X) \to \mathbb{Z}$ is *degree data* given by

$$\mu(\beta) = \sum \text{ degree } (f_\alpha : M_\alpha \to X_\beta),$$

the sum running through all α such that $\check{f}\alpha = \beta$ and $\dim M_\alpha = \dim X_\beta$ (compare [**DP1**, (1.12), p. 5]).

Frequently it suffices to work with a portion of the indexing data. As in Section II.2, let $\mathcal{H}$ be a nonempty family of subgroups of G that is closed under conjugation and passage to subgroups, and for each compact locally linear G-manifold Y let $\pi(Y;\mathcal{H})$ denote the set of all closed substrata in $\pi(Y)$ that are components of fixed point sets of subgroups in $\mathcal{H}$. If $f : M \to X$ is a map such that $\check{f} : \pi(M) \to \pi(X)$ sends $\pi(M;\mathcal{H})$ to $\pi(X;\mathcal{H})$, then the *$\mathcal{H}$-restricted functional indexing data* $\lambda(f;\mathcal{H})$ will consist of the G-posets $\pi(M;\mathcal{H})$, $\pi(X;\mathcal{H})$, the restrictions of the dimension, normal slice, and orientation data to these G-posets, the map

$$\check{f}_{\mathcal{H}} : \pi(M;\mathcal{H}) \to \pi(X;\mathcal{H})$$

determined by $\check{f}$, and the restriction $\mu_{\mathcal{H}}$ of the degree data to $\pi(X;\mathcal{H})$ and $\pi(M;\mathcal{H})$; the latter is defined by

$$\mu_{\mathcal{H}}(\beta) = \sum_{\mathcal{H}} \text{degree } (f_\alpha : M_\alpha \to X_\beta),$$

where one only adds degrees over closed substrata in $\pi(M;\mathcal{H})$.

DEFINITION: If $f : M \to X$ and $\mathcal{H}$ are as above, we shall say that f is **$\mathcal{H}$-isogeneric** if for each $\alpha \in \pi(M;\mathcal{H})$ we have $G^{\check{f}\alpha} = G^\alpha$, where G^γ denotes the generic isotropy subgroup of the closed stratum γ. An $\mathcal{H}$-isogeneric map clearly satisfies the condition needed to construct $\lambda(f;\mathcal{H})$.

If $\mathcal{H}$ is the family $\mathfrak{P}_G$ of all subgroups that are trivial or p-groups (where p can be any prime dividing the order of G), we shall generally use **weakly isogeneric** to denote $(\mathcal{H} = \mathfrak{P}_G)$-isogeneric.

REMARKS: 1. If $\check{f} : \pi(M) \to \pi(X)$ is an isomorphism, then f is $\mathcal{A}\ell\ell$-isogeneric where $\mathcal{A}\ell\ell$ is the family of all subgroups (this follows from [**DP1**, Section 1]). It follows that f is also $\mathcal{H}$-isogeneric for every smaller family $\mathcal{H}$.

2. One always has $G^{\check{f}\alpha} \supseteq G^\alpha$. Here is an example where proper containment holds: If a and b are positive integers, let $t^a \oplus t^b$ be the 2-dimensional unitary S^1-representation $z(x,y) = (z^a x, z^b y)$, and let $S(t^a \oplus t^b)$ be the associated unit sphere. If we also assume that a and b are relatively prime, then there is an S^1-equivariant map

$$f : S(t^a \oplus t^b \oplus \mathbf{R}) \to S(t^{ab} \oplus t^1 \oplus \mathbf{R})$$

that arises repeatedly in the work of Petrie (compare [**MP**, Section 2]). We may view f as an equivariant map of $\mathbf{Z}_{ab}$-manifolds by restricting to $\mathbf{Z}_{ab} \subset S^1$. Then the associated map $\check{f} : \pi(S(t^a \oplus t^b \oplus \mathbf{R})) \to \pi(S(t^{ab} \oplus t^1 \oplus \mathbf{R}))$ is given as follows: The fixed sets of $\mathbf{Z}_a$ and $\mathbf{Z}_b$ in $S(t^a \oplus t^b \oplus \mathbf{R})$ are distinct connected submanifolds, and $\check{f}$ takes both into the fixed set of $\mathbf{Z}_{ab}$ in $S(t^{ab} \oplus t^1 \oplus \mathbf{R})$. Therefore $G^{\check{f}\alpha} = \mathbf{Z}_{ab}$ and $G^\alpha = \mathbf{Z}_a$ or $\mathbf{Z}_b$ in this example. Incidentally, the degrees of the maps $f_\alpha : M_\alpha \to X_{\check{f}\alpha}$ in this example are integers q and r such that $qa + rb = 1$.

In [**Do2**, (2.8)] two additional pieces of data (γ, δ) are imposed in order to construct a sum operation on surgery problems with indexing data $(\lambda; \gamma, \delta)$. The invariants γ and δ assigns to each closed substratum β some virtual representations γ_β and $\delta_\beta \in RO(G^\beta)$

(see [**Do2**, 4.7, p. 254]; also see [**Do1**, p. 27]). To describe δ, we first recall that an equivariant surgery problem $f : M \to X$ generally involves something like a stable G–vector bundle isomorphism $T(M) \cong f^*\xi$ for some G-vector bundle ξ over X. Given a closed substratum M_β and a point $y \in M_\beta$ with generic isotropy subgroup G^β, the representation δ_β is simply the local representation of G^β in the fiber of ξ over y. By connectedness this is independent of the choice of y. This construction is compatible with stabilization, for if we replace ξ by $\xi \oplus W$ (where W is some G-module), then δ_β is replaced by $\delta_\beta \oplus W$. It will be convenient for us to view two objects as interchangeable if the vector bundle data in one is interchangeable with the vector bundle data in the other, and therefore we shall replace δ by a *stabilized* analog ∇.

Specifically, consider the difference bundle

$$\xi \oplus \nu_{(X,\mathbf{E})} \qquad (\nu = \text{equivariant normal bundle of some smooth } G\text{-embedding of } X \text{ in a } G\text{-module } \mathbf{E}).$$

The virtual representation ∇_β of G^β is then defined by the formula

$$\nabla_\beta = (\xi \oplus \nu_{(X,\mathbf{E})})|_\beta - \operatorname{res} \mathbf{E} \qquad (\operatorname{res} = \text{restriction to } G^\beta).$$

The identity $\delta_\beta = \nabla_\beta \oplus T(X)_\beta$ allows us to pass back and forth between δ and ∇. One can show directly that the stabilization map from $I(-,\delta)$ to $I(-,\delta \oplus W)$ is an isomorphism (of groups, if a group structure is present), and therefore we shall usually write $I(-,\nabla)$ instead of $I(-,\delta)$.

Finally, we describe γ. An equivariant surgery problem $f : M \to X$ usually involves a system of *unstable* vector bundle isomorphisms $\Pi(T(M)) \cong \Pi(f^*\eta)$ where $\Pi(T(M))$ is the system of equivariant normal bundles defined by the inclusions of the closed substrata of M in each other, and η is an appropriate system of vector bundles $\{\eta_\beta\}$ over the closed substrata X_β (these systems are also called Π-*bundles*). If $y \in X_\beta$ has generic isotropy subgroup G^β, then γ_β is the representation of G^β on the fiber $\eta_\beta|_y$. When f is strongly isogeneric the invariant γ_β can be recovered from s_M.

Several notational conventions for indexing data were listed at the end of Section I.2. We shall use the same conventions for the functional indexing data described in this section.

Normal maps

Throughout this discussion I^a will refer to the theory I^h or one of the theories $I^{h(R)}$, and the appropriate family of such groups will be called $\mathcal{H}^a$; by definition $I^h = I^{h(\mathbb{Z})}$ and $\mathcal{H}^h = \mathcal{H}^{h(\mathbb{Z})}$. As indicated in (II.2.5) the family $\mathcal{H}^{h(R)}$ is all subgroups whose orders are 1 or powers of primes that are not units in R. The appropriate family of manifolds

$$\mathcal{F}^a \qquad (a = h \text{ or } h(R))$$

will be all manifolds Y such that the Gap Hypothesis holds and all closed substrata in $\pi(Y;\mathcal{H}^a)$ are simply connected and at least 5-dimensional.

REMARK: In [**DP1-2**] the family $\mathcal{H}^h$ is called $\mathfrak{P}_G$. We shall generally use $\mathfrak{P}_G$ for the sake of consistency with these references.

In order to apply the theory I^a to a degree 1 map $f : M \to X$ of connected manifolds it is necessary to assume five additional conditions on the indexing data $\lambda(f;\mathcal{H}^a)$. The first four are easy to state.

(1.0.A) *The map f is $\mathcal{H}^a$-isogeneric.*

(1.0.B) *For each $\alpha \in \pi(M,\mathcal{H}^a)$ one has* $\dim M_\alpha = \dim X_{f(\alpha)}$.

(1.0.C) *For each $\beta \in \pi(X;\mathcal{H}^a)$ the degree $\mu(\beta)$ is relatively prime to the order of G^β (recall the latter is a prime power).*

(1.0.D) *For each $\beta \in \pi(X;\mathcal{H}^a)$ the (simply connected) manifold X_β is $n(\lambda)$-connected, where $n(\lambda)$ is defined as in* [**DP1**, p. 19, between Thm. 2.A and Thm. 2.11].

REMARK: The simple connectivity assumption on X_β follows from the restrictions on the family $\mathcal{F}^a$.

A description of the integer $n(\lambda)$ would require a very lengthy digression; fortunately, for our purposes it will suffice to know that the condition can be ignored in many important special cases. Ultimately we shall work with groups that are nilpotent and have odd order, and for these groups the final condition is necessary:

COMPLEMENT TO (1.0.D). *If G is a (finite) nilpotent group, then $n(\lambda) \leq 1$.*

As noted in [**DP1**, p. 20, line 4] this follows from results of Oliver and Petrie (see [**OP**, 5.13–14] and [**DP1**, Cor. 2.12, p. 19] for the case of abelian groups).

The fifth (and last) condition is significantly more difficult to formulate; it concerns restrictions on the Euler characteristics of various components of fixed point sets in the mapping cone (see [**DP1**, (3.13), p. 25]). Results of Oliver and Petrie [**OP**] show that surgery up to pseudoequivalence is impossible if the Euler characteristics do not satisfy suitable restrictions, and therefore it is reasonable to consider only maps that satisfy the given conditions. In order to minimize the notational requirements we shall only give the explicit definitions of normal maps for groups of odd order. However, the whole discussion extends directly to even order groups if condition (1.0.F) below is replaced by a more elaborate statement; we shall discuss the necessary changes at the end of this section.

Unfortunately, the discussion of Euler characteristic restrictions requires additional terminology.

DEFINITION: Let X be a separable metric G-space, where G is a finite group. The **complete G-poset** $\Pi(X)$ is the set

$$\bigcup_{H\subset G} \pi_0(X^H) \times \{H\},$$

with

(i) partial ordering $(C,H) \leq (C',H')$ if $C \subseteq C'$ and $H \supseteq H'$,
(ii) G-action given by $g(C,H) = (gC,\ gHg^{-1})$.

The **essential elements** of $\Pi(X)$ are the pairs (C, H) such that H is the isotropy subgroup for all points in some open subset of C. It follows directly from the definitions that G acts order-preservingly on $\Pi(X)$ and the set of essential elements is G-invariant. If X is a compact smooth G-manifold, then the essential elements of $\Pi(X)$ form a set isomorphic to $\pi(X)$ and the G-action on essential elements coincides with the previously defined G-action on $\pi(X)$.

Let $(Z, \{z_0\})$ be a pair consisting of a finite G-CW complex and an invariant subcomplex $\{z_0\}$ (= the base point). If X is a finite G-CW complex, then a $\Pi(X)$-**complex structure** on Z is a surjective map

$$\psi : \Pi(Z - \{z_0\}) \to \Pi(X)$$

which is equivariant, monotonic, and subgroup-preserving; *i.e.*, $\psi(C, H) = (C', H)$ for suitable C'. If $\alpha \in \Pi(X)$, with second coordinate H, then set

$$Z_\alpha = \{z_0\} \cup \ (\cup_{B \in \Psi(\alpha)} B)$$

where

$$\Psi(\alpha) := \{B \in \pi_0((Z - \{z_0\})^H) \mid \psi(B) \subset C_\alpha\}.$$

The definition differs slightly from [**DP1**, (2.1), page 14], but the two formulations are equivalent.

If $\alpha = (C, H) \in \Pi(X)$, we shall say that Z_α is an **essential subcomplex** if some point of Z_α has isotropy group H, and if $\mathcal{F} \subset \Pi(X)$ we shall say that $(Z, \{z_0\}; \psi)$ is an $\mathcal{F}$-*complex* if all essential subcomplexes lie in $\mathcal{F}$ (compare [**DP1**, (2.2), page 14]).

EXAMPLES: 1. The complex X itself does not have a $\Pi(X)$-structure, but the augmented complex $X_+ = X \amalg \{x_0\}$ does have such a structure, where $(X_+)_\alpha = C \amalg \{x_0\}$ for $\alpha = (C, H)$.

2. The wedge of two $\mathcal{F}$-complexes is an $\mathcal{F}$-complex.

3. If $f : M \to X$ is an equivariant map of finite G-CW complexes, for $\alpha = (C, H) \in \Pi(X)$ define

$$(\ddagger) \qquad M_\alpha := \{y \in M^H \mid f(y) \in C\}.$$

Then M_α is a subcomplex of M^H, and f induces a map $f_\alpha : M_\alpha \to C$. Let $M(f)$ be the mapping cone of f with base point given by the cone vertex. It follows that $M(f)$ has a $\Pi(X)$-structure with $M(f)_\alpha = M(f_\alpha)$. These concepts are discussed in more detail on pages 14–15 of [**DP1**].

Given two $\mathcal{F}$-complexes Z and W, we say Z and W are $\mathbf{\Omega}(G, \mathcal{F})$-*equivalent* if $\chi(Z_\alpha) = \chi(W_\alpha)$ for all $\alpha \in \Pi(X)$. The corresponding equivalence classes form an abelian group denoted by $\mathbf{\Omega}(G, \mathcal{F})$; the addition corresponds to wedge sum. By construction, if $\mathcal{F}' \subset \mathcal{F}$ then $\mathbf{\Omega}(G, \mathcal{F}')$ is a subgroup of $\mathbf{\Omega}(G, \mathcal{F})$. The groups $\mathbf{\Omega}(G, \mathcal{F})$ can be viewed as variants of the usual Burnside ring $\mathbf{\Omega}(G)$ as defined in (say) [**tD2**]. In fact, $\mathbf{\Omega}(G)$ is isomorphic to $\mathbf{\Omega}(G, \mathcal{O}_+)$, where $\mathcal{O}_+(G)$ is the G-poset consisting of coset spaces G/H ordered by

$G/H \leq G/K \Leftrightarrow K \subset H$ together with an extra minimal "basepoint class" $*_G$; the isomorphism is given by sending a class in $\mathbf{\Omega}(G)$ represented by a finite G-CW complex K to the class of $K \amalg \{*_G\}$.

We shall need two more facts from [**DP1**]. First, $\Pi(X)$ is functorial for G-maps, and given $f : M \to X$ we denote the induced map of G-posets by $\widetilde{f}$. Second, we shall need the set $\Delta(G, \mathcal{F}) \subset \mathbf{\Omega}(G, \mathcal{F})$ as defined in [**DP1**, Section 2] or [**OP**]. For our purposes it suffices to view $\Delta(G, \mathcal{F})$ as the subset of elements representable by (nonequivariantly!) contractible $\mathcal{F}$-complexes (compare [**OP**, Proposition 1.6] or [**DP1**, Theorem 2.10]).

The following Euler characteristic condition for I^h and $I^{h(R)}$ normal maps is used throughout [**DP1**]:

$$(1.0.E) \qquad [M(f)] \in \Delta\Big(G, \pi(X) \cup \widetilde{f}(\pi(M))\Big) + \mathbf{\Omega}(G; \pi(X)).$$

This condition allows one to prove an equivariant version of the $\pi - \pi$ Theorem [**DP1**, Section 7], but it is not quite what one needs to obtain well-defined stepwise surgery obstructions as in [**DP2**]. Therefore we also need a variant of (1.0.E):

$$(1.0.F) \qquad [M(f)] \in \Delta\Big(G, \pi(X) \cup \widetilde{f}(\pi(M))\Big) + (1 + (-1)^{\dim M})\,\mathbf{\Omega}(G; \pi(X)).$$

(Less formally, one replaces $\mathbf{\Omega}$ with $2\mathbf{\Omega}$ if $\dim M$ is even and with zero if $\dim M$ is odd.)

The bundle data for the theories in this chapter are very similar to the bundle data introduced in Chapter I; the main difference is that unstable bundle data are needed only over the closed substrata in $\pi(X; \mathcal{H}^a)$. Specifically, if $f : M \to X$ is an equivariant degree 1 map of closed smooth manifolds satisfying (1.0.A)–(1.0.C) above, then a *system of equivariant bundle data of type I^a for f* will be a triple (b_0, c, φ) where

(i) ξ is a stable G-vector bundle over M,
(ii) $b_0 : T(N) \to f^*\xi$ is a stable G-vector bundle isomorphism,
(iii) η is a vector bundle system over $\pi(X; \mathcal{H}^a)$,
(iv) $c : \Pi(T(N); \mathcal{H}^a) \to f^*\eta$ is an isomorphism of vector bundle systems over $\mathcal{H}^a$,
(v) $\varphi : \eta \oplus \Pi(M \times V; \mathcal{H}^a) \to \Pi(\xi; \mathcal{H}^a)$ is a stable isomorphism of vector bundle systems over $\mathcal{H}^a$ such that $f^*\varphi \circ (c \oplus V) = \Pi(b_0; \mathcal{H}^a)$.

REMARKS: Vector bundle systems over subsets $\Pi \subset \pi(X)$ are defined in Section I.3, and conditions (i)–(v) are straightforward extensions of the conditions for bundle data in Section I.4 (*i.e.*, the five paragraphs preceding Thm. I.4.4).

A **G-normal map of type I^a** will consist of

(1) an equivariant degree 1 map between manifolds in $\mathcal{F}^a$ such that (1.0.A)–(1.0.D) and (1.0.F) hold,
(2) a system of equivariant bundle data of type I^a for f.

A **G-normal cobordism of type I^a** will be a triad object

$$((F; f_0, f_1), \text{ other data})$$

such that $(f_0, -)$ and $(f_1, -)$ are G-normal maps of type I^a, the map F satisfies (1.0.A)–(1.0.E), and the inclusions induce isomorphisms $\lambda(f_i; \mathcal{H}^a) \cong \lambda(F; \mathcal{H}^a)$ for $i = 0, 1$.

One slightly anomalous feature of these definitions seems worth mentioning explicitly: *The Euler characteristic condition for I^a normal maps is more restrictive than the condition for an I^a normal cobordism.*

If $\lambda = \lambda(f; \mathcal{H}^a)$ it is now fairly straightforward to define the surgery obstruction sets $I^a(G; \lambda)$ as normal bordism classes of normal maps (see [**DP1**, pp. 26–29]). Similarly, if all closed substrata in $\pi(X; \mathcal{H}^a)$ and $\pi(M; \mathcal{H}^a)$ are at least 5-dimensional, and the Gap Hypothesis holds, then one also has restricted surgery obstruction sets $I^a(G; \lambda; \gamma, \delta)$ as in [**Do1-2**]. Furthermore, the results of [**Do1-2**] show that these sets have canonical abelian group structures if $n(\lambda) \leq 1$.

The following special cases will play an important role in this chapter.

PROPOSITION 1.1. *If G is nilpotent, the data $\lambda = \lambda(f; \mathcal{H}^a), \gamma$, and δ are as above, the dimensions of all closed substrata are ≥ 3, the dimensions of all closed substrata in $\pi(-; \mathcal{H}^a)$ are ≥ 5, and the Gap Hypothesis holds, then there is a geometrically defined abelian group structure on $I^a(G; \lambda; \gamma, \delta)$.*

This follows from the previously mentioned results of [**Do1-2**] because $n(\lambda) \leq 1$ if G is nilpotent (see the complement to (1.0.D)).■

REMARKS: 1. Earlier in this section we mentioned that the information carried by δ was also contained in a stabilized object ∇; one can show directly that the stabilization map from $I(-, \delta)$ to $I(-, \delta \oplus W)$ is an isomorphism (of groups, if group structures are present), and therefore we shall usually write $I(-, \nabla)$ instead of $I(-, \delta)$.

2. The sets $I^a(G; \lambda)$ are related to the groups $I^a(G; \lambda; \gamma, \nabla)$ by a disjoint union formula

$$I^a(G; \lambda) \cong \coprod_{(\gamma, \nabla)} I^a(G; \lambda; \gamma, \nabla), \tag{1.2}$$

where the disjoint union runs through all possible choices of γ and ∇.

REMARK: It was not necessary to introduce γ and ∇ for the other theories studied in this book because the information carried by these invariants can be recovered from the data specified in the other cases.

Extension to even order groups

Although our definitions of normal maps and normal bordisms are only formulated for groups of odd order, the definitions themselves probably do not suggest any reasons for this restriction. In fact, the crucial property of odd order groups is that the dimensions of all closed substrata are congruent mod 2. This implies that property (1.0.F) is invariant under I^a normal cobordism. In contrast, it is easy to construct examples of orientation-preserving even order group actions that are smooth and have closed substrata in both even and odd dimensions; the simplest example is the one point compactification of the regular representation of $\mathbf{Z}_2 \times \mathbf{Z}_2$. For such group actions it is necessary to replace (1.0.F) by a more complicated relation on Euler characteristics that reflects the mod 2 dimensions of the various closed substrata. With this change the entire discussion above extends to actions of even order groups.

2. The main results

In Section III.1 we showed that many equivariant surgery theories have product homomorphisms given by taking the direct product of an equivariant surgery problem $(f, -)$ with the identity on a suitable G-manifold Y. Although the existence and additivity of the product maps were not difficult to verify, the proofs were less trivial than in the nonequivariant case. Since the existence and additivity of products requires considerably more work for the theories $I^a = I^h$ or $I^{h(R)}$ we shall begin by stating two theorems on products.

THEOREM 2.1. *Let G be an odd order nilpotent group. Suppose that $\lambda = \lambda(f; \mathcal{H}^a)$ satisfies the basic conditions for defining $I^a(G;\lambda)$, let Y be a closed smooth G-manifold with simply connected substrata satisfying the relevant conditions from* (1.0.A)–(1.0.D), *and assume that $\lambda(f) \times \text{data}(Y)$ satisfies the Gap Hypothesis. Then there is a well-defined product map $\mathbf{P}_Y : I^a(G;\lambda) \to I^a(G; \lambda \times \text{data}(Y))$ that takes the class of* $(f : M \to X,\ \text{data})$ *to* $(f \times \mathbf{id}_Y : M \times Y \to X \times Y, \text{other data})$.

THEOREM 2.2. *In the setting of Theorem 2.1 also assume that $I^a(G; \lambda; \gamma, \nabla)$ satisfies the conditions for defining addition in* [**Do2**, 6.1, p. 266] *including the Gap Hypothesis (and $n(\lambda) \leq 1$). Then $\lambda \times \text{data}(Y)$ also satisfies the conditions for defining addition, and the restriction of the product map $\mathbf{P}_Y$ to $I^a(G;\lambda;\gamma,\nabla)$ defines a map*

$$\mathbf{P}_Y(\lambda;\gamma,\nabla) : I^a(G;\lambda;\gamma,\nabla) \to I^a(G;\lambda \times \text{data}(Y); \gamma + \gamma_Y, \nabla)$$

that is a homomorphism with respect to the abelian group structures of [**Do1-2**].

In principle these are straightforward results, but some effort is needed to show that $f \times \mathbf{id}_Y$ satisfies the conditions in (1.0.A)–(1.0.D) and that the Euler characteristic conditions (1.0.E)–(1.0.F) are preserved. The additivity property can be established by the methods of Section 1. Proofs of both theorems will be given in Section 8.

REMARKS: 1. If one defines I^a normal maps and normal bordisms for even order group actions by modifying (1.0.F) as suggested at the end of Section 1, then Theorems 2.1 and 2.2 will extend directly to actions of even order groups.

2. The restriction to nilpotent groups in Theorems 2.1 and 2.2 deserves some explanation. Our main examples of periodicity manifolds have been $\mathbf{CP}^{2n}$ and $\mathbf{CP}^{2n}\uparrow G$; the closed substrata of these manifolds are 1-connected but not 2-connected. Also, the discussion of addition requires the condition $n(\lambda) \leq 1$. Therefore we are forced to restrict attention to situations where $n(\lambda) \leq 1$. In order to ensure this, we assume G is nilpotent; as indicated in the Complement to (1.0.1)) the condition $n(\lambda) \leq 1$ is automatic for such groups.

If G is nilpotent of odd order then there are generalizations of the main Periodicity Theorems III.2.7–9 to the equivariant surgery theories I^h and $I^{h(R)}$. In order to keep the exposition relatively simple, we shall formulate the theorems and proofs only for the theory I^h. There are extensions of the periodicity theorems to $I^{h(R)}$ and other variants of I^h; these will be discussed at the end of Section 8.

STANDING HYPOTHESIS. *For the rest of this section we shall assume G is nilpotent and has odd order.*

We shall begin by stating the analog of Periodicity Theorem III.2.7.

THEOREM 2.3. *Let $\lambda = \lambda(f)$ satisfy the conditions for $I^h(G;\lambda(f))$ to be defined and for an abelian group structure on $I^h(G;\lambda(f);\gamma,\nabla)$, and assume that $\lambda(f)$ is weakly isogeneric. Let Y be a periodicity manifold in the sense of Section III.2, and assume that the following hold:*

(i) *$\lambda(f) \times \mathrm{data}(Y)$ satisfies the Gap Hypothesis.*
(ii) *$\lambda(f)$ is Y-preperiodic in the sense of Section III.2.*
(iii) *The set $I^h(G;\mathbf{S}^{-2}\lambda;\gamma,\nabla)$ is nonempty (where $\mathbf{S}^k\lambda$ is defined as in Section I.2).*

Then the product map

$$\mathbf{P}_Y : I^h(G;\lambda;\gamma,\nabla) \to I^h(G;\lambda\times\mathrm{data}(Y),\gamma\oplus\gamma_Y,\nabla)$$

defines an isomorphism of abelian groups.

We shall prove this in Section 8. In analogy with Chapter III, the following two results are essentially special cases of 2.3.

THEOREM 2.4. *Suppose that $\lambda(f)$ satisfies the conditions for an abelian group structure on $I^h(G;\lambda;\gamma,\nabla)$ as in 2.3, and assume further that*

(i) *λ is weakly isogeneric,*
(ii) *$\mathbf{S}^{4n}\lambda$ satisfies the Gap Hypothesis,*
(iii) *$I^h(G;\mathbf{S}^{-2}\lambda;\gamma,\nabla) \neq \varnothing$.*

Then $\mathbf{P}_Y$ defines an isomorphism if $Y = \mathbf{CP}^{2n}$ (with trivial G-action). If (iii) *is not true but $\mathbf{S}^{8n}\lambda$ satisfies the Gap Hypothesis, then*

$$\mathbf{P}_Y : I^h(G;\lambda\times\mathrm{data}(\mathbf{CP}^{2n});\gamma,\nabla) \to I^h(G;\lambda\times\mathrm{data}((\mathbf{CP}^{2n})^2);\gamma\oplus\gamma_Y,\nabla)$$

is an isomorphism of abelian groups.

THEOREM 2.5. *Suppose that $\lambda(f)$ satisfies the conditions for a group operation on $I^h(G;\lambda;\gamma,\nabla)$. Let $Y = \mathbf{CP}^{2n}\uparrow G$ be defined as in Chapter III. Then the product maps*

$$\mathbf{P}_Y : I^h(G;\lambda \times \mathrm{data}(Y^r);\gamma,\nabla) \to I^h(G;\lambda \times \mathrm{data}(Y^{r+1});\gamma \oplus \gamma_Y,\nabla)$$

are isomorphisms for all $r \geq 0$.

In the next five sections we shall develop the machinery needed to prove these results; we shall complete the proofs in Section 8.

In analogy with Complement III.2.10 and the discussions following the statements of Theorems IV.6.2 and IV.6.4, the results of [Ya] yield the analog of Theorem 2.5 if $Y = \mathbf{CP}^2\uparrow S$ where S is a finite G-set, X is $\mathbf{CP}^2\uparrow S$-preperiodic, and all relevant indexing data satisfy the Gap Hypothesis. and all relevant indexing data satisfy the Gap Hypothesis.

The nilpotence condition

Obviously one would like to extend the periodicity theorems for I^h and $I^{h(R)}$ to all odd order groups. In order to do this it is necessary to deal with the assumption $n(\lambda) \leq 1$. There are some conditions besides nilpotence that imply $n(\lambda) \leq 1$ (see [**DP1**, p. 19]), and in such cases the proofs of the periodicity theorems extend word for word. The sum construction from [**Do1-2**] requires $n(\lambda) \leq 1$, but in many cases it should be possible to circumvent this. If so, then one could use the quaternionic and Cayley projective spaces $\mathbf{KP}^{2n}$ and $\mathbf{CayP}^2$ and the associated products $(\mathbf{KP}^{2n})\uparrow G$, $(\mathbf{CayP}^2)\uparrow G$ to treat cases where $n(\lambda) \leq 3$ or $n(\lambda) \leq 7$. In order to find similar adaptations for even larger values of $n(\lambda)$, it would be necessary to have an affirmative answer to the following question:

PROBLEM 2.6. *Given $k > 0$ is there a k-connected closed smooth manifold P^{4n} with odd signature?*

If such a manifold exists, one could attempt to use $P\uparrow G$ as a replacement for $\mathbf{CP}^2\uparrow G$ throughout; the main complication is the possibility of torsion in $H^*(P;\mathbf{Z})$. Questions similar to 2.6 have arisen in other contexts; for example, D. Burghelea has asked if such manifolds can exist with $k \geq \frac{4}{3}n$. Presumably the results of R. Stong [**St**] on highly connected fiberings of BSO would be needed in any study of Problem 2.6.

3. Stepwise obstructions and addition

The proofs of the periodicity theorems in Chapters III and IV depend on two quantitative principles. First of all, global equivariant surgery obstruction groups can be broken into pieces arising from ordinary Wall groups. Second, product maps behave nicely on these pieces. In Chapters III and IV we formulated these principles in terms of adjusted surgery obstruction groups. It is possible to set up a similar formal apparatus for the theories I^h and $I^{h(R)}$. However, in order to avoid undue complications we shall adopt an approach that does not require adjusted groups; the concept of adjustability will still be important, but in a less formal manner. As in Section 2 we shall only deal explicitly with the theory I^h. The results of this section generalize to the theories $I^{h(R)}$ and other variants of I^h; further information on such extensions appears at the end of Section 8.

Except for the final paragraph, we assume throughout this section that G is an odd order nilpotent group. In the final paragraph we shall discuss extensions to other finite groups.

Recall that $\mathfrak{P}_G$ is the family of subgroups of G that are either trivial or p-groups and that $\mathfrak{P}_G$ is the distinguished family of subgroups $\mathcal{H}^h$ for the theory I^h.

If $Y_\alpha \in \pi(Y, \mathfrak{P}_G)$, let $\ell(\alpha)$ denote the product of all primes p such that Y_α is a component of Y^P with P a p-group, and define a subring $\mathbb{Z}_{(\alpha)} \subset \mathbb{Q}$ by

$$\mathbb{Z}_{(\alpha)} := \mathbb{Z}_{(\ell(\alpha))} = \{x \in \mathbb{Q} |\ x = a/b \text{ with } a, b \in \mathbb{Z} \text{ and g.c.d. } (b, \ell(\alpha)) = 1\}.$$

The next definition provides the standard framework for handling equivariant surgery problems in I^h as sequences of ordinary surgery problems.

DEFINITION: Let ($f : M \to X$, other data) represent a class in $I^h(G; \lambda; \gamma, \nabla)$, and let $\Sigma \subset \pi(M)$ be a *closed* G-invariant set (*i.e.*, $M_\gamma \subset M_\alpha$ and $\alpha \in \Sigma \Rightarrow \gamma \in \Sigma$). In the case of pseudoequivalence surgery, also assume that $\Sigma \subset \pi(M, \mathfrak{P}_G)$ and is closed in the corresponding sense. We shall say that f is **Σ-adjusted** (called **Σ-good** in [**DP1-2**]) if for each $\alpha \in \Sigma$ the map $f_\alpha : M_\alpha \to X_\beta$ (where $\beta = \check{f}(\alpha)$) is (1) a $\mathbb{Z}_{(\alpha)}$-homology equivalence if α is not a maximal element of $\pi(X)$, or (2) a homotopy equivalence if α is not a maximal element of $\pi(X)$.

We shall say that f is **Σ-adjustable** if it is equivalent to a Σ-adjusted map; by abuse of language, we shall also say that an equivalence class with a Σ-adjusted representative is Σ-adjustable. The set of all Σ-adjustable classes in $I^h(G; \lambda; \gamma, \nabla)$ will be denoted by

$$K^h_\Sigma(G; \lambda; \gamma, \nabla)$$

or simply by $K^h_\Sigma(\lambda)$ or K^h_Σ if no chance of ambiguity seems likely.

PROPOSITION 3.1. *$K^h_\Sigma(G; \lambda; \gamma, \nabla)$ is a subgroup of $I^h(G; \lambda; \gamma, \nabla)$.*

PROOF: Since inverses are given by reversing the orientations of everything in sight, it is obvious that the inverse of a Σ-adjustable class is Σ-adjustable. In order to show that K_Σ^h is closed under addition, it is necessary to recall some important features of the addition construction in [**Do2**, Section 5]:

(3.2) *Given representatives* ($f_i : M_i \to X_i$, other data) *of classes in* I^h *for* $i = 1$ *or* 2, *there is a map of cobordisms* ($F : W \to Z$, other data) *such that*

(i) $\partial_+ F = f_1 \amalg f_2$,

(ii) $\partial_- F$ *and the corresponding data determine a representative for an element of* I^h,

(iii) *the inclusion maps* $M_i \subset W$, $X_i \subset Z$, $\partial_- W \subset W$, $\partial_- Z \subset Z$ *all induce isomorphisms of G-posets,*

(iv) (F, other data) *is a* $(G; \mathbf{S}\lambda; \gamma, \nabla)$*-normal map in the sense of* [**DP1**, Section 3], *but not a normal cobordism (because* $\partial_+ F$ *is not an* I^h *normal map),*

(v) *the class of* ($-\partial_- F$, other data) *represents the sum* $(f_1, -) + (f_2, -)$. ■

There is a very useful converse to (3.2v) that characterizes the sum construction:

PROPERTY (3.2vi). *If* (F^*, other data) *satisfies conditions* (3.2i)–(3.2iv), *then* ($-\partial_- F^*$, other data) *represents* $(f_1, -) + (f_2, -)$.

To see this, attach $-F^*$ to F along the upper boundaries. This almost yields a $(G; \lambda; \gamma, \nabla)$-normal cobordism between $-\partial_- F^*$ and $\partial_- F$ (with the appropriate extra data); the most obvious problem is that the cobordisms are not simply connected (compare Figure I below).

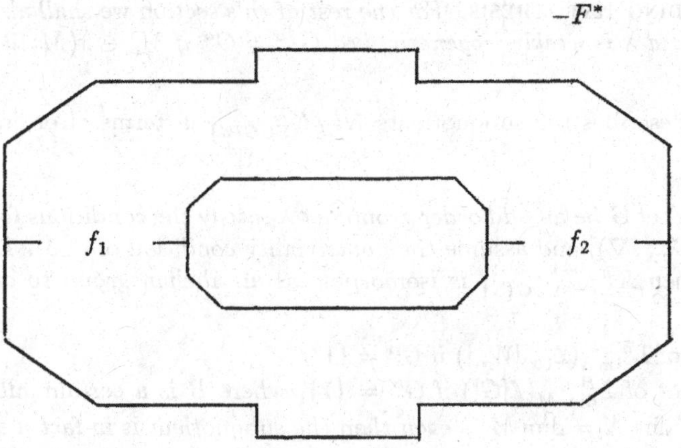

Figure I

However, it is possible to proceed as in the first two paragraphs of [**Do2**, page 267] and modify $F \cup -F^*$ to a $(G; \lambda; \gamma, \nabla)$-normal cobordism between $-\partial_- F^*$ and $\partial_- F$. Therefore $-\partial_- F^*$ and $-\partial_- F$ represent the same class in $I^h(G; \lambda; \gamma, \nabla)$. □

CONTINUATION OF PROOF OF (3.1): Suppose that $(f_1, -)$ and $(f_2, -)$ are Σ-adjusted, and let $(F, -)$ be as in (3.2). The inductive argument used to prove the $\pi - \pi$ theorem of [**DP1**, Section 7] may be applied over all the closed substrata in Σ (this follows by a direct examination of the argument in [**DP1**, Thm. 7.2, pp. 74–75]). If this is done one obtains a cobordism rel $M_0 \amalg M_1$ from $(F; -)$ to to $(F' : W' \to Z', -)$ such that F' is Σ-adjusted, $(\partial_- F, -)$ is $(G; \lambda; \gamma, \nabla)$-normally cobordant to $(\partial_- F', -)$ and $\partial_- F'$ is Σ-adjusted. Furthermore, we have $\lambda(F') \cong \lambda(F)$ and $\lambda(\partial_- F') \cong \lambda(\partial_- F)$. All these statements are direct analogs of [**DP1**, Cor. 7.3, p. 75].

Combining this with (3.2) we see that $(-\partial_- F', -)$ is a Σ-adjusted representative of $(-\partial F, -)$, where the latter represents $(f_1, -) + (f_2, -)$. Therefore $(f_1, -) + (f_2, -)$ is Σ-adjustable and accordingly K_Σ is closed under addition.■

Suppose now that $(f, -)$ is Σ-adjusted, and suppose that M_α is a minimal closed substratum in $\pi(M; \mathfrak{P}_G) - \Sigma$. As in previous chapters, the basic inductive problem is to find necessary and sufficient conditions for $(f, -)$ to be $\Sigma \cup G\{\alpha\}$-adjustable. In our situation, this requires a description of the subquotient of I^h given by

$$K^h_\Sigma / K^h_{\Sigma \cup G\{\alpha\}}.$$

At this point we must assume

$$(3.3) \qquad I^h(G; \mathbf{S}^{-1}\lambda; \gamma, \nabla) \neq \emptyset.$$

ADDITIONAL STANDING HYPOTHESIS. *For the rest of this section we shall also assume that the indexing data λ is weakly isogeneric; i.e., $G^{f\alpha} = G^\alpha$ if $M_\alpha \in \pi(M; \mathfrak{P}_G)$.*

Our next result describes the subquotients $K^h_\Sigma / K^h_{\Sigma \cup G\{\alpha\}}$ in terms of ordinary Wall groups:

PROPOSITION 3.4. *Let G be an odd order group, let λ satisfy the conditions for a group structure on $I^h(G; \lambda; \gamma, \nabla)$, and assume the nontriviality condition of (3.3). Let Σ and α be as above. Then $K^h_\Sigma / K^h_{\Sigma \cup G\{\alpha\}}$ is isomorphic as an abelian group to one of the following:*

(i) *A subgroup of $L^h_{\dim \alpha}(\mathbb{Z}_{(\alpha)}[W_\alpha])$ if $G^\alpha \neq \{1\}$.*
(ii) *A subquotient of $L^B_{\dim M}(\mathbb{Z}[G])$ if $G^\alpha = \{1\}$, where B is a certain subgroup of $\widetilde{K}_0(\mathbb{Z}[G])$. If $\dim X = \dim M$ is even then the subquotient is in fact a subgroup.*

Ultimately we shall need a more precise description of the subquotients and subgroups in (i)-(ii). In fact, in (ii) the subquotients turn out to be quotients (and if $\dim M$ is even the subgroup is all of L^B), and there is a canonical relationship between the subgroups and quotient groups in (i) and (ii). This will be discussed in Section 6.

REMARK: The orientation homomorphisms for the Wall groups in 3.4 are trivial (the groups under consideration have odd order), and therefore we have not mentioned these maps explicitly. We shall also use this notational convention at other points in this chapter.

PROOF OF 3.4: The first step is to define a *stepwise surgery obstruction*

$$\sigma_\alpha : K^h_\Sigma \to L^b_{\dim(\alpha)}(R_\alpha[W_\alpha])/\mathcal{Z}_\Sigma(\lambda)$$

where the L-group in the codomain is defined by (i), or (ii) as appropriate and $\mathcal{Z}_\Sigma(\lambda)$ is an appropriate subgroup of this L-group. This subgroup will be $\{0\}$ **except** perhaps in case (ii) when $\dim M = \dim X$ is odd. The basic idea for defining σ_α has already been discussed fairly thoroughly in [**DP2**, (5.1–5.3)], so we shall be somewhat sketchy here.

Case 1. Even-dimensional surgery. Since the Gap Hypothesis holds, we can do surgery up to the middle dimension on f_α away from the lower strata as in [**DP2**, Section 5] and extend the resulting cobordism to obtain a new map f' that is $(G;\lambda;\gamma,\nabla)$-normally cobordant to f with f'_α highly connected. As usual $\widetilde{H}_*(\mathrm{Cone}\,(f'_\alpha);\mathbf{Z}_{(\alpha)})$ is concentrated in the middle dimension and has a nonsingular Hermitian form of the standard type over the ring $\mathbf{Z}_{(\alpha)}[W_\alpha]$. Furthermore, the middle dimensional homology is a finitely generated projective $\mathbf{Z}_{(\alpha)}[W_\alpha]$-module.

If $G^\alpha \neq \{1\}$ then one can use further considerations as in [**OP**] and [**DP1**, Section 6] to conclude that the middle dimensional homology is in fact stably free, and therefore one can replace f' with a $(G;\lambda;\gamma,\nabla)$-normally cobordant map f'' such that f'' is Σ-adjusted, f''_α is connected up to the middle dimension, and $\widetilde{H}_*(\mathrm{Cone}\,(f''_\alpha))$ is a free $\mathbf{Z}_{(\alpha)}[W_\alpha]$-module. Thus one obtains an element $\sigma(f''_\alpha) \in L^h_{\dim\alpha}(\mathbf{Z}_{(\alpha)}[W_\alpha])$.

We claim this element does not depend upon the representative f''. It suffices to show that if f''' is also $(G;\lambda;\gamma,\nabla)$-normally cobordant to f with the same properties as f'', then one can modify the resulting cobordism F from f'' to f''', holding the ends fixed, to a Σ-adjusted normal cobordism. Then one can prove that $\sigma(f''_\alpha) = \sigma(f'''_\alpha)$ as in [**DP2**, Section 5]; problems with finding stably free modules make the proof more difficult than in [**W1**, Section 5]. In order to prove that a normal cobordism from f'' to f''' can be made Σ-adjusted, it suffices to show that a sequence of obstructions in

$$L^h_{\dim\beta+1}(\mathbf{Z}_{(\beta)}[W_\beta]) \qquad \beta \in \Sigma$$

must vanish; again, the proof is more complicated because of considerations involving projective modules, and the basic argument is outlined in [**DP1**, Section 6]. Granting this, we need the following vanishing result for Wall groups:

(3.5) *If H has odd order and R is a subring of the rationals, then*

$$L^h_{\mathrm{odd}}(R[H]) = 0.$$

A proof of this result is given in Section 9; special cases of this have been known for some time, and the general result will follow directly from previous work in the area.

If $G^\alpha = \{1\}$ so that $M_\alpha = M$, where $\dim M$ is even, then surgery below the middle dimension is again possible, and thus a Σ-adjusted map f is $(G;\lambda;\gamma,\nabla)$-normally cobordant to a map f' which is Σ-adjusted and connected up to the middle dimension on M_α. The surgery kernel $\widetilde{H}_*(\mathrm{Cone}\,(f'))$ is projective, but unlike the case $G^\alpha \neq 1$ it may not be possible to modify the problem to obtain a free module (compare [**Pe1**]). However, one can say that the class of $\widetilde{H}_*(\mathrm{Cone}\,(f'))$ in $\widetilde{K}_0(\mathbf{Z}[G])$ lies in a subgroup called

$$B_0(G,\mathcal{F}(\lambda)) \qquad \text{(see [DP1, Section 2])}.$$

Therefore the form on $\widetilde{H}_*(\mathrm{Cone}\,(f))$ defines a class in a surgery obstruction group

$$L^B_{\dim\alpha}(\mathbf{Z}[G]),$$

defined by forms on projective modules whose classes lie in $B = B_0(G,\mathcal{F}(\lambda))$. The proof that $\sigma(f')$ does not depend upon the highly connected modification f' of f is parallel to the argument in [**DP2**, Section 5].

NOTATIONAL CONVENTION: We shall often abbreviate $B_0(G,\mathcal{F}(\lambda))$ to $B_0(G,\lambda)$ or, when no confusion seems likely, simply by B.

Case 2. Odd-dimensional surgery. Let Ω be the set of all closed **nonmaximal** $\mathfrak{P}_G$ strata (*i.e.*, exclude the closed substratum X). We claim that every map f is Ω-adjustable. Let Σ and α be as before, and assume by induction that f is Σ-adjusted. One proceeds to modify f into a map f' with f'_α connected up to the middle dimension. Then there is a surgery obstruction

$$\sigma_\alpha(f') \in L^h_{\dim\alpha}(\mathbf{Z}_{(\alpha)}[W_\alpha]),$$

and the results of [**DP2**] say that f' (and therefore f) is $(\Sigma \cup G\{\alpha\})$-adjustable if $\sigma_\alpha(f') = 0$. But by (3.5) the L-group in question is zero, and thus f' is indeed $\Sigma \cup G\{\alpha\}$-adjustable. By induction it follows that f and f' are Ω-adjustable.

The above argument of course implies that if $\Sigma \subsetneqq \Omega$ then

$$K^h_\Sigma / K^h_{\Sigma\cup G\{\alpha\}} \cong L^h_{\dim\alpha}(\mathbf{Z}_{(\alpha)}[W_{\alpha'}])$$

since both groups are zero. Thus 3.4(ii) is trivially true if $\dim M = \dim X$ is odd.

Now assume that $\Sigma = \Omega$, so that $M_\alpha = M$ and $G^\alpha = \{1\}$. Suppose that f is Ω-adjusted, and let f' be an Ω-adjusted modification of f that is highly connected. The results in [**DP2**, Section 6] show that f' determines a class $\sigma(f') \in L^B_{\dim M}(\mathbf{Z}[G])$.

Unfortunately, unlike the even-dimensional case there is no simple observation to imply that $\sigma(f')$ only depends upon the class of f in $I^h(G;\lambda;\gamma,\nabla)$. The element $\sigma(f')$ may well depend upon the way in which one has chosen a highly-connected Ω-adjusted representative. In order to obtain a well-defined image we must factor out by a suitable indeterminancy. The following results supply the needed information:

Lemma 3.6. *Let $\mathcal{Z}(\lambda) \subset L^B_{\dim M}(\mathbb{Z}[G])$ be the set of all classes $\sigma(f')$ for f' equivalent to zero in $I^h(G;\lambda;\gamma,\nabla)$. Then $\mathcal{Z}(\lambda)$ is a subgroup.*

Lemma 3.7. *Suppose that f_1 and f_2 are highly connected Ω-adjusted representatives for the same class in I^h. Then $\sigma(f_1)-\sigma(f_2)$ lies in $\mathcal{Z}(\lambda)$.*

These results imply that a class in $I^h(G;\lambda;\gamma,\nabla)$ determines a well-defined element of $L^B_{\dim M}(\mathbb{Z}[G])/\mathcal{Z}(\lambda)$.

Proof of 3.6: We first show $\mathcal{Z}(\lambda)$ is closed under sums. Let f_1 and f_2 be highly connected Ω-adjusted maps that are cobordant to pseudo-equivalences. Then there exist cobordisms

$$h_i : V_i \to Y_i \qquad (i=1,2)$$

such that $\partial_+ h_i = f_i$ and $\partial_- h_i$ is a homotopy equivalence. Let

$$F : W \to U$$

be the sum cobordism for f_1 and f_2 constructed by (3.2). By construction and the $\pi-\pi$ theorem we may assume F is Ω-adjusted, but the h_i are not necessarily Ω-adjusted.

We may make $\partial_- F$ and F highly connected by surgery away from f_1, f_2, and the singular set. This will yield a new cobordism F' such that $-\partial_- F'$ is still a representative for $[f_1]+[f_2]$ (compare (3.2)). We claim the following:

(a) $\sigma(\partial_- F') = \sigma(f_1)+\sigma(f_2)$.
(b) $-\partial_- F'$ represents zero in $I^h(G;\lambda;\gamma,\nabla)$.

These combine to show that $\mathcal{Z}(\lambda)$ is closed under sums.

To prove (a), consider the formation (= odd-dimensional surgery obstruction) determined by $f_1 \cup f_2 \cup -\partial_- F'$. Since the union of the three maps in question bounds F' and all of the maps $f_1, f_2, -\partial_- F'$, F' are highly connected, one can proceed as in [**Ra**, Section 5] to show that the algebraic invariants of F' yield an algebraic trivialization of

$$\sigma(f_1 \cup f_2 \cup -\partial_- F') = \sigma(f_1)+\sigma(f_2)+\sigma(-\partial_- F').$$

This implies (a).

On the other hand, (b) is immediate from the basic results on sums in [**Do2**].

Finally, $\mathcal{Z}(\lambda)$ is closed under inverses because if f_1 represents zero in $I^h(G;\lambda;\gamma,\nabla)$, then $-f_1$ (= f_1 with reversed orientations) also represents zero and $\sigma(-f_1) = -\sigma(f_1)$.■

Proof of 3.7: Apply the sum construction to f_1 and $-f_2$, obtaining F_0. By construction F_0 may be assumed to be Ω-adjusted, and once again by equivariant surgery away from the singular set we may assume F_0 and $\partial_- F_0$ are highly connected. Exactly as in 3.6 we see that $\sigma(f_1)-\sigma(f_2) = \sigma(-\partial_- F_0)$. It suffices to prove that $-\partial_- F_0$ represents zero in $I^h(G;\lambda;\gamma,\nabla)$.

Since f_1 and f_2 represent the same class in I^h, there is a cobordism $h : V \to Y$ between them. Attach h to F_0 along $f_1 \amalg -f_2$. Then we have a new cobordism F' with

$\partial F' = \partial_- F_0$. Furthermore, by construction $\lambda(\partial F') \cong \lambda(F')$. The cobordism F' fails to be a normal map in only one respect; namely, the fundamental groups of the closed substrata of the domains and codomains of F' are all infinite cyclic, and in fact all the canonical maps induce isomorphisms of fundamental groups. Standard methods may be used to kill these homotopy classes for the closed substrata in $\pi(\text{Domain } F', \mathfrak{P}_G)$ as in [**Do2**] or Chapter IV, and this yields a normal map F with $\lambda(F') = \lambda(F)$ and $\partial F' = \partial F$. Therefore the $\pi - \pi$ theorem again implies that $\partial F = \partial_- F_0$ represents zero, and consequently $\sigma(-\partial_- F_0) \in \mathcal{Z}(\lambda)$.■

Proof of 3.4 *(Resumed)*

ADDITIVITY OF STEPWISE OBSTRUCTIONS: If $\dim M = \dim X$ is odd and we are considering pseudoequivalence surgery, then the arguments used in 3.6 and 3.7 may be modified directly to yield

$$\sigma(-\partial_- F_1) = \sigma(f_1) + \sigma(f_2)$$

if f_1 and f_2 are arbitrarily Ω-adjusted normal maps and F_1 is the Ω-adjusted sum construction obtained from (3.2) by killing homotopy and homology in the appropriate fashion. Since this is the only nontrivial case when $\dim M = \dim X$ is odd, it will suffice to dispose of the even-dimensional case.

Now suppose $\dim M = \dim X$ is even and $f_1, f_2, F_0, -\partial_- F_0$ are as before, where everything in sight is Σ-adjusted. As usual let $\alpha \notin \Sigma$ be minimal. By preliminary surgery we may assume $(f_{0\alpha}, -)$ and $(F_\alpha, -)$ are connected up to the middle dimension. If we are dealing with surgery up to equivariant homotopy equivalence, then the surgery kernels

$$K(i) = \widetilde{H}_*(\text{Cone}\,(f_i)_\alpha) \qquad i = 0, 1, 2$$

are all (stably) free, and as in [**W1**, Section 5] one can use the homology sequence of the pair

$$(\text{Cone}\,(F)_\alpha, \text{Cone}\,(\partial F)_\alpha)$$

to construct a free subkernel in

$$K(1) \oplus K(2) \oplus -K(0).$$

But $K(i)$ with its associated form represents $\sigma_\alpha(f_i, -)$ for $i = 0, 1, 2$; therefore we have that

$$\sigma_\alpha(f_0) = \sigma_\alpha(f_1) + \sigma_\alpha(f_2).$$

But by construction $[f_0] = [f_1] + [f_2]$, and therefore σ_α is additive.□

REMARK: It is also possible to prove additivity using the very general study of additivity for stepwise obstructions in [**Do1**, Section 9]. The (unpublished) results of that work say that additivity holds if $W_\alpha \cong W_{\bar{f}(\alpha)}$. This is true in our case because f induces a $1-1$ correspondence $\pi_0(M^P) \cong \pi_0(X^P)$ for each $P \in \mathfrak{P}_G$ (hence $N(M_\alpha) = N(X_{\bar{f}(\alpha)})$ if $a \in \pi(M, \mathfrak{P}_G)$), while $G^{\bar{f}(\alpha)} = G^\alpha$ holds by hypothesis. Since $W_\gamma = N(Y_\gamma)/G^\gamma$, the relation $W_\alpha \cong W_{\bar{f}(\alpha)}$ follows.

THE KERNEL OF σ_α: Suppose first that $\dim M = \dim X$ is **even**. The main result on stepwise surgery obstructions is that $\sigma_\alpha(f) = 0$ if and only if f is $\Sigma \cup G\{\alpha\}$-adjustable (see [**DP2**, Lemma 5.2]). This implies that σ_α induces a monomorphism

$$K^h_\Sigma / K^h_{\Sigma\cup G\{\alpha\}} \to L^b_{\dim \alpha}(R_\alpha[W_\alpha]).$$

Suppose now that $\dim M = \dim X$ is **odd**, so that $\Sigma = \Omega$. Let f be a highly connected, Ω-adjusted normal map such that $\sigma(f)$ lies in $\mathcal{Z}(\lambda)$. We claim that f represents the zero element of $I^h(G;\lambda;\gamma,\nabla)$; *i.e.*, f is cobordant to a pseudoequivalence.

Since $\sigma(f)$ lies in $\mathcal{Z}(\lambda)$, there is a second highly connected, Ω-adjusted normal map f^* such that f^* is cobordant to a pseudoequivalence and $\sigma(f^*) = \sigma(f)$. Let F be the sum construction for f and $-f^*$ as in (3.2), and let h be the cobordism between f^* and a pseudoequivalence e. As before we have that F and $\partial_- F$ are Ω-adjusted, and we can also make them highly connected by surgery. It follows that

$$\sigma(-\partial_- F) = \sigma(f) - \sigma(f^*) = 0.$$

Therefore by the results of [**DP2**] we know that $-\partial_- F$ can be surgered to a pseudoequivalence. Let h' denote the cobordism, and let e' be the pseudoequivalence. Consider the cobordism

$$E = F \cup h \cup h'.$$

By construction $\partial E = f \cup e \cup e'$ where e and e' are pseudoequivalences. Furthermore, $\lambda(f) \cong \lambda(E)$ by construction, and therefore one can perform surgery on E, leaving e and e' untouched, to obtain a pseudoequivalence. If D represents the cobordism given by surgery on E, then a piece of the boundary of D will give an appropriate cobordism from f to a pseudoequivalence.□

THE IMAGE OF σ_α: The image of σ_α is a subgroup by the additivity of σ_α.■

We shall need the following result on the image of σ_0 later in this chapter.

PROPOSITION 3.8. *Let λ be a set of data appropriate for surgery up to pseudoequivalence, and let Ω denote the proper closed substrata in $\pi(X;\mathfrak{P}_G)$. Assume that $\dim M = \dim X$ is even and $I^h(G;\mathbf{S}^{-1}\lambda;\gamma,\nabla)$ is nonempty. Then the image of the surgery obstruction*

$$\sigma_0 : I^h(G;\lambda;\gamma,\nabla) \to L^B_{\dim X}(\mathbf{Z}[G])$$

contains the image of the forgetful map from L^h to L^B.

NOTE: If $\dim M = \dim X$ is odd there is nothing to prove because $L^h = 0$.

PROOF: (*Sketch*) Let $u \in L^h_{\dim X}(\mathbf{Z}[G])$. Since $I^h(G;\mathbf{S}^{-1}\lambda;\gamma,\nabla)$ is nonempty the results of [**DP1**] show there is a zero class represented by an equivalence $f_0 : M_0 \to X_0$. One can add handles to $M_0 \times I$ away from the nonfree orbits to obtain an equivariant, Ω-adjusted normal map $F : W \to X \times I$ such that the surgery kernel $K_*(F)$ is $\mathbf{Z}[G]$-free and concentrated in the middle dimension, and the intersection form on $K_*(F)$ represents u. It follows immediately that the stepwise obstruction of F is just the image of u in $L^B_{\dim X}(\mathbf{Z}[G])$.■

REMARKS: 1. It is possible to formulate an analog of 3.4 without assuming that λ is weakly isogeneric. In general, if $\alpha \in \pi(M; \mathfrak{P}_G)$ then there is a canonical surjection from W_α to $W_{\dot{f}\alpha}$, and the codomain of the stepwise obstruction will be a group

$$\Gamma_{\dim \alpha}\left(\mathbb{Z}_{(\alpha)}[W_\alpha] \rightarrow \mathbb{Z}_{(\alpha)}[W_{\dot{f}\alpha}]\right)$$

of the type considered by Cappell and Shaneson [**CS**].

2. If $D(G) \subset \widetilde{K}_0(\mathbb{Z}[G])$ is the kernel of the map $\widetilde{K}_0(\mathbb{Z}[G]) \rightarrow \widetilde{K}_0(\mathcal{M})$ induced by passage to a maximal order $\mathcal{M}$, then results of R. Oliver [**OL**] show that the subgroups $B \subset \widetilde{K}_0(\mathbb{Z}[G])$ that arise in 3.4 are actually subgroups of $D(G)$. Furthermore, if X has strongly saturated orbit structure (in the sense of Section III.2), then the results of [**OL**] imply that $B = D(G)$.

Extensions to other finite groups

Much of the preceding discussion is also valid in other situations. For example, the only role of the nilpotence condition on G is to ensure that $n(\lambda) \leq 1$, and the entire discussion of this section applies whenever the latter condition holds and G has odd order. If G has even order and $n(\lambda) \leq 1$ it is still possible to define stepwise obstructions like those of Proposition 3.4. However, one needs a special property like (3.5) to conclude that stepwise obstructions are well defined, and in general one must factor out an indeterminacy coming from stepwise obstructions on lower strata. A special case with a nontrivial indeterminacy is treated in [**Do3**]; analogous phenomena for actions of compact connected classical groups were also observed in work of M. Davis, W.-C. Hsiang, and J. Morgan [**DH,DHM**].

4. Restriction morphisms

We shall need algebraic maps on $I^h(G; \lambda; \gamma, \nabla)$ that roughly correspond to taking fixed point sets of various subgroups. More precisely, if $\alpha \in \pi(X; \mathfrak{P}_G)$, we would like to discuss the restriction of a normal map to the closed substratum with index α. Throughout this section α will represent a fixed *nonmaximal* (important!) closed substratum as above. As in Section 3 we shall assume $G^\alpha = G^{\dot{f}\alpha}$. We shall **also** assume that Σ is a closed G-invariant subset of $\pi(X; \mathfrak{P}_G)$ such that α is a minimal element of $\pi(-, \mathfrak{P}_G)$ not in Σ. Furthermore, the entire setting degenerates to statements about zero groups or homomorphisms if $\dim M = \dim X$ is odd, and therefore *we shall assume* $\dim M = \dim X$ *is even.*

Formal constructions associated to restrictions

The obstruction group that is appropriate to restriction will be denoted by

$$I^{h|\alpha}(W_\alpha; \lambda|\alpha; \gamma|\alpha, \nabla|\alpha).$$

Of course, this deserves some explanation. The subquotient W_α of G is defined in Section I.2; the important point is that W_α acts effectively on M_α and X_α. The data in $\lambda|\alpha$, $\gamma|\alpha$, and $\nabla|\alpha$ is given by the induced W_α-equivariant data for the restriction of ($f : M \to X$, other data) to the α-substratum, and one is interested in equivariant surgery up to $\mathbf{Z}_{(\alpha)}$-*localized pseudoequivalence.* In other words, the goal is an equivariant map $f : N \to X_\alpha$ which is a (nonequivariant) homotopy equivalence after localization at $\ell(\alpha)$. We recall from Section 3 that $\ell(\alpha)$ is the product of all primes p such that X_α is a component of X^P for some p-subgroup $P \subset G$. If $\ell(\alpha)$ is simply a prime (e.g., this happens if X has strongly saturated orbit structure), then the definition of $I^{h|\alpha}$ is given in Appendix A of [**DP1**].

The entire setting generalizes directly to arbitrary $\ell(\alpha)$ with suitable choices of definitions. First of all, the distinguished family of subgroups corresponding to $\mathcal{H}^{h|\alpha}$ (compare [**DP1**, (A1.2), p. 93]) is given by the set of all p-subgroups of W_α where p divides $\ell(\alpha)$. If $\Pi(X)$ is the complete G-poset introduced in Section 1 and $\mathcal{F} \subset \Pi(X)$ is a closed G-invariant subset, then $\mathcal{F}|\alpha$ will denote the corresponding subset of the complete W_α-poset $\Pi(X_\alpha)$ obtained by restricting to X_α with its associated W_α-action. The following properties can be verified directly by elementary arguments:

(4.1) *Let X, α, $\mathcal{F}$, and $\mathcal{F}|\alpha$ be as above, and suppose that Y is an $\mathcal{F}$-complex. Then the associated subcomplex Y_α is an $\mathcal{F}|\alpha$-complex.*■

(4.2) *If $f : M \to X$ is an equivariant map satisfying* (1.0.A)–(1.0.E), *the map $f_\alpha : M_\alpha \to X_\alpha$ is defined by identity* (‡) *in Section* I.1, *and*

$$\mathcal{F}_f := \pi(X) \cup \widetilde{f}(\pi(M)))$$

is the set considered in (1.0.E) *then $\mathcal{F}_f|\alpha$ is equal to $\pi(X_\alpha) \cup \widetilde{f}_\alpha(\pi(M_\alpha))$.*■

We shall also need a set of nicely representable elements $\Delta^{h|\alpha}(W_\alpha, \mathcal{F}|\alpha)$ in the generalized Burnside ring $\mathbf{\Omega}(W_\alpha, \mathcal{F})$ analogous to [**DP1**, Definition A1.3, p. 94]. The definition of the set $\Delta^{h|\alpha}(\cdots)$ requires some preparation. As in [**DP1**, Appendix A], for each prime p dividing $|G|$ let $\mathcal{G}_1^p(G)$ be the set of subgroups K of G that fit into short exact sequences of the form

$$1 \to P \to K \to C \to 1$$

where P is a p-group and C is a cyclic group of order prime to p. We then define

$$\begin{aligned}\Delta^{h|\alpha}(W_\alpha, \mathcal{F}_f|\alpha) = \{Y \mid\ & Y \text{ is an } \mathcal{F}-\text{complex and} \\ & \chi(Y_\beta) = 1 \text{ for all } \beta \in \Pi(X_\alpha) \text{ with} \\ & G^\beta \in \mathcal{G}_1^p(W_\alpha) \text{ for some } p \text{ dividing } \ell(\beta)\ \}.\end{aligned}$$

The next result shows that Condition (1.0.E) is preserved under restriction to the α-substratum; the verification of this is a routine exercise:

PROPOSITION 4.3. *If f, $\mathcal{F}_f, \ldots$ are as in (4.1) – (4.2) and Z represents a class in the subset $\Delta\left(G, \pi(X) \cup \widetilde{f}(\pi(M))\right) + \Omega(G; \pi(X))$, then the subcomplex Z_α represents a class in the subset $\Delta\left(W_\alpha, \pi(X_\alpha) \cup \widetilde{f}_\alpha(\pi(M_\alpha))\right) + \Omega(W_\alpha; \pi(X_\alpha))$.*■

By construction and the preceding discussion, it follows directly that there is a restriction map

$$\text{res}_\alpha : I^h(G; \lambda; \gamma, \nabla) \to I^{h|\alpha}(W_\alpha; \lambda|\alpha; \gamma|\alpha, \nabla|\alpha) \tag{4.4}$$

that sends Σ-adjustable classes to $\Sigma|\alpha$-adjustable classes and $(\Sigma \cup G\{\alpha\})$-adjustable classes to a zero element in the sense of [**DP1**]. The following formal properties are a straightforward consequence of the corresponding techniques used to study I^h.

(4.5) *The appropriate analog of the $\pi - \pi$ theorem* [**DP1**, Section 7] *holds for $I^{h|\alpha}$.*■ (Results of this type are discussed in [**DP1**, Section A2].)

(4.6) *There is a natural abelian group structure on $I^{h|\alpha}$ constructible by the methods of* [**Do1-2**].■

(4.7) *The set of $\Sigma|\alpha$-adjustable classes $K_{\Sigma|\alpha}$ in $I^{h|\alpha}$ is a subgroup with respect to the sum defined by* (4.6).■

(4.8) *The map res_α is a group homomorphism.*

The proofs of (4.6)–(4.8) follow directly from the same methods used in [**Do1-2**].■

Restrictions and stepwise obstructions

There is also a good theory of stepwise surgery obstructions for $I^{h|\alpha}$. For our purposes it will suffice to consider $\Sigma|\alpha$-adjusted classes; by the hypotheses introduced at the beginning of this section, α is a minimal element in $\pi - \Sigma$. In this case the only closed substratum of X_α on which the map is not adjusted is the maximal closed substratum X_α itself. The following is then immediate from the discussion above, the methods of Section 3, and the foundational material in [**DP1-2**]:

PROPOSITION 4.9. *There is a commutative diagram of group monomorphisms as follows:*

$$\begin{array}{ccc} K^h_\Sigma / K^h_{\Sigma \cup G\{\alpha\}} & \xrightarrow{\text{res}'_\alpha} & K^{h|\alpha}_{\Sigma|\alpha} \\ \sigma_\alpha \big\downarrow & & \sigma'_\alpha \big\downarrow \\ L^h_{\dim \alpha}(\mathbb{Z}_{(\alpha)}[W_\alpha]) & = & L^h_{\dim \alpha}(\mathbb{Z}_{(\alpha)}[W_\alpha]). \end{array}$$

Furthermore, the map σ'_α is surjective if $I^h(G; \mathbf{S}^{-1}\lambda) \neq \varnothing$.

PROOF: (*Sketch*) The existence of σ'_α was mentioned before, and one can prove this is a group monomorphism as in Section 3. Commutativity is immediate since the definition of σ_α only involves data for the closed substrata associated to α. The restriction map res'_α is a group homomorphism because res_α is a group homomorphism, and furthermore res'_α is a monomorphism because σ_α is a monomorphism.

It remains to show that σ'_α is a surjection. At this point we need $I^h(G; \mathbf{S}^{-1}\lambda) \neq \emptyset$. Let $h : N \to Y$ be an equivalence representing a null class in $I^h(G; \mathbf{S}^{-1}\lambda)$. Then h_α will be an equivariant $\mathbb{Z}_{(\alpha)}$-local pseudoequivalence. Let Sing N_α denote the subsets of points on singular orbits of N_α. Then $N_\alpha -$ Sing N_α is equivariantly the interior of a compact free smooth W_α-manifold with boundary. Let $C \times (0, \infty)$ be an open collar on the end of $N_\alpha -$ Sing N_α, and let N^*_α be $(N_\alpha - \mathrm{Sing}\ N_\alpha) - C \times (1, \infty)$, so that $\partial N^*_\alpha = C$.

Let u be an arbitrary element of $L^h_{\dim \alpha}(\mathbb{Z}_{(\alpha)}[W_\alpha])$. As in [**DP1**, Sections 5–6] one can add handles to $(N^*_\alpha \times I)/W_\alpha$ over the complement of C/W_α to obtain a normal map

$$F_\alpha : U_\alpha \to N^*_\alpha \times [0,1]/W_\alpha$$

that is an equivalence on the boundary and has surgery obstruction u. We may in fact do this so that F_α is a diffeomorphism on

$$F_\alpha^{-1}(N^*_\alpha \times [0,\varepsilon]/W_\alpha \cup C \times (0,1) \times [0,1]/W_\alpha).$$

Let $\widetilde{F}_\alpha$ be the induced map of regular W_α-covering spaces. Then one can use $\widetilde{F}_\alpha$ and f_0 to define an equivariant $h|\alpha$-normal map

$$F^* : \widetilde{U}_\alpha \cup (N_\alpha - \mathrm{Int}\ N^*_\alpha) \times [0,1] \to X_\alpha \times [0,1]$$

such that $\partial_0 F^*$ is the original equivalence f_0, the map $\partial_1 F^*$ is also an equivalence, and F^* is $\Sigma|\alpha$-adjusted. From the construction it follows that (F^*, other data) represents a class in $K^{h|\alpha}_{\Sigma|\alpha}(\lambda)$, and it also follows that $\sigma'_\alpha(F^*$, other data$) = u$. Therefore σ'_α is onto as claimed.■

5. Projective class group obstructions

STANDING HYPOTHESIS. *In this section G will always denote an odd order nilpotent group.*

Let X be a compact smooth G-manifold in the class $\mathcal{F}^h$ defined in Section 1. As in Section 3, Ω will denote all the proper closed substrata X_α that are expressible as components of X^P for some nontrivial p-subgroup $P \subset G$ (where p is some prime); in symbols, we require $\alpha \in \pi(X, \mathfrak{P}_G)$-(maximal closed substratum).

As noted in Theorem 3.4, the stepwise surgery obstruction for $K^h_\Omega \subset I^h$ is an exceptional case; specifically, this obstruction lies in a Wall group $L^B_{\dim \alpha}(\mathbb{Z}[G])$ defined using projective G-modules representing the elements of some subgroup $B \subset \widetilde{K}_0(\mathbb{Z}[G])$. This leads to complications that cannot be handled by a straightforward extension of the techniques in Wall's book [**W1**]; one major goal of the G-surgery papers [**DP1-2**] is the development of machinery to deal with surgery kernels that are projective but might not be stably free. In this section we shall prove some additional results in this direction. One motivation is the need to prove the following result:

THEOREM 5.1. *If $I^h(G; \mathbf{S}^{-1}\lambda)$ is nontrivial (and the other assumptions of 3.4 also hold), then the stepwise surgery obstruction map*

$$\sigma_0 : K^h_\Omega(\lambda) \to L^B_{\dim X}(\mathbf{Z}[G])/\mathcal{Z}(\lambda)$$

is surjective (and thus by 3.4 a group isomorphism).

The proof will be given at the end of this section.

The key to understanding $L^B_*(\mathbf{Z}[G])$ geometrically is the Ranicki exact sequence **[Ra]**:

$$\to L^h_m(\mathbf{Z}[G]) \to L^B_m(\mathbf{Z}[G]) \to H_{m-1}(\mathbf{Z}_2; B) \to L^h_{m-1}(\mathbf{Z}[G]) \to$$

If the group G has odd order, then the vanishing of $L^h_{\text{odd}}(\mathbf{Z}[G])$ yields two important conclusions.

(5.2) *There are exact sequences*

$$L^h_{2q}(\mathbf{Z}[G]) \to L^B_{2q}(\mathbf{Z}[G]) \xrightarrow{\text{CLS}} H_{2q-1}(\mathbf{Z}_2; B) \to 0$$

$$0 \to L^B_{2q-1}(\mathbf{Z}[G]) \xrightarrow{\text{CLS}} H_{2q-2}(\mathbf{Z}_2; B),$$

where CLS *is the map that takes* (i) *a Hermitian form to the class of the underlying projective module in the even-dimensional case,* (ii) *a formation to the difference of the projective classes of the two lagrangians (or subkernels) in the odd-dimensional case.*■

Since the codomain of the class map does not look very much like anything reasonably constructible from $\widetilde{K}_0(\mathbf{Z}[G])$, we shall recall the precise relationship. There is a standard conjugation $(\cdots)^*$ on $\widetilde{K}_0(\mathbf{Z}[G])$ given by taking P to $P^* = \text{Hom}_{\mathbf{Z}}(P, \mathbf{Z})$. The subgroup $B = B_0(G, \mathcal{F}(\lambda))$ is invariant under conjugation, and

$$H_k(\mathbf{Z}_2; B) \cong \frac{\{x \in B \mid x^* = (-1)^{k+1}x\}}{\{y - (-1)^k y^* \mid y \in B\}}$$

Since the classes described in (5.2) are $(-1)^{k+1}$-symmetric and lie in B, it is clear that any representative form or formation for an element of L^B_k will determine an element in $H_{k-1}(\mathbf{Z}_2; B)$; in fact, the image turns out to be independent of the choice of representative.

The following result provides a useful means for computing the composite CLS $\circ \sigma_0$ in the even-dimensional case.

PROPOSITION 5.3. *Let* ($f : M \to X$, other data) *be an Ω-adjusted normal map representing a class in $I^h(G; \lambda)$. Assume that f is at least 2-connected,* $\dim M = \dim X \geq 6$ *is even, and the homology groups $\widetilde{H}_i(\text{Cone}(f), \mathbf{Z})$ are all finitely generated projective $\mathbf{Z}[G]$-modules. Then* CLS $(\sigma_0(f, \dots))$ *is given up to sign by the projective class*

$$\chi(K_*(f)) = \Sigma(-1)^i[\widetilde{H}_i(\text{Cone}(f); \mathbf{Z})],$$

where $[\cdots]$ *denotes the stable equivalence class of a finitely generated projective module in* $\widetilde{K}_0(\mathbf{Z}[G])$.

PROOF: Following standard practice we write $K_i(f) = \widetilde{H}_{i+1}(\mathrm{Cone}(f); \mathbf{Z})$, so that the expression in question is simply $\pm\chi(K_*(f))$. Suppose that f is k-connected, where $k \leq q = (\dim X)/2$. If $k = q$ then K_q is the only nonzero group, and in this case the result is an immediate consequence of (5.2). On the other hand, if $k < q$ then we may kill k-dimensional homotopy classes to obtain a $(k+1)$-connected representative f'. The crucial point is to obtain a relationship between $\chi(K_*(f))$ and $\chi(K_*(f'))$.

If $k+1 < i \leq q$ then killing k-dimensional homotopy will leave K_i intact. On the other hand, if $k+1 < q$, then killing $K_k(f)$ will add a summand Q to $K_{k+1}(f')$; specifically if $\varphi : F \to K_k(f)$ is a surjection from a free module, then Q is a projective module stably equivalent to the kernel of φ. It follows that the projective classes satisfy

$$[K_{i+1}(f')] = [K_{i+1}(f)] - [K_i(f)].$$

Since $[K_{2q-i}] = [K_i]^*$ by Poincaré duality, it follows that $\chi(K_*(f')) = \chi(K_*(f))$ if $k+1 < q$. Finally, if $k+1 = q$, then the same sort of argument shows that

$$[K_q(f')] = [K_q(f)] - ([K_{q-1}(f)] + [K_{q+1}(f)]).$$

Thus $\chi(K_*(f')) = \chi(K_*(f))$ in this case also.■

The idea in the proof of 5.3 goes back to Wall's original work on finiteness obstructions [W0]. The odd-dimensional analog of 5.3 is less straightforward and is based upon the following:

DEFINITION: Given an Ω-adjusted, 2-connected I^h-normal map ($f : M \to X$, other data), a **geometric formation** on f is a decomposition of f into a map of triads

$$f = f_0 \cup_\partial f_1 : M_0 \cup_\partial M_1 \to X_0 \cup_\partial X_1,$$

where the M_i and X_i are smooth G-manifolds with boundary, the maps f_0, f_1, and $\partial f_0 = \partial f_1$ are 2-connected, G acts freely on M_1 and X_1, the surgery kernels $K_*(f_1)$ and $K_*(\partial f_1 = \partial f_0)$ are free in each dimension, and the surgery kernel $K_*(f_0)$ is projective in each dimension.

The motivation for this definition is the following result.

PROPOSITION 5.4. *Let* (f, other data) *be a* $(2q+1)$*-dimensional* Ω*-adjusted* h*-normal map, and assume* f *is* q*-connected. Then there is a geometric formation on* f *such that* f_0, f_1, *and* $\partial f_0 = \partial f_1$ *are all* q*-connected and* $\sigma_0(f)$ *is entirely determined by the class of* $\chi(K_*(f_0))$ *in* $\widetilde{K}_0(\mathbf{Z}[G])$.

PROOF: Take a finite set of equivariant generators of $K_q(M)$ and represent these classes by pairwise disjointly embedded spheres. The Gap Hypothesis ensures that these embeddings may be chosen to miss the singular orbits. Exactly as in Section 6 of Wall's book, this defines a splitting of f, and the connectivity assertions follow from the construction.

The surgery obstruction is defined by the formation with module $K_q(\partial f_0 = \partial f_1)$ and subkernels

(i) $K_{q+1}(f_0, \partial f_0) \cong K_q(f_1)^*$,
(ii) $K_{q+1}(f_1, \partial f_1) \cong K_1(f_0)^*$.

By (5.2) the surgery obstruction is entirely determined by the difference between (i) and (ii) in the projective class group; since (i) is free by construction, the obstruction is carried entirely by $K_q(f_0)$. To prove the assertion in the proposition, it suffices to show that $K_i(f_0) = 0$ for $i \neq q$. But $K_i(f_0) = 0$ for $i < q$ by the connectivity condition, and if $i > q$ then one can use a duality argument as in [**DP1**, Section 6] to show that $K_i(f_0) = 0$.■

We shall need an analog of the formula in 5.4 when the geometric formation is k-connected for some k with $2 \leq k \leq q$.

PROPOSITION 5.5. *Let* (f, other data) *be a* $(2q+1)$*-dimensional* Ω*-adjusted* h*-normal map, and let* $(f_0, f_1, \partial f_0 = \partial f_1)$ *be a geometric formation on* f*. Then the obstruction* $\sigma_0(f)$ *is entirely determined by the class of* $\chi(K_*(f_0))$ *in* $\widetilde{K}_0(\mathbb{Z}[G])$.

PROOF: We shall first outline the idea behind the argument. Suppose that the maps f_1, f_1, and $\partial f_0 = \partial f_1$ are all known to be k-connected, where $2 \leq k \leq q-1$; by definition this is true at least in the case $k = 2$. The crucial first step is to show that one can kill $K_k(f_0)$, $K_k(f_1)$, and $K_{k+1}(\partial f_0 = \partial f_1)$ in a prescribed fashion to obtain a new geometric formation $(f_0', f_1', \partial f_0' = \partial f_j')$ such that the relevant maps are all $(k+1)$-connected and $\chi(K_*(f_0')) = \chi(K_*(f_0))$. This will yield a geometric formation which is connected up to the middle dimensions and has the same projective invariant. The next step is to construct the geometric formation of 5.4 in terms of the given, highly connected geometric formation and to verify that the projective invariants for both geometric formations are the same. Since the projective invariant of the geometric formation in 5.4 carries the entire surgery obstruction by 5.2, it follows that the original invariant $\chi(K_*(f_0))$ determines the surgery obstruction of f.

Given an integer k satisfying $2 \leq k \leq q-1$ and a geometric formation $(f_0, f_1, \partial f_0 = \partial f_1)$ with all maps k connected, the first step is to kill $K_k(\partial M_0 = \partial M_1)$. This kernel is a free $\mathbb{Z}[G]$-module, and surgery on a set of free generators yields a cobordism Q from $\partial M_0 = \partial M_1$ to some N. By construction Q is equal to $\partial M_0 = \partial M_1$ with $(k+1)$-cells attached to kill the free generators of K_k; an elementary duality and exactness argument shows that $K_i(N) = 0$ for $i \leq k$ or $i \geq 2q-k$, while $K_i(N) \cong K_i(\partial M_0 = \partial M_1)$ for $i < k < 2q-k$. By the standard normal cobordism extension argument, it is possible to extend Q to a normal cobordism Q^* from ($f: M \to X$, other data) to some $f': M' \to X$, other data) such that one has a map of triples $(\partial_0 f', \partial_1 f', f'|N)$. The corresponding splitting $M' = M_0' \cup M_1'$ has $M_j' = M_i \cup Q$. We claim that, at least after some stabilization by adding trivial handles, *the new map of triples is a geometric formation on* f'.

Since M_j' is formed from M_j by adding $(k+1)$-handles, it follows that $K_i(M_j') = K_i(M_j)$ unless $i = k, k+1$. In these cases we have an exact sequence

$$0 \to K_{k+1}(M_j) \to K_{k+1}(M_j') \to F \to K_k(M_j) \to K_k(M_j') \to 0,$$

where F is a free $\mathbf{Z}[G]$-module generated by a suitable collection of $(k+1)$-cells. We claim that the $K_*(M_j')$ are projective $\mathbf{Z}[G]$-modules. It is immediate from the construction and the exact sequence of $(M_j, \partial M_j)$ that $K_k(M_j, \partial M_j) \cong K_k(M_j')$. However, by duality we have that $K_k(M_j, \partial M_j) \cong K^{2q+1-k}(M_j)$, and the latter is isomorphic to $K_{2q+1k-k}(M_j)^*$ by the Universal Coefficient Theorem (recall that all $K_i(M_j)$ are projective and hence torsion free). For $j = 1$ we in fact have that $K_k(M_1')$ is free. It follows from the exact sequence that $K_{k+1}(M_j')$ is also projective, and in fact its stable equivalence class is

$$[K_{k+1}(M_j)] - [K_k(M_j)] + [K_k(M_j')] \in \widetilde{K}_0(\mathbf{Z}[G]).$$

The latter implies that $\chi(K_*(f_0')) = \chi(K_*(f_0))$ and that $K_{k+1}(M_1')$ is stably free. As in Wall [**W1**, Sections 5 and 6] we may add trivial $(k+1)$ handles to Int M_2' to make $K_{k+1}(M_1')$ free. Thus we not only have a new geometric formation, but also we have that its projective invariant equals that of the original formation.

The next step is to kill $K_k(M_0')$ and $K_k(M_1')$ by surgery on suitably embedded k-spheres in the interiors of M_0' and M_1'. This yields a new normal map of triads $(f_0'', f_1'', f'|N)$ with all maps $(k+1)$-connected and since $k \leq q-1$ the following sequence is exact:

(5.6) $$0 \to K_{k+1}(M_j') \to K_{k+1}(M_j'') \to F' \to K_k(M_j') \to 0$$

(F' is again a free $\mathbf{Z}[G]$-module.)

It follows as before that $K_{k+1}(M_1'')$ is stably free and $K_{k+1}(M_0'')$ is projective and represents

$$[K_{k+1}(M_j')] - [K_k(M_0')] \in \widetilde{K}_0(\mathbf{Z}[G]).$$

Further analysis shows that $K_i(M_j'') = K_i(M_j')$ unless $i = k,\ k+1,\ 2q-k-1$ or $2q-k$. By construction $K_k = 0$, and the discussion above yields K_{k+1}. Furthermore, by duality, the Universal Coefficient Theorem, and exactness we have

$$0 = K_k(M_j'') \cong K_k(M_j'', \partial M_j'') \cong K^k(M_j'', \partial M_j'') \cong K_{2q-k+1}(M_j''),$$

and a similar argument shows that

$$K_{2q-k}(M_j'') \cong K_{k+1}(M_j'', \partial M_j'').$$

But $\partial M_j'' = \partial M_j' = N$, and because of this one has an exact sequence analogous to (5.6) with $(M_j'', \partial M_j'')$ and $(M_j', \partial M_j')$ replacing M_j'' and M_j'. By the same argument we have already used several times, it follows that $K_{2q-k}(M_1'')$ is stably free (and thus can be made free after a suitable stabilization), while $K_{2q-k}(M_0'')$ is projective and has class $[K_{2q-k}(M_0')] - [K_{2q-k+1}(M_0')]$. Therefore we have shown that $(f_0'', f_1'', f'|N)$ is a geometric formation with all maps $(k+1)$-connected and $\chi(K_*(f_0'')) = \chi(K_*(f_0))$. This completes the proof of the first step in the argument.

For the final step we may assume that we have a geometric formation $(f_0, f_1, \partial f_0 = \partial f_1)$ with all maps q-connected; recall that $\dim M = \dim X = 2q+1$. Construct disjointly embedded q-spheres in Int M_1'' which represent a set of free generators for $K_q(M_1'')$, we may assume these embedded subspheres and their G-translates are pairwise disjoint. Let $U_1 \subset$ Int M_1'' be a well-chosen closed tubular neighborhood of these embedded spheres and their translates. Similarly construct embedded q-spheres representing $\mathbf{Z}[G]$-generators for $K_q(M_0'')$; we may assume these embedded spheres and their G-translates are pairwise disjoint, and by the Gap Hypothesis we may also assume that the embedded spheres do not meet the singular set. Let $U_0 \subset$ Int M_0 be a well-chosen closed tubular neighborhood. By construction (see 5.4), *the geometric formation associated to the splitting* $M = \{M - \mathrm{Int}\,(U_0 \cup U_1)\} \cup \{U_0 \cup U_1\}$ *determines the surgery obstruction of* f.

Let $M_j' = M_j - \mathrm{Int}\, U_j$, let $M_0^* = M_0' \cup M_1$, let $N = \partial M_0 = \partial M_1$, and let $U = U_0 \cup U_1$. It will suffice to show that $\chi(K_*(M_0^*)) = \chi(K_*(M_0))$.

First of all we compute $K_*(M_0 \cup M_1')$. By exactness and excision we have

$$K_*(M_0 \cup M_1', M_0) \cong K_*(M_1', N),$$

and by duality and excision the latter is isomorphic to

$$K^*(M_1', \partial U_1) \cong K^*(M_1', U_1).$$

But by construction we have that $K_i(M_1', U_1) = 0$ if $i \neq q+1$ and $K_{q+1}(M_1', U_1)$ is the free module $K_{q+1}(M_1')$. Therefore by the Universal Coefficient Theorem and a diagram chase we have that $K_i(M_0 \cup M_1') \cong K_i(M_0)$ if $i \neq q$ and $K_q(M_0 \cup M_1') \cong K_q(M_0) \oplus K_{q+1}(U_1)$. In other words, from the viewpoint of projective modules and stable equivalences the submanifolds M_0 and $M_0 \cup M_1'$ have the same invariants.

We must now relate the classes

$$\chi(K_*(M_0)) = \chi(K_*(M_0 \cup M_1')) \text{ and } \chi(K_*(M_0^*)).$$

By construction (compare [**W1**, Section 6] and [**DP1**, Sections 5 and 6]) we know that $K_i(M_0^*) = 0$ for $i \neq q$ and $K_q(M_0^*)$ is $\mathbf{Z}[G]$-projective. Consider the exact sequence associated to the pair $(M_0 \cup M_1', M_0^*)$. By excision we have

$$K_*(M_0 \cup M_1', M_0^*) \cong K_*(U_0, \partial U_0)$$

Therefore the relative groups are zero in all dimensions except $q+1$, and in this dimension the group is a free $\mathbf{Z}[G]$-module. Furthermore, the groups $K_i(M_0^*)$ all vanish unless $i = q$ or $q+1$, and therefore the exact sequence of the pair reduces to the following:

$$0 = K_{q+1}(M_0^*) \to K_{q+1}(M_0 \cup M_1') \to F \to K_q(M_0^*) \to K_q(M_0 \cup M_1') \to 0.$$

(where F is a free $\mathbf{Z}[G]$-module).

From this sequence the relation

$$[K_q(M_0^*)] = [K_q(M_0 \cup M_1')] - [K_{q+1}(M_0 \cup M_1')] \in \widetilde{K}_0(\mathbf{Z}[G])$$

is immediate. Since the remaining K_i for M_0^* and $M_0 \cup M_1'$ all vanish, it follows that

$$\chi(K_*(M_0^*)) = \chi(K_*(M_0 \cup M_1')) = \chi(K_*(M_0)).$$

But we know that $\chi(K_*(M_0^*))$ carries the surgery obstruction of f, and therefore the same is true of $\chi(K_*(M_0))$.■

The preceding two results are useful in proving product formulas for equivariant surgery obstructions; in particular, we shall establish a formula of this type in Section 7 (see Theorem 7.10).

PROOF OF THEOREM 5.1: **Case 1.** $\dim M = \dim X = 2q$.—Since $I(G; \mathbf{S}^{-1}\lambda)$ is nonempty, there is an equivalence ($f_0 : M_0 \to X_0$, other data) with $\dim M_0 = \dim X_0 = \dim X - 1$. Let $\alpha \in B \subset \widetilde{K}_0(\mathbf{Z}[G])$ satisfy $\alpha - \alpha^* = 0$. Then by the construction of [**DP1**, Section 6], it is possible to perform equivariant surgery on f_0 and obtain an Ω-adjusted normal cobordism $F : M' \to X_0 \times [0,1]$ such that

(1) $\partial M' = M_0 \cup \partial_+ M \cup \partial M_0 \times [0,1]$,
(2) $F|M_0 \cup \partial M_0 \times [0,1] = f_0 \cup \partial f_0 \times \mathbf{id}_{[0,1]}$,
(3) $K_*(F)$ is projective and $\chi(K_*(F)) = \alpha$.

In fact, this can be done so that F and $(F, \partial_+ F)$ become q-connected and $\partial_+ F$ becomes $(q-1)$-connected. We note two important points:

(i) The surgeries done to form M' will necessarily involve the singular orbits (compare [**DP1**, Section 6]).
(ii) The map F comes very close to representing a class in I^h with projective invariant α. In fact, the only substantial difficulty is that $\partial_+ F$ might not be a pseudo-equivalence.

In view of (ii), we would like to show that $\partial_+ F$ can be surgered to a pseudo-equivalence without changing the projective obstruction. This will be the case if all surgeries are q-dimensional and one can do all surgeries entirely in the free part of $\partial_+ M'$, and thus it suffices to show (5.6) *the surgery obstruction for $\partial_+ F$ is trivial.*

Recall that the surgery obstruction is determined by a projective $\mathbf{Z}[G]$-module; specifically, if U is a well-chosen tubular neighborhood for an G-invariant set of embedded $(q-1)$-spheres that generate $K_{q-1}(\partial_+ M')$, then $K_q(\partial_+ M', U) \cong K^{q-1}(\partial_+ M' - \text{Int } U) = K_{q-1}(\partial_+ M - \text{Int } U)^*$ is a projective module which completely determines the surgery obstruction. Consider the exact sequence of the triple $U \subset \partial_+ M' \subset M'$. The nontrivial terms are contained in the short exact sequence

$$0 \to K_q(\partial_+ M', U) \to K_q(M, U) \to K_q(M', \partial_+ M') \to 0.$$

On the other hand, from (M, U) we obtain another exact sequence

$$0 \to K_q(M) \to K_q(M, U) \to K_{q-1}(U) \to 0.$$

All modules in these short exact sequences are projective, and therefore it follows that

$$[K_q(\partial_+M', U)] = [K_q(M)] - [K_q(M, \partial_+M')] = \alpha - \alpha^*;$$

by the assumptions of the first paragraph of the proof, the right hand side is equal to zero. Therefore the subkernel $K_{q-1}(\partial_+M' - \mathrm{Int}\ U)$ is stably free (and hence can be made free), and accordingly one can do q-dimensional surgeries in the free part of M' to kill the remaining surgery kernels for ∂_+M.

The argument above shows that the composite

$$\mathrm{CLS}\, \sigma_0 : I^h_{\Omega}(\lambda) \to H_{2q-1}(\mathbb{Z}_2; B)$$

is onto. On the other hand, by 3.8 the image of σ_0 contains the image of the forgetful map $L^h_{2q} \to L^B_{2q}$. Therefore the Ranicki sequence 5.2 implies that σ_0 is onto.

Case 2. $\dim M = \dim X = 2q+1$.—In this case we assume $I(G; \mathbf{S}^{-1}\lambda)$ is nonempty, so let $f_0 : M_0 \to X_0$ represent an equivalence in $I(\mathbf{S}^{-1}\lambda)$. Let $\alpha \in B$ satisfy $\alpha + \alpha^* = 0$, and let $\mathsf{H}(\alpha) \in L^h_{2q}(\mathbb{Z}[G])$ be a hyperbolic form with subkernel represented by α. Since $\alpha + \alpha^*$ is stably free, the hyperbolic form does define an element of L^h, but this element is not necessarily trivial (compare results of Bak [**Ba2**] when $G = \mathbb{Z}_p$). However, if $\alpha = \mathrm{CLS}\,(v)$ for some element $v \in L^B_{2q+1}(\mathbb{Z}[G])$, then $\mathsf{H}(\alpha) = 0$ by the Ranicki exact sequence discussed in (5.2).

Once again by the methods of ([**W1**, Chapter 6]) there is a cobordism

$$F' : W \to X \times [0,1]$$

such that

(i) $\partial_- F' = f$,
(ii) $K_*(F')$ is projective with $\chi(K_*(F')) = \mathrm{CLS}\,(v)$, where $v \in L^B_{2q+1}(\mathbb{Z}[G])$,
(iii) $K_*(F')$ is Ω-adjusted, where Ω is defined as in Section 3.

In fact, after surgery away from the singular set we may assume F' is q-connected. It will follow that ∂_+F' is also q-connected, and in fact the surgery obstruction ($K_q(\partial_+F')$, intersection form) is given by $H(\mathrm{CLS}\,(v))$. Since this class vanishes, we may do surgery away from the singular set to make ∂_+F' a pseudoequivalence. Suppose that $F'' : W' \to X \times [1,2]$ is the associated cobordism from $\partial_-F'' = \partial_+F'$ to the equivalence ∂_+F''. By construction $K_i(F'') = 0$ if $i \neq q+1$ and is $\mathbb{Z}[G]$-free if $i = q+1$. Thus the map $F = F' \cup F''$ will once again have projective homology with $\chi(K_*(F)) = \chi(K_*(F')) = \mathrm{CLS}\,(v)$. Furthermore, the restrictions $\partial_-F = \partial_-F' = f$ and $\partial_+F = \partial_+F''$ are both pseudoequivalences, and therefore F represents an element of $I^h(G; \lambda)$. It is immediate from (5.2) and the construction that

$$\mathrm{CLS}\,(\sigma_0([F])) = \chi(K_*(F)) = \mathrm{CLS}\,(v).$$

Since the class map CLS is monic, it follows that $\sigma_0[F] = v$. ■

6. An exact sequence

STANDING HYPOTHESIS: Throughout this section G denotes a nilpotent group of odd order.

In this section we describe the cokernels of the stepwise surgery obstructions σ_α defined in Section 3. Throughout this section we deal with surgery up to pseudoequivalence, we assume that $\dim M = \dim X$ is even, and we also assume that $I^h(G; \mathbf{S}^{-2}\lambda) \neq \emptyset$ satisfies the conditions for the existence of a group structure.

DEFINITION: Let $\Sigma \subset \pi(X; \mathfrak{P}_G)$ be closed and G-invariant. The set

$$\mathcal{Z}_\Sigma(\mathbf{S}^{-1}\lambda) \subset L^B_{\dim X-1}(\mathbf{Z}[G])$$

consists of all obstructions realizable as $\sigma(f)$, where (f, other data) represents a class in $I^h(G; \mathbf{S}^{-1}\lambda)$ that is Σ-adjustedly normally cobordant to a pseudoequivalence.

By the $\pi - \pi$ theorem for I^h [**DP1**, Section 7], this is equivalent to the set of all obstructions $\sigma(F_0)$ associated to normal maps $F_0 : W_0 \to X_0$ that extend to cobordism $(F; F_0, F_1)$ such that

$$\lambda(F_0) \to \lambda(F)$$

is an isomorphism, F_1 is a pseudoequivalence, F is Σ-adjusted, and F satisfies most of the conditions for a normal map in [**DP1**] (specifically, see [**DP1**, (7.1), p. 73]; the crucial omission is that F_0 need not be a pseudoequivalence).

NOTATION: A cobordism $(F; F_0, F_1)$ satisfying the conditions of the preceding paragraph will be called an **equivariant $\pi - \pi$ trivialization** of F_0.

EXAMPLES: 1. If $\Sigma = \Omega = \pi(X; \mathfrak{P}_G)$-{maximal stratum of $\pi(X)$}, then one can show that $\mathcal{Z}_\Omega(\mathbf{S}^{-1}\lambda) = \{0\}$. The reason is fairly simple; if a highly connected Ω-adjustable map f is Ω-adjustably bordant to a pseudoequivalence, then the surgery obstruction of f in $L^B_{2q-1}(\mathbf{Z}[G])$ is zero by the same sort of argument used in Section 3 for the even-dimensional case.

2. If $\Sigma = \emptyset$, then $\mathcal{Z}_\emptyset(\mathbf{S}^{-1}\lambda)$ is precisely the subgroup $\mathcal{Z}(\mathbf{S}^{-1}\lambda)$ defined in Section 3.

The following is an immediate consequence of the definitions:

(6.0) *If $\Sigma \subset \Sigma'$, then $\mathcal{Z}_{\Sigma'}(\mathbf{S}^{-1}\lambda)$ is contained in $\mathcal{Z}_\Sigma(\mathbf{S}^{-1}\lambda)$.*■

The methods of Section 3 also yield the following basic fact:

(6.1) *The set $\mathcal{Z}_\Sigma(\mathbf{S}^{-1}\lambda)$ is a subgroup of $L^B_{\dim X-1}$.*■

Suppose now that α is a minimal closed substratum of $\pi(X; \mathfrak{P}_G)$ not in Σ; assume also that X_α is not the maximal closed stratum. We shall define an **obstruction homomorphism**

$$\theta_\alpha : L^h_{\dim\alpha}(\mathbf{Z}_{(\alpha)}[W_\alpha]) \to L^B_{\dim X-1}(\mathbf{Z}[G])/\mathcal{Z}_{\Sigma\cup G\{\alpha\}}(\mathbf{S}^{-1}\lambda)$$

such that the kernel of θ_α is the image of the stepwise surgery obstruction map σ_α of 3.4. The definition of θ_α requires the choice of a pseudoequivalence $f : M_0 \to X_0$ representing $0 \in I(S^{-1}\lambda)$ and a form Φ representing $u \in L^h$. Using Φ one can proceed as in 3.4 or [**W1**, Section 5], to add handles to $M_0 \times I$ and obtain a G-normal cobordism

$$F : W \to X_0 \times I$$

such that $F|\partial_- W = f_0$, the map F is Σ-adjusted, the restriction $F|\partial_+ W$ is $\Sigma \cup G\{\alpha\}$-adjusted, and the stepwise surgery obstruction for F_α is u. We can in fact assume that $F|\partial_+ W$ is Ω-adjusted (by the vanishing of L^h_{odd}; compare Section 9 and 3.5), and we define $\theta_\alpha(u)$ to be the class of $\sigma_0(F|\partial_+ W)$ in the quotient of $L^B_{\dim X-1}$.

By construction, this definition depends on the choice of representatives Φ and f_0, and less obviously it depends upon the specific manner in which handles are added and the way in which the $\Sigma \cup G\{\alpha\}$-adjusted map $F|\partial_+ W$ is made Ω-adjusted. However, the following result shows that θ_α is in fact independent of all choices.

PROPOSITION 6.2. *Let F_0 and F_0' be Ω-adjusted representatives of classes in the set $I^h(G; \mathbf{S}^{-1}\lambda)$, and let F and F' be Σ-adjusted equivariant $\pi - \pi$ trivializations of F_0 and F_0'. Assume that $\sigma_\alpha(F) = u = \sigma_\alpha(F')$, while $\sigma_0(F_0) = v$ and $\sigma_0(F_0') = v'$. Then*

$$v - v' \in \mathcal{Z}_{\Sigma \cup G\{\alpha\}}(\mathbf{S}^{-1}\lambda).$$

PROOF: Perform a sum construction on F_0 and $-F_0'$, holding ∂F_0 and $\partial F_0'$ fixed throughout the surgery process. The result will be a cobordism F^* from the sum F_0'' to $F_0 \amalg -F_0$, where $\partial F_0'' = \partial F_0 \amalg -\partial F_0'$ and

$$\lambda(F_0'') \cong \lambda(F^*),$$
$$\lambda(F_0) \cong \lambda(F^*),$$
$$\lambda(F_0') \cong \lambda(F^*).$$

Since F_0 and F_0' are Ω-adjusted, the construction and the proof of the $\pi - \pi$ theorem will yield Ω-adjusted maps F_0'' and F^*. The surgery obstruction of F_0'' will be $v - v'$. Set

$$F'' = F^* \cup (F \amalg -F').$$

Then F'' will be a Σ-adjusted equivariant $\pi - \pi$ trivialization of F_0'' with

$$\sigma_\alpha(F'') = \sigma_\alpha(F') - \sigma_\alpha(F) = u - u = 0.$$

In other words, F'' is $(\Sigma \cup G\{\alpha\})$-adjustable. Let F''' be a $(\Sigma \cup G\{\alpha\})$-adjusted map obtained by surgery on F'' away from the boundary. Then F''' will define a $(\Sigma \cup G\{\alpha\})$-adjusted equivariant $\pi - \pi$ trivialization of F_0''. But this means that

$$\sigma(F_0'') = v - v' \in \mathcal{Z}_{\Sigma \cup G\{\alpha\}}(\mathbf{S}^{-1}\lambda). \blacksquare$$

A similar argument proves another property of θ_α.

PROPOSITION 6.3. *The map θ_α is a homomorphism.*

PROOF: Let F_0 and F_0' represent classes in $I^h(G;\mathbf{S}^{-1}\lambda)$, assume these maps are Ω-adjusted, and let F and F' be Σ-adjusted $\pi-\pi$ trivializations. Assume that $\sigma_\alpha(F)=u$ and $\sigma_\alpha(F')=u'$. Perform a sum construction on F_0 and F_0' as in 6.2, and let F_0'', F^*, and F'' be defined exactly as in 6.2. (**Caution:** In this case the orientation of F_0' has not been reversed!) Then F'' will be a Σ-adjusted $\pi-\pi$ trivialization of F_0'', and the stepwise surgery obstructions will satisfy

$$\begin{aligned}\sigma(F_0'') &= -(v+v') \quad (\text{since } -[F_0''] = [F_0]+[F_0']),\\ \sigma(F'') &= -(u+u').\end{aligned}$$

In view of 6.2 this implies

$$\theta_\alpha(u+u') = v+v' = \theta_\alpha(u)+\theta_\alpha(u'),$$

and therefore θ_α is additive.■

This brings us to the main result:

THEOREM 6.4. *There is an exact sequence of abelian groups:*

$$\begin{aligned}&0 \to K_{\Sigma\cup G\{\alpha\}}(\lambda) \to K_\Sigma(\lambda) \xrightarrow{\sigma_\alpha} L^h_{\dim\alpha}(\mathbf{Z}_{(\alpha)}[W_\alpha])\\ &\xrightarrow{\theta_\alpha} L^B_{\dim X-1}\mathbf{Z}[G]/\mathcal{Z}_{\Sigma\cup G\{\alpha\}}(\mathbf{S}^{-1}\lambda) \to L^B_{\dim X-1}(\mathbf{Z}[G])/\mathcal{Z}_\Sigma(\mathbf{S}^{-1}\lambda) \to 0.\end{aligned}$$

(Recall that $\dim X$ is even.)

PROOF: Exactness at K was established in the proof of 3.4, and therefore the proof of 6.4 reduces to verifying

(i) Kernel $\theta_\alpha \subset$ Image σ_α,
(ii) Image $\sigma_\alpha \subset$ Kernel θ_α,
(iii) Image $\theta_\alpha =$ Image $\mathcal{Z}_\Sigma(\mathbf{S}^{-1}\lambda)$.

The arguments are variations of the proof of 6.2.

(i) Suppose that $u \in$ Kernel θ_α. Let $(F;F_0,F_1)$ be a Σ-adjusted G-normal cobordism from the Ω-adjusted map F_0 to the pseudoequivalence F_1, where $\sigma_\alpha(F)=u$. Then the surgery obstruction $v=\sigma_\alpha(F_0)$ represents $\theta_\alpha(u)$. Since $u\in$ Kernel θ_α, it follows that

$$v \in \mathcal{Z}_{\Sigma\cup G\{\alpha\}}(\mathbf{S}^{-1}\lambda).$$

In other words, there is a $(\Sigma\cup G\{\alpha\})$-adjusted G-normal cobordism $(F';F_0',F_1')$ with F_1' a pseudoequivalence and $\sigma_0(F_0')=v$. Apply the sum obstruction to F_0 and $-F_0'$ as in 6.2; the proof of the latter then yields a new $\pi-\pi$ trivialization F'' such that

(a) $\partial F'' = F_0'' \cup (F_1 \amalg -F_1')$,
(b) F'' is Σ-adjusted,

(c) $\sigma_\alpha(F'') = u$,
(d) F_0'' is Ω-adjusted with obstruction $v - v = 0$.

We may therefore perform surgery on F_0'' away from the singular orbits to obtain a pseudoequivalence F_0'''. Let F''' be the corresponding $\pi - \pi$ trivialization obtained by adding handles to F'' along F_0''. Then F''' satisfies the analogs of (a)–(c), and in addition F_0''' is a pseudoequivalence. Therefore F''' represents a class ω in $I^h(G;\lambda)$. By construction the class ω is Σ-adjustable and has stepwise obstruction $\sigma_\alpha\omega = u$.□

(ii) Suppose now that $u = \sigma_\alpha[F]$, where F is a Σ-adjusted G-normal map with ∂F a pseudoequivalence. Let F' be a Σ-adjusted $\pi - \pi$ trivialization of some F_0' with stepwise obstruction u, and suppose that F_0' is Ω-adjusted with stepwise obstruction v (which then represents $\sigma_\alpha(u)$). Perform the sum construction on F' and $-F$. This will yield a Σ-adjusted map F'' that is a $\pi - \pi$ trivialization of F_0' (for $\partial F'' = (F_0' \cup F_1') \amalg -F$), and the stepwise obstruction $\sigma_\alpha(F'')$ of F'' is zero. Therefore we can modify F'' away from the boundary to obtain a $\Sigma \cup G\{\alpha\}$-adjusted $\pi - \pi$ trivialization of F_0'. By the alternate definition this implies that

$$v = \sigma_0(F_0') \in \mathcal{Z}_{\Sigma \cup G\{\alpha\}}(\mathbf{S}^{-1}\lambda),$$

or, equivalently, $\theta_\alpha(u) = 0$.□

(iii) By construction it is immediate that Image $\theta_\alpha \subset$ Image $\mathcal{Z}_\Sigma(\mathbf{S}^{-1}\lambda)$, for $\sigma_\alpha(u)$ is represented by $\sigma_0(F_0)$, where $(F; F_0, F_1)$ is a Σ-adjusted normal cobordism with F_1 a pseudoequivalence. Conversely, if $v \in \mathcal{Z}_\Sigma(\mathbf{S}^{-1}\lambda)$ then there is a Σ-adjusted $\pi - \pi$ trivialization $(F; F_0, F_1)$ of some F_0 with $v = \sigma_0(F_0)$; let $\sigma_\alpha(F) = u$. Clearly we would like to say that v represents $\theta_\alpha(u)$. In any case we may construct a representative v' for $\theta_\alpha(u)$ by taking a pseudoequivalence F_1' representing $0 \in I(\mathbf{S}^{-1}\lambda)$ and adding handles to form a Σ-adjusted G-normal cobordism $(F'; F_0', F_1')$ with $\sigma(F') = u$ and $v' = \sigma(F_0')$ representing $\theta_\alpha(u)$. But Proposition 6.2 now implies that $v - v' \in \mathcal{Z}_{\Sigma \cup G\{\alpha\}}(\mathbf{S}^{-1}\lambda)$, and therefore v does represent $\theta_\alpha(u)$ as claimed.■

The exact sequence implies that σ_α comes close to being surjective.

COROLLARY 6.5. *The cokernel of σ_α is a finite 2-group (possibly trivial).*

Corollary 6.5 has a noteworthy consequence: *For many choices of λ the groups $I^h(G;\lambda)$ are not finitely generated.* Suppose that $\alpha \in \pi(X, \mathfrak{P}_G)$ is such that X_α is a proper closed substratum and $\dim \alpha \equiv 0 \mod 4$; then the codomain of the stepwise surgery obstruction σ_α is $L_0^h(\mathbb{Z}_{(\alpha)}[W_\alpha])$. This group is infinitely generated, for it contains $L_0^h(\mathbb{Z}_{(\alpha)})$ as a direct summand, and the latter is well-known to be infinitely generated (compare [**Pa1**]). On the other hand, the image of σ_α has finite index by 6.5, and thus the image of σ_α must be infinitely generated. Since Image σ_α is a subquotient of $I^h(G;\lambda)$, it is immediate that the latter is also infinitely generated.

PROOF OF 6.5: As noted previously the class map from $L^B_{\mathrm{odd}}(\mathbb{Z}[G])$ to $H_*(\mathbb{Z}_2; B)$ is injective. Since the relevant homology group of $\mathbb{Z}_2$ has exponent dividing 2, it follows that $2 \cdot L^B_{\mathrm{odd}}(\mathbb{Z}[G]) = 0$. Therefore it suffices to show that $L^B_{\mathrm{odd}}(\mathbb{Z}[G])$ is also finite.

Using the class map once again, we see that it suffices to prove that B is finite. Since B is a subgroup of $\widetilde{K}_0(\mathbb{Z}[G])$, it suffices to observe that the latter is finite by a fundamental result of R. Swan (see [**Sw**, Section 9]).■

7. Products and stepwise obstructions

STANDING HYPOTHESIS. *Throughout this section G denotes an odd order nilpotent group.*

The usefulness of the groups $I^h(G;\lambda)$ depends strongly on their description in terms of standard Wall groups via the stepwise surgery obstructions. Similarly, the value of a product map $\mathbf{P}_Y$, depends on its describability in terms of the decomposition of $I^h(G;\lambda)$ by standard Wall groups. If G acts freely on M and X, then $I^h(G;\lambda) = L_n^h(\mathbb{Z}[G])$, where $n = \dim M = \dim X$, and the map $\mathbf{P}_Y$ is precisely the twisted product of [**Yo**]. In Chapters III and IV we showed that the products in many equivariant surgery theories could be described by twisted products in ordinary surgery. The purpose of this section is to prove similar results for I^h using the machinery of the preceding sections and the methods of Chapters III and IV.

(7.1) SIMPLIFYING HYPOTHESIS: *We shall assume all fixed point sets Y^H are connected and that Y is λ-preperiodic.*

It is possible to work in greater generality, but the results become considerably more difficult to state.

The first result is a straightforward extension of the results in Sections III.4 and IV.6.

THEOREM 7.2. *Let λ satisfy the hypotheses in Theorems 2.1 and 2.2, so that $\mathbf{P}_Y$ is a group homomorphism. Assume that $I^h(G;\mathbf{S}^{-1}\lambda) \neq \varnothing$ and λ is weakly isogeneric. In addition, assume Simplifying Hypothesis (7.1), and identify λ with $\lambda \times \mathrm{data}(Y)$ using preperiodicity. Then $\mathbf{P}_Y$ sends the Σ-adjustable elements $K_\Sigma^h(\lambda)$ to $K_\Sigma^h(\lambda \times \mathrm{data}(Y))$. If $\alpha \in \pi(X, \mathfrak{P}_G)$ represents a minimal closed substratum not in Σ and α is not the maximal closed substratum of M and X, then there is a commutative diagram*

$$\begin{array}{ccc} K_\Sigma^h(\lambda) & \xrightarrow{\sigma_\alpha} & L^h_{\dim \alpha}(R_\alpha[W_\alpha]) \\ \Big\downarrow{\scriptstyle \mathbf{P}_Y} & & \Big\downarrow{\scriptstyle \mu_\alpha} \\ K_\Sigma^h(\lambda \times \mathrm{data}(Y)) & \xrightarrow{\sigma'_\alpha} & L^h_{\dim \alpha + \dim Y_\alpha}(R_\alpha[W_\alpha]), \end{array}$$

where the W_α-manifold $Y_\alpha = \mathrm{Fix}(Y, G^\alpha)$ is defined as in Chapters III–IV and R_α equals $\mathbb{Z}_{(\alpha)}$.

NOTE: As in Section 3 the integer $\ell(\alpha)$ is the product of all primes p such that X_α is a component of X^P for some p-group $P \subset G$.

PROOF: The assertion that $\mathbf{P}_Y$ preserves Σ-adjustability follows from the fact that the product of a homology equivalence with the identity is a homology equivalence.

The main step in constructing the commutative diagram is to use the geometric restriction maps of Section 4 to form the following commutative diagram:

$$\begin{array}{ccccc}
K^h_\Sigma(\lambda) & \xrightarrow{\text{res }\alpha} & I^{h|\alpha}_{\Sigma|\alpha}(\lambda|\alpha) & \underset{\cong}{\xrightarrow{\tau_\alpha}} & L^h_{\dim\alpha}(R_\alpha[W_\alpha]) \\
\mathbf{P}_Y \downarrow & & \mathbf{P}_{Y_\alpha}\downarrow & & \\
K^h_\Sigma(\lambda\times \text{data}(Y)) & \xrightarrow{\text{res }\alpha} & I^{h|\alpha}_{\Sigma|\alpha}((\lambda\times\text{data}(Y))|a) & \underset{\cong}{\xrightarrow{\tau'_\alpha}} & L^h_{\dim\alpha+\dim Y_\alpha}(R_\alpha[W_\alpha])
\end{array}$$

The horizontal composites are the maps σ_α and σ'_α by 4.9. Notice that the left square commutes by definition of the geometric restriction maps. Thus it suffices to show that $\mu_\alpha\tau_\alpha = \tau'_\alpha\mathbf{P}_{Y_\alpha}$. But this follows by the same sort of argument used in Theorem III.4.2.■

We shall also need an analog of 7.2 for the stepwise surgery obstruction on the maximal closed substratum. This is more difficult because standard surgery techniques only work well for elements of $L^B_n(\mathbf{Z}[G])$ that lie in the image of $L^h_n(\mathbf{Z}[G])$. In particular, we need effective means for dealing with the portion of the G-surgery obstruction involving the projective class group $\widetilde{K}_0(\mathbf{Z}[G])$.

In order to describe the projective factor of the G-surgery obstruction, it is necessary to use the standard description of $\widetilde{K}_0(\mathbf{Z}[G])$ as a module over the Burnside ring (compare [tD2]). One way of defining the module structure is to stipulate that the coset $G/H \in \Omega(G)$ sends a projective module P to the projective module $\mathbf{Z}[G]\otimes_{\mathbf{Z}[H]}P \cong \mathbf{Z}[G/H]\otimes_{\mathbf{Z}}P$, where $a\in G$ sends $[gH]\otimes y$ in the latter to $[agH]\otimes ay$. For our purposes it is important to know that the subgroups $B\subset \widetilde{K}_0\mathbf{Z}[G]$ arising in equivariant surgery are mapped to themselves by certain elements of the Burnside ring.

PROPOSITION 7.3. *Let λ be as above, and assume Y is λ-preperiodic. Furthermore, assume that $H^*(Y;\mathbf{Z})$ is torsion free in each dimension and each $H^i(Y;\mathbf{Z})$ is a direct sum of G-permutation modules. Let $\chi_G(Y)\in\Omega(G)$ be the equivariant Euler characteristic of Y. Then $\chi_G(Y)$ maps the subset $B_0(G,\lambda)$ defined in* [**DP1**, (2.6)] *to itself.*

PROOF: By construction $\chi_G(Y)$ maps $B_0(G,\lambda)$ into $B_0(G,\lambda\times\text{data}(Y))$. However the latter are isomorphic by λ-preperiodicity.■

COROLLARY 7.4. *Under the assumptions above, multiplication by $\chi_G(Y)$ induces a linear endomorphism $\chi_G(Y)_*$ of $H_*(\mathbf{Z}_2;B_0(G,\lambda))$.*

PROOF: Multiplication by elements of $\Omega(G)$ is compatible with the involution; specifically, for all $H\subset G$ and all projective $\mathbf{Z}[G]$-modules P we have $([G/H]\cdot P)^*\cong [G/H]\cdot(P^*)$.■

The maps $\chi_G(Y)_*$ in 7.4 completely describe the effect of $\mathbf{P}_Y$ on the projective class group factor of the final stepwise surgery obstruction.

THEOREM 7.5. *Let λ and Y satisfy all the conditions in* 7.2 *and* 7.3, *assume that* $\dim X$ *and* $\dim Y$ *are even, and let $\Omega = \pi(X,\mathfrak{P}_G)-\{X_{\text{maximal}}\}$. Let $\sigma_0 : I^h_\Omega(G;\lambda)\to$*

$L^B_{\dim X}(\mathbf{Z}[G])$ be the stepwise surgery obstruction, let B be defined as in Section 3, and let $\mathrm{CLS} : L^B_{2n}(\mathbf{Z}[G]) \to H_{2n-1}(\mathbf{Z}_2; B)$ be the projective class invariant, where $B = B_0(G, \lambda)$. Then the following diagram is commutative:

$$\begin{array}{ccccc} I^h_\Omega(\lambda) & \xrightarrow[\cong]{\sigma_0} & L^B_{2n}(\mathbf{Z}[G]) & \xrightarrow{\mathrm{CLS}} & H_{2n-1}(\mathbf{Z}_2; B) \\ \downarrow{\scriptstyle \mathbf{P}_Y} & & & & \downarrow{\scriptstyle \chi_G(Y)_*} \\ I^h_\Omega(\lambda \times \mathrm{data}\,(Y)) & \xrightarrow{\sigma_0'} & L^B_{2n+2m}(\mathbf{Z}[G]) & \xrightarrow{\mathrm{CLS}} & H_{2n-1}(\mathbf{Z}_2; B) \end{array}$$

$$(2n = \dim X,\ 2m = \dim Y)$$

PROOF: Let $(f$, other data) represent a class in $I^h_\Omega(G; \lambda)$, and assume f is highly connected. Then the surgery kernel $K_*(f)$ is projective and concentrated in one dimension, and by 5.3 we have

$$\mathrm{CLS}\,\sigma_0(f) = \text{ class of } \chi(K_*(f)) \text{ in } H_{2n-1}(\mathbf{Z}_2; B).$$

Since the homology group has exponent 2, we need not worry about signs.

By the Künneth formula the surgery kernel $K_*(f \times \mathbf{id}_Y)$ is just the tensor product $K_*(f) \otimes H_*(Y)$, and this graded module is projective in every dimension. Therefore by 5.3 we have

$$\begin{aligned} &\mathrm{CLS}\,(\sigma_0'(\mathbf{P}_Y[f])) = \chi(K_*(f \times \mathbf{id}_Y)) = \\ &\chi\,(K_*(f) \otimes H_*(Y)) = \chi_G(Y)_*\chi(K_*(f)) = \\ &\chi_G(Y)_*\mathrm{CLS}\,(\sigma_0[f]), \end{aligned}$$

which is the statement to be proved.■

The next result describes the effect of $\mathbf{P}_Y$ on the final stepwise obstruction more completely.

THEOREM 7.6. *Let λ, Y, B, etc., be as in 7.5, let $\dim X = 2q$, and assume that $I^h(G; \mathbf{S}^{-1}\lambda) \neq \varnothing$. Then there is the following commutative diagram:*

$$\begin{array}{ccc} L^h_{2q}(\mathbf{Z}[G]) & \xrightarrow{\mu_Y} & L^h_{2q+2m}(\mathbf{Z}[G]) \\ \downarrow{\scriptstyle \gamma_0} & & \downarrow{\scriptstyle \gamma_0'} \\ I^h_\Omega(G; \lambda) & \xrightarrow{\mathbf{P}_Y} & I^h_\Omega(\lambda \times \mathrm{data}\,(Y)) \\ \downarrow{\scriptstyle \sigma_0} & & \downarrow{\scriptstyle \sigma_0'} \\ L^B_{2q}(\mathbf{Z}[G]) & & L^B_{2q+2m}(\mathbf{Z}[G]) \\ \downarrow{\scriptstyle \mathrm{CLS}} & & \downarrow{\scriptstyle \mathrm{CLS}} \\ H_{2q-1}(\mathbf{Z}_2; B) & \xrightarrow{\chi_G(Y)_*} & H_{2q-1}(\mathbf{Z}_2; B) \end{array}$$

The map μ_Y is a twisted product homomorphism, the maps γ_0 and γ_0' are cut-and-paste constructions, the maps $\mathbf{P}_Y$ represent taking products with Y, the maps σ and σ' are ordinary surgery obstructions, and the composites $\sigma_0\gamma_0$ and $\sigma_0'\gamma_0'$ are the forgetful homomorphisms.

NOTES: 1. Kernel CLS = Image σ_0 (or σ_0') by the exactness of the Ranicki sequence.

2. The results of Chapters III and IV suggest that the diagram in Theorem 7.6 should also commute if one inserts a twisted product map

$$\mu_Y^B : L_{2q}^B(\mathbb{Z}[G]) \to L_{2q+2m}^B(\mathbb{Z}[G])$$

as defined in Section IV.4; however, we have not verified this.

PROOF OF THEOREM 7.6: (*Sketch*) Let $j_* : L_*^h(\mathbb{Z}[G]) \to L_*^B(\mathbb{Z}[G])$ denote the forgetful homomorphsisms. Then the relations $\sigma_0\gamma_0 = j_{2q}$, $\sigma_0'\gamma_0' = j_{2q+2m}$, and $\gamma_0'\mu_Y = \mathbf{P}_Y\gamma_0$ follow exactly as in the proof of Theorem 4.2. Furthermore, the relation $\chi_G(Y)_*\mathrm{CLS}\,\sigma_0 = \mathrm{CLS}\,\sigma_0'\mathbf{P}_Y$ was established in 7.5.■

We turn next to the odd-dimensional case. Since CLS induces an injection from $L_{2q+1}^B(\mathbb{Z}[G])$ to $H_{2q}(\mathbb{Z}_2; B)$, the projective obstruction carries all the information about the surgery obstruction. One complication is that the surgery obstruction in $L_{2q+1}^B(\mathbb{Z}[G])$ is only defined up to an indeterminancy given by the subgroup $\mathcal{Z}(\lambda)$. The first order of business is to show that the map $\chi_G(Y)_*$ in 7.4 preserves this indeterminancy.

PROPOSITION 7.7. *Let $\lambda, Y, B, \ldots$ be as before, but assume* $\dim X = 2q+1$, $\dim Y = 2m$, *and $I^h(G; \mathbf{S}^{-1}\lambda) \neq \varnothing$. Then $\chi_G(Y)_*$ sends* $\mathrm{CLS}\,(\mathcal{Z}(\lambda)) \subset H_{2q}(\mathbb{Z}_2; B)$ *to* $\mathrm{CLS}\,(\mathcal{Z}(\lambda \times \mathrm{data}\,(Y))) \subset H_{2q+2m}(\mathbb{Z}_2; B)$.

The proofs of this proposition and many other facts about odd-dimensional products are based upon the following consequence of 5.5. We assume the setting of 7.7.

LEMMA 7.8. *Let $f : M \to X$ have indexing data λ, and assume f is highly connected and Ω-adjusted. Let $u \in L_{2q+1}^B(\mathbb{Z}[G])$ be the surgery obstruction of f. Then $f \times \mathbf{id}_Y$ is also Ω-adjusted, and one can do surgery on $f \times \mathbf{id}_Y$ away from the singular set to obtain a highly connected Ω-adjusted map f^* with surgery obstruction satisfying*

$$\mathrm{CLS}\,(\sigma_0(f^*)) = \chi_G(Y)_*\mathrm{CLS}\,(u).$$

PROOF: By 5.1 there is a geometric formation $(f; f_0, f_1)$ on f with projective surgery kernel $K_*(f_0)$ and

$$\mathrm{CLS}\,(u) = \chi(K_*(f_0)).$$

Since $H_*(Y)$ is a sum of permutation modules, it follows that the tensor product of $H_*(Y)$ with a graded projective (resp., free) $\mathbb{Z}[G]$-module will again be projective (resp., free). Therefore it follows that

$$(f \times \mathbf{id}_Y; f_0 \times \mathbf{id}_Y, f_1 \times \mathbf{id}_Y)$$

is again a geometric formation, and in fact

$$\chi(K_*(f_0 \times \mathbf{id}_Y)) = \chi(K_*(f_0) \otimes H_*(Y)) =$$
$$\chi_G(Y)_*(\chi(K_*(f_0))) = \chi_G(Y)_* \text{CLS}\,(u).$$

But the proof of 5.5 shows that $f \times \mathbf{id}_Y$ can be equivariantly surgered away from the singular set to yield an Ω-adjusted map f^* with

$$\text{CLS}\,(\sigma_0(f^*)) = \chi(K_*(f_0 \times \mathbf{id}_Y)),$$

and we have already shown that the latter is $\chi_G(Y)_* \text{CLS}\,(u)$.■

PROOF OF 7.7: If $u \in \mathcal{Z}(\lambda)$ there is a $\pi - \pi$ trivialization $(F; \partial_+ F, \partial_- F)$ such that $\partial_- F$ is a pseudoequivalence, $\partial_+ F$ is Ω-adjusted, and highly connected, and $\sigma_0(\partial_+ F) = u$. Add handles to $\partial_+ F \times \mathbf{id}_Y \times [1,2]$ away from the singular set to form a highly connected Ω-adjusted map f^*, and let $(F'; f^*, \partial_+ F \times \mathbf{id}_Y)$ be the associated map of triads. Join F' and $F \times \mathbf{id}_Y$ together along $\partial_+ F \times \mathbf{id}_Y$, and call the resulting object $(F''; f^*, \partial_- F \times \mathbf{id}_Y)$. By construction this is a $\pi - \pi$ trivialization, and by 7.8 we know that $\text{CLS}\,\sigma(f^*) = \chi_G(Y)_* \text{CLS}\,(u)$.■

The same argument proves somewhat more:

COROLLARY 7.9. *The conclusion of 7.7 remains valid if one replaces $\mathcal{Z}(\lambda)$ by $\mathcal{Z}_\Sigma(\lambda)$ and $\mathcal{Z}(\lambda \times \text{data}\,(Y))$ by $\mathcal{Z}_\Sigma(\lambda \times \text{data}\,(Y))$ for arbitrary Σ.*

PROOF: (*Sketch*) In this case $(F; \partial_+ F, \partial_- F)$ is assumed Σ-adjusted; this remains true if one takes products with $\mathbf{id}_Y$. Furthermore, the cobordism $(F', f^*, \partial_+ F \times \mathbf{id}_Y)$ is Ω-adjusted (hence Σ-adjusted) by construction, and therefore $(F''; f^*, \partial_- F \times \mathbf{id}_Y)$ will be Σ-adjusted.■

The preceding results and 3.4 yield a complete description of $\mathbf{P}_Y$ in the odd-dimensional case.

THEOREM 7.10. *In the setting of 7.7 there is a commutative diagram*

$$\begin{array}{ccc} I^h(G;\lambda) & \xrightarrow{\text{CLS}^{\#}\sigma_0} & H_{2q}(\mathbb{Z}_2;B)/\text{CLS}\,(\mathcal{Z}(\lambda)) \\ \mathbf{P}_Y \big\downarrow & & \big\downarrow \chi_G(Y)_* \\ I^h(G;\lambda \times \text{data}\,(Y)) & \xrightarrow{\text{CLS}^{\#}\sigma_0'} & H_{2q}(\mathbb{Z}_2;B)/\text{CLS}\,(\mathcal{Z}(\lambda \times \text{data}\,(Y))) \end{array}$$

in which $\text{CLS}^{\#}\sigma_0$ and $\text{CLS}^{\#}\sigma_0'$ are injective. The map $\text{CLS}^{\#}$ is obtained from CLS by passage to quotients.

PROOF: The injectivity of $\text{CLS}^{\#}\sigma_0$ and $\text{CLS}^{\#}\sigma_0'$ follows directly from 3.4 and 5.2.

Given an element of $I^h(G;\lambda)$, choose an Ω-adjusted highly connected representative f, and let $u = \text{CLS}\,\sigma_0(f)$ be the surgery obstruction of this map. By 7.8 there is an Ω-adjusted highly connected representative f^* for $\mathbf{P}_Y(f)$ such that $\text{CLS}\,\sigma_0'(f^*) = \chi_G(Y)_* u$. It follows that

$$\chi_G(Y)_*\text{CLS}\,{}^\#\sigma_0[f] = \chi_G(Y)_*(u) = \text{CLS}\,\sigma_0'(f^*) =$$
$$\text{CLS}\,{}^\#\sigma_0'[f^*] = \text{CLS}\,{}^\#\sigma_0'\mathbf{P}_Y[f],$$

and thus the diagram commutes.■

Finally, we need to describe the behavior of products with respect to the obstruction homomorphisms

$$\theta_\alpha : L^h_{\dim\alpha}(\mathbf{Z}_{(\alpha)}[W_\alpha]) \to L^B_{2q-1}(\mathbf{Z}[G])/\mathcal{Z}_{\Sigma\cup G\{\alpha\}}(\lambda)$$

introduced in Section 6.

PROPOSITION 7.11. *Assume the setting of 7.6, and also assume that $I^h(G;\mathbf{S}^{-2}\lambda) \neq \emptyset$. Then the following diagram is commutative:*

$$\begin{array}{ccc} L^h_{\dim\alpha}(\mathbf{Z}_{(\alpha)}[W_\alpha]) & \xrightarrow{\theta_\alpha} & L^B_{2q-1}(\mathbf{Z}[G])/\mathcal{Z}_{\Sigma\cup G\{\alpha\}}(\lambda) \\ \downarrow \mu_\alpha & & \downarrow \chi_G(Y)_* \\ L^h_{\dim\alpha+\dim Y_\alpha}(\mathbf{Z}_{(\alpha)}[W_\alpha]) & \xrightarrow{\theta'_\alpha} & L^B_{2q-1}(\mathbf{Z}[G])/\mathcal{Z}_{\Sigma\cup G\{\alpha\}}(\lambda\times\text{data}\,(Y)). \end{array}$$

PROOF: Recall that θ_α is defined by taking a map of triads $(F;\partial_+F,\partial_-F)$ with indexing data $\mathbf{S}^{-1}\lambda$ such that ∂_-F is a pseudoequivalence (with indexing data $\mathbf{S}^{-1}\lambda$), F is Σ-adjusted with stepwise obstruction $u \in L^h_{\dim\alpha}(\mathbf{Z}_{(\alpha)}[W_\alpha])$, and ∂_+F is an Ω-adjusted highly connected normal map representing a class in $I^h(G;\mathbf{S}^{-1}\lambda)$. The class $\theta_\alpha(u)$ is the image of the obstruction $\sigma_0(\partial_+F)$ in the quotient $L^B_{2q-1}/\mathcal{Z}_{\Sigma\cup G\{\alpha\}}$.

Consider the map of triads $(F\times\mathbf{id}_Y;\partial_+F\times\mathbf{id}_Y,\partial_-F\times\mathbf{id}_Y)$. This is again Σ-adjusted, and the stepwise surgery obstruction is $\mu_\alpha(u)$ by 7.2. Furthermore, $\partial_-F\times\mathbf{id}_Y$ is a pseudoequivalence and $\partial_+F\times\mathbf{id}_Y$ is Ω-adjusted. If we perform surgery on $\partial_+F\times\mathbf{id}_Y$ away from the singular set to obtain a highly connected map, we obtain a new map of triads $(F';f^*,\partial_-F\times\mathbf{id}_Y)$ such that F' is Σ-adjusted with the same stepwise obstruction $\mu_\alpha(u)$, the map f^* is Ω-adjusted and highly connected, and $\sigma_0(f^*) = \chi_G(Y)_*\sigma_0(\partial_+F)$ by 7.8. In other words, we have $\theta'_\alpha(\mu_\alpha(u)) = \chi_G(Y)_*(\sigma_0(\partial_+F)) = \chi_G(Y)_*\theta_\alpha(u)$, and therefore the diagram under consideration commutes as claimed.■

8. Proofs of main results

In the first half of this section we shall verify Theorems 2.1 and 2.2 on the existence and additivity of products in the theories I^h and $I^{h(R)}$, and in the second half we shall use the results of Sections 3–7 to prove the periodicity theorems for I^h stated in Section 2.

Proofs of Theorems 2.1 – 2.2

We shall begin by describing our basic setting.

STANDING HYPOTHESES. *The finite group G is nilpotent, the smooth G-manifold Y satisfies the Codimension ≥ 2 Gap Hypothesis, and $f : M \to X$ is an I^h-normal map with indexing data λ.*

We now go through the conditions that must hold if $f \times 1_Y$ is to be an I^h-normal map. The first five conditions are very straightforward, so we shall merely describe the main ideal in these cases and leave the details to the reader.

(8.1) *All the closed substrata in $\pi(M \times Y; \mathfrak{P}_G)$ and $\pi(X \times Y; \mathfrak{P}_G)$ are simply connected.*

PROOF: *(Sketch)* This holds because closed substrata in $X \times Y$ and $M \times Y$ have the form $C_1 \times C_2$ where C_1 and C_2 are components of X^H (resp., M^H) and Y^H for some $H \in \mathfrak{P}_G$.■

(8.2) *For each $H \in \mathfrak{P}_G$, the map $f \times 1_Y$ sends a component of $(M \times Y)^H$ to a component of $(X \times Y)^H$ with the same dimension.*

PROOF: *(Sketch)* A component of $(M \times Y)^H$ has the form $C \times B$ where C and B are components of M^H and Y^H respectively; if f maps C to the component C' of X^H, then $f \times 1_Y$ sends $C \times B$ to $C' \times B$ and $\dim C = \dim C'$ implies $C \times B = \dim C' \times B$.■

(8.3) *For each $H \in \mathfrak{P}_G$ the map $f \times 1_Y$ induces an isomorphism between the components of $(M \times Y)^H$ and $(X \times Y)^H$.*

PROOF: *(Sketch)* This follows from the description of components in (8.1) and (8.2).■

(8.4) *For each $H \in \mathfrak{P}_G$ and each component Q of $(M \times Y)^H$ the induced map from Q to a component of $(X \times Y)^H$ has degree prime to the order of H.*

PROOF: *(Sketch)* Q has the form $C \times B$ and $f \times 1_Y$ maps $C \times B$ to some product $C' \times B$. By hypothesis the degree of f is prime to the order of H, so the conclusion follows because $\deg(f \times 1_Y) = \deg f$.■

(8.5) *There is a system of bundle data for $f \times 1_Y$.*

PROOF: *(Sketch)* If (b_0, c, φ) are bundle data for f, then the bundle map $b_0 \times 1_{T(Y)}$, the map vector bundle systems determined by $c \times \Pi(T(Y))$ (where Π is the construction from vector bundles to vector bundle systems in Section I.3), and the compatibility relation determined by $\varphi \times 1_{\Pi(T(Y))}$ form a system of bundle data for $f \times 1_Y$.■

The final condition for $f \times 1_Y$ is considerably less elementary than (8.1)–(8.5) and involves the Euler characteristic restrictions. All the notation that appears below is explained in Section 1.

Let X be a finite G-CW complex. If Y is a second finite G-CW complex and Z is a $\Pi(X)$-complex, then the smash product $Z \wedge (Y_+)$ has a natural $\Pi(X \times Y)$-complex structure. Furthermore, it is elementary to verify that smash product with Y_+ induces a homomorphism

$$\text{(8.6)} \qquad \mathbf{P}(Y_+) : \Omega(G, \Pi(X)) \to \Omega(G, \Pi(X \times Y)).$$

As noted in Section 1, the basic condition on Euler characteristics are given by

(8.7-d) $$[M(f)] \in \Delta\Big(G, \pi(X) \cup \widetilde{f}(\pi(M)\Big) + d\boldsymbol{\Omega}(G, \pi(X))$$

where $d = 1$ if f comes from an I^a normal cobordism and $d = 0$ or 2 if f comes from a map of odd- or even-dimensional manifolds respectively. We need to show that (8.7-d) implies the corresponding Euler characteristic condition for $f \times \mathbf{id}_Y$. Since $M(f \times \mathbf{id}_Y) \cong M(f) \wedge (Y_+)$, the first step in the proof can be expressed as follows:

PROPOSITION 8.8. *(i) If K is a $\pi(X)$-complex* (**sic**), *then $K \wedge (Y_+)$ is a $\pi(X \times Y)$-complex.*

(ii) If K is a $\pi(X) \cup \widetilde{f}\pi(M)$-complex, then $K \wedge (Y_+)$ is a $\pi(X \times Y) \cup (f \times \mathbf{id}_Y)^{\sim} \pi(M \times Y)$-complex.

Before proving this result we shall use it to establish the Euler characteristic condition for $f \times \mathbf{id}_Y$:

(8.9-d). *If f satisfies* (8.7-d), *then*

$$[M(f \times \mathbf{id}_Y)] \in \Delta\Big(G, \pi(X \times Y) \cup (f \times \mathbf{id}_Y)^{\sim} \pi(M \times Y)\Big) + d\boldsymbol{\Omega}(G, \pi(X \times Y)).$$

PROOF THAT 8.8 IMPLIES 8.9-d: Write the equivalence class of $M(f)$ in $\boldsymbol{\Omega}(G, \Pi(X))$ as $[K_1] + d[K_2]$, where K_1 represents an elements in the first summand of (8.7-d) and K_2 represents an element in $\boldsymbol{\Omega}(G, \pi(X))$; we may assume K_1 is nonequivariantly contractible. Then by (8.8) we know that $K_1 \wedge Y_+$ is a $\{\pi(X \times Y) \cup (f \times \mathbf{id}_Y)^{\sim} \pi(M \times Y)\}$-complex and $K_2 \wedge Y_+$ is a $\pi(X \times Y)$ complex. All that remains is to note that $K_1 \wedge Y_+$ is contractible if K_1 is contractible.■

PROOF OF PROPOSITION 8.8: (i) We must show that the essential subcomplexes of $K \wedge Y_+$ all correspond to elements of $\pi(X \times Y) \subset \Pi(X \times Y)$. An essential subcomplex of $K \wedge Y_+$ has the form $K_\alpha \wedge (Y_{\beta+})$ where $\alpha = (X_\alpha, H)$, $\beta = (Y_\beta, H)$, and some point $(u, y) \in K_\alpha \wedge Y_{\beta+}$ has isotropy subgroup precisely H. Let $G_1 = G_u$ and $G_2 = G_y$, so that

$$H = G_1 \cap G_2.$$

Then there are essential subcomplexes $K_{\alpha'}, Y_{\beta'}$ such that $u \in K_{\alpha'}, y \in Y_{\beta'}, G^{\alpha'} = G_1$, and $G^{\beta'} = G_2$.

Since K is a $\pi(X)$-complex, it follows that $X_{\alpha'+}$ is an essential subcomplex of X_+ (*i.e.*, $X_{\alpha'}$ is a closed substratum). Therefore the subcomplex $X_{\alpha'} \times Y_{\beta'}$ has generic isotropy subgroup H. But $X_{\alpha'}$ and $Y_{\beta'}$ are components of X^{G_1} and Y^{G_2} respectively, and $G_1 \cap G_2 = H$ implies that $X_{\alpha'} \subset X_\alpha$ and $Y_{\beta'} \subset Y_\beta$. Therefore the component $X_\alpha \times Y_\beta$ of $(X \times Y)^H$ has a point with isotropy subgroup H, so that $(X_\alpha \times Y_\beta, H) \in \Pi(X \times Y)$ is an essential element and thus lies in $\pi(X \times Y)$.

(ii) The first paragraph in the proof of (i) also goes through in this case, so we shall use the notation from that part freely. However, in this case we are assuming K is a $\pi(X) \cup \widetilde{f}\pi(M)$-complex. Therefore **either** (a) $X_{\alpha'+}$ is an essential subcomplex of

X_+ **or else** (b) there is a closed substratum $M_{\gamma'}$ of M such that $G(\gamma) = G_1$ and $f(M_{\gamma'}) \subset X_{\alpha'}$. If (a) holds then $(X_\alpha \times Y_\beta, H) \in \pi(X \times Y)$ by the argument in part (i), so assume that (b) holds henceforth. Let M_γ be the component of M^H containing $M_{\gamma'}$. Then $M_\gamma \times Y_\beta$ is a component of $(M \times Y)^H$ and has a point with isotropy subgroup H, so that $M_\gamma \times Y_\beta \in \Pi(M \times Y)$. But now we have that $f \times \mathbf{id}_Y$ maps $M_\gamma \times Y_\beta$ into $X_\alpha \times Y_\beta$. For X_α is the unique component of X^H containing $X_{\alpha'}$, M_γ is the unique component of M^H containing $M_{\gamma'}$, and $f(M_{\gamma'}) \subset X_{\alpha'}$, so that $f(M_\gamma) \subset X_\alpha$ by continuity considerations. Therefore we have

$$(X_\alpha \times Y_\beta, H) = (f \times \mathbf{id}_Y)^{\sim}(M_\gamma \times Y_\beta, H),$$

where $(M_\gamma \times Y_\beta, H)$ is an essential element of $\Pi(M \times Y)$. In other words, if (b) holds then $(X_\alpha \times Y_{\beta'} H) \in (f \times \mathbf{id}_Y)^{\sim} \pi(M \times Y)$.■

Having established that the product construction $\mathbf{P}_Y$ is always defined, we show next that it is additive.

PROOF OF 2.2: (*Additivity of* $\mathbf{P}_Y$) Let $(f_1, -)$ and $(f_2, -)$ represent classes in $I^h(\lambda)$, and construct the sum cobordism $(F; f_1 \amalg f_2, \partial_- F)$ as in [**Do2**]. The most important features are that $\lambda(\partial_- F) \cong \lambda(F)$ and $\lambda(f_i)' \cong \lambda(F)$, for these determine the sum uniquely (compare 3.2). Consider the cobordism

$$(F \times \mathbf{id}_Y; f_1 \times \mathbf{id}_Y \amalg f_2 \times \mathbf{id}_Y, \partial_- F \times \mathbf{id}_Y). \tag{8.10}$$

By the characterization of sums in 3.2 it follows that

$$[-\partial_- F \times \mathbf{id}_Y] = [f_1 \times \mathbf{id}_Y] + [f_2 \times \mathbf{id}_Y],$$

which translates to

$$\mathbf{P}_Y[-\partial_- F] = \mathbf{P}_Y[f_1] + \mathbf{P}_Y[f_2],$$

which in turn implies

$$\mathbf{P}_Y([f_1] + [f_2]) = \mathbf{P}_Y[f_1] + \mathbf{P}_Y[f_2],$$

so that $\mathbf{P}_Y$ is in fact a homomorphism.■

Proofs of periodicity theorems

Theorems 2.4 and 2.5 can be derived from Theorem 2.3 in the same way that Theorems III.2.8 and III.2.9 are derived from Theorem III.2.7. Therefore it will suffice to prove Theorem 2.3.

PROOF OF THEOREM 2.3: We shall assume that the indexing data $\lambda(f)$ has $\dim M = \dim X = 2q$, and we shall work simultaneously on proving 2.7 for $I(\lambda)$ and $I(\mathbf{S}^{-1}\lambda)$. If $\lambda(f)$ involves odd-dimensional manifolds we can take $\lambda_1 = S\lambda$ and go through the entire proof for λ_1 (noting at the end that $\lambda = \mathbf{S}^{-1}\lambda_1$).

STEP 1. *Proof that* $\mathbf{P}_Y$ *induces an isomorphism from* $I^h_\Omega(\lambda)$ *to* $I^h_\Omega(\lambda \times \text{data}(Y))$ *and that* $\chi_G(Y)_*$ *induces an isomorphism on* $L^B_{2q-1}(\mathbf{Z}[G])$.

We shall prove the statement about $L^B_*(\mathbf{Z}[G])$ first. The crucial insight is a consequence of the following facts.

(8.10). *If G has odd order and K is a finite G-CW complex such that $\chi(K^H)$ is even for all $H \subset G$, then $\chi_G(K)$ is divisible by 2 in the Burnside ring.*

(8.11). *Let G be a group of odd order, and let A_2 be the 2-primary component of the finite group $\widetilde{K}_0(\mathbb{Z}[G])$. If Y is a periodicity manifold, the multiplication by the element $\chi_G(Y)$ in the Burnside ring $\mathbf{\Omega}(G)$ induces an isomorphism of A_2.*

PROOFS OF (8.10) AND (8.11): To prove the first statement, note that it suffices to consider finite virtual G-sets; for such sets the result follows by a simple counting argument (this uses the fact that $|G|$ is odd). To prove the second statement, note that $\chi(Y^H)$ is always odd because the signature of Y^H is always 1. Therefore $\chi_G(Y) = 1+2z$ for some $z \in \mathbf{\Omega}(G)$. Since any endomorphism of $\widetilde{K}_0(\mathbb{Z}[G])$ maps A_2 into itself, it follows that $z(A_2) \subset A_2$. But $2z$ is then a nilpotent endomorphism of A_2, and therefore $\chi_G(Y) = 1 + 2z$ must be an isomorphism.∎

FIRST STEP IN THE PROOF OF THEOREM 2.3. (CONTINUED): By preperiodicity we know that $\chi_G(Y)$ maps $B = B_0(G, \lambda) \subset \widetilde{K}_0(\mathbb{Z}[G])$ into itself, and therefore we also know that $\chi_G(Y)$ maps the 2-primary component $B_2 = A_2 \cap B$ into itself. Since $\chi_G(Y)$ is bijective on A_2, the induced self-map of B_2 is automatically injective. However, the finiteness of B_2 now implies that $\chi_G(Y)$ must also be bijective on B_2.

Next observe that

$$H_*(\mathbb{Z}_2; B_2) \cong H_*(\mathbb{Z}_2; B)$$

because $H_*(\mathbb{Z}_2; C) = 0$ if C has odd order. It follows that $\chi_G(Y)_*$ induces an automorphism of $H_*(\mathbb{Z}_2; B)$. Since the map CLS : $L^B_{2q+1}(\mathbb{Z}[G]) \to H_{2q}(\mathbb{Z}_2; B)$ is a monomorphism and $\chi_G(Y)_*(\text{Image CLS}) \subset \text{Image CLS}$ by 7.5 and the proof of 5.1, it follows that $\chi_G(Y)_*$ induces an injective self-map of L^B_{2q+1}; since the latter is finite, the self-map must also be surjective.

It remains to prove that $\mathbf{P}_Y$ is bijective on $I^h_\Omega(G; \lambda)$. For this we must use the commutative diagram of Theorem 7.6. Let

$$\mathsf{H} : H_{2m}(\mathbb{Z}_2; B) \to L^h_{2m}(\mathbb{Z}[G])$$

be the hyperbolization map of [**Ba2**, p. 1397]. By the Ranicki sequence the image of H is the kernel of the forgetful map ι_{hB} from $L^h_{2m} \to L^B_{2m}$.

We prove first that $\mathbf{P}_Y$ is injective. Suppose that $\mathbf{P}_Y(z) = 0$. Then

$$0 = \text{CLS}\, \sigma_0' \mathbf{P}_Y(z) = \chi_G(Y)_* \text{CLS}\, \sigma_0(z),$$

and since $\chi_G(Y)_*$ is bijective by (8.11) this means that CLS $\sigma_0(z) = 0$. By the exactness of the sequence in Theorem 7.6 it follows that $z = \gamma_0(w)$ for some $w \in L^h_{2q}(\mathbb{Z}[G])$. Therefore

$$0 = \sigma_0' \mathbf{P}_Y(z) = \sigma_0' \mathbf{P}_Y \gamma_0(w) = \sigma_0' \gamma_0' \mu_Y(w) = \sigma_0 \mu_Y(w)$$

so that $\mu_Y(w) = \mathsf{H}(v)$ for some v. But μ_Y is essentially the identity map because $\text{Witt}(G, Y) = 1$ and Yoshida's result implies that μ_Y depends only on the Witt class. Therefore $w = \mathsf{H}(v)$, so that $0 = \iota_{hB}(w) = \sigma_0 \gamma_0(w) = \sigma_0(z)$, and therefore $z = 0$ because σ_0 is injective.

To prove $\mathbf{P}_Y$ is surjective, first notice that a diagram chase based upon the diagram in 7.6 and the bijectivity of μ_Y shows that $\sigma_0'(\text{Image } \mathbf{P}_Y)$ contains Image ι'. Therefore it suffices to show that $\text{CLS}\,\sigma_0'\mathbf{P}_Y = \chi_G(Y)_*\text{CLS}\,\sigma_0$ is surjective. But $\chi_G(Y)_*$ is surjective by (8.11), the map CLS is surjective by (5.2), and σ_0 is bijective by 3.4 and 5.1. Since a composite of surjective maps is surjective, it follows that $\text{CLS}\,\sigma_0'\mathbf{P}_Y$ is surjective, and therefore $\mathbf{P}_Y$ is also surjective.■

SECOND STEP IN THE PROOF OF THEOREM 2.3: We shall prove the following inductive statement:

Let Σ be closed and G-invariant, and let α be minimal in $\pi(X; \mathfrak{P}_G) - \Sigma$. Assume that α is not the maximal closed stratum. Suppose that $\mathbf{P}_Y$ induces an isomorphism on $K^h_{\Sigma \cup G\{\alpha\}}(\lambda)$ and $\chi_G(Y)_*$ induces an isomorphism on $\mathcal{Z}_{\Sigma\cup\{\alpha\}}(\lambda)$. Then $\mathbf{P}_Y$ induces an isomorphism on $K^h_\Sigma(\lambda)$ and $\chi_G(Y)_*$ induces an isomorphism on $\mathcal{Z}_\Sigma(\lambda)$.

Theorem 2.3 will be an immediate consequence of Step 1, Step 2, and induction on the number of elements of $\pi(X; \mathfrak{P}_G) - \Sigma$ (after a finite number of inductive steps one reaches the case $\Sigma = \varnothing$).

The statement regarding the subgroups $\mathcal{Z}_Q(\lambda)$ is really not an inductive argument. By 7.9 we know that $\chi_G(Y)_*$ maps $\mathcal{Z}_Q(\lambda)$ into itself, and this map must be injective because $\chi_G(Y)_*$ is a monomorphism by (8.11) and $\mathcal{Z}_Q(\lambda)$ is finite for all Q.

To prove the assertion regarding K^h_Σ, we combine 3.4, the exact sequence of 6.4, and the results on Section 7 to obtain the following diagram with exact columns:

$$
\begin{array}{ccc}
0 & & 0 \\
\downarrow & & \downarrow \\
K^h_{\Sigma \cup G\{\alpha\}}(\lambda) & \xrightarrow{\mathbf{P}_Y(\Sigma \cup G\{\alpha\})} & K^h_{\Sigma \cup G\{\alpha\}}(\lambda \times \mathrm{data}\,(Y)) \\
\downarrow & & \downarrow \\
K^h_{\Sigma}(\lambda) & \xrightarrow{\mathbf{P}_Y(\Sigma))} & K^h_{\Sigma}(\lambda \times \mathrm{data}\,(Y)) \\
\sigma_\alpha \downarrow & & \sigma'_\alpha \downarrow \\
L^h_{\dim \alpha}(\mathbb{Z}_{(\alpha)}[W_\alpha]) & \xrightarrow{\mu_\alpha} & L^h_{\dim \alpha + \dim Y^H}(\mathbb{Z}_{(\alpha)}[W_\alpha]) \\
\theta_\alpha \downarrow & & \theta'_\alpha \downarrow \\
L^B_{2q-1}(\mathbb{Z}[G])/\mathcal{Z}_{\Sigma \cup G\{\alpha\}}(\lambda) & \xrightarrow{\chi_G(Y)_*} & L^B_{2q+\dim Y-1}(\mathbb{Z}[G])/\mathcal{Z}_{\Sigma \cup G\{\alpha\}}(\lambda \times \mathrm{data}\,(Y)) \\
\downarrow & & \downarrow \\
L^B_{2q-1}(\mathbb{Z}[G])/\mathcal{Z}_{\Sigma}(\lambda) & \xrightarrow{\chi_G(Y)_*} & L^B_{2q+\dim Y-1}(\mathbb{Z}[G])/\mathcal{Z}_{\Sigma}(\lambda \times \mathrm{data}\,(Y)) \\
\downarrow & & \downarrow \\
0 & & 0
\end{array}
$$

We know that the map $\mathbf{P}_Y(\Sigma \cup G\{\alpha\})$ is an isomorphism and the maps $\chi_G(Y)_*$ are isomorphisms. Once again the Witt invariant of Y^H is the unit in $Witt_+(\mathbb{Z}[W_\alpha])$ because Y is a periodicity manifold, and therefore the results on twisted products in [**Yo**] and Chapters III–IV imply as before that the maps μ_α are isomorphisms. Thus all the vertical arrows in the diagram except $\mathbf{P}_Y(\Sigma)$ are known to represent isomorphisms. By the Five Lemma, the maps $\mathbf{P}_Y(\Sigma)$ must also be isomorphisms.■

Remarks on other theories

There are analogs of the periodicity theorems for several theories related to I^h. For the record we shall mention some examples; a reader who wishes to verify the periodicity results in these cases should be able to do so by modifying the discussion of this chapter at the appropriate points.

The theories $I^{h(R)}$ for equivariant surgery up to R-local pseudoequivalence as discussed in [**DP1**, Appendix A]. These theories were discussed in (II.2.5) and Subsection II.3D, and special cases were considered more specifically in Section 4 of this chapter. Theorems 2.3–2.5 remain valid if I^h is replaced by $I^{h(R)}$, and in fact the methods of this chapter go through with only minimal changes.

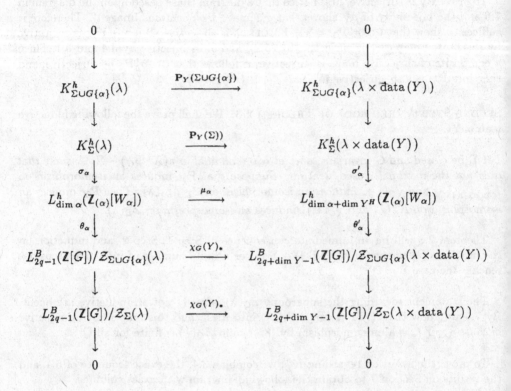

The theory $I^{h/Sm}$ with Smith bundle data instead of bundle data as in Section 1. This theory is introduced in [**PR**], and the results of that paper show that I^h and $I^{h/Sm}$ are very similar in many respects. The main difference is that the stable G-vector bundle map $b_0 : T(M) \to E(f^*\xi)$ is replaced by a so-called **completed** stable vector bundle map $\hat{b}_0 : T(M) \times_G EG \to E(\hat{f}^*\xi)$ where EG is the total space of the universal principal G-bundle over the classifying space BG, ξ is a vector bundle over the Borel construction $X \times_G EG$, and $\hat{f} := f \times_G \mathbf{id}_{EG}$. The techniques of [**PR**] and this chapter show that the stepwise surgery obstructions for the theories $I^{h/Sm}$ and I^h are the same, so periodicity theorems in this case follow immediately.

A variant $I^{h\Gamma}$ of I^h with a weaker notion of equivalence as in [**DS**]. The theory $I^{h\Gamma}$ is analogous to I^h but only requires homology surgery on the nonmaximal closed substrata. More precisely, one wants a map $f : M \to X$ such that the maps $f_\alpha : M_\alpha \to X_\alpha$ induce isomorphisms of homology with local coefficients in $R_\alpha[W_\alpha]$ for all *nonmaximal* closed substrata and f itself is a homotopy equivalence (but, as usual, not necessarily a G-homotopy equivalence). Two advantages of this theory are (i) it applies to simply connected G-manifolds whose nonmaximal closed substrata are not necessarily simply connected, (ii) it might be useful in situations where 3-dimensional closed substrata are present because homology surgery works in dimension 3. In such a theory one encounters stepwise obstructions in Cappell-Shaneson Γ-groups on the nonmaximal closed substrata and a final obstruction in $L_*^B(\mathbf{Z}[G])$ on the maximal closed substratum. Special cases of such a theory for the group $\mathbf{Z}_2$ are considered in [**DS**]. Periodicity theorems for the theory $I^{h\Gamma}$ can be obtained by combining the methods of this chapter with the results of Section IV.5 on twisted products for Cappell-Shaneson Γ-groups.

9. Appendix: A result on Wall groups

The purpose of this section is to prove the following result:

THEOREM 9.1. *Let G be an odd order group, and let R be a subring of the rationals. Then $L^h_{2q+1}(R[G]) = 0$ for all q.*

Special cases of Theorem 9.1 have been known for twenty years. A basic and well known theorem on Wall groups states that $L^h_{2q+1}(\mathbf{Z}[G]) = L^s_{2q+1}(\mathbf{Z}[G]) = 0$ if G has odd order (compare [**Ba1, W2**]). Numerous other results of this sort are also in print. For example, if G is a finite group then $L^h_1(\mathbf{Q}[\pi]) = 0$ (compare Connolly [**Co**]). Finally, if G is a p-group, then the vanishing of $L^h_{2q+1}(\mathbf{Z}_{(p)}[G])$ is shown in [**DP2**, Section 3]. We shall derive 9.1 as an elementary consequence of $L^h_{2q+1}(\mathbf{Z}[G]) = 0$, Connolly's results on $L^h_{2q+1}(\mathbf{Q}[G])$ in [**Co**], and the long exact sequence associated to the change of coefficients map $R[G] \to \mathbf{Q}[G]$; this localization sequence is discussed in papers of W. Pardon [**Pa1-2**], and G. Carlsson and R. J. Milgram [**CM**].

PROOF OF 9.1: The localization sequence for $\mathbf{Z}[G] \to \mathbf{Q}[G]$ contains an exact subsequence of the form

$$\bigoplus_{p \text{ prime}} T_p \to L^h_{2q+1}(\mathbf{Z}[G]) \to L^h_{2q+1}(\mathbf{Q}[G]), \tag{9.2}$$

where $\bigoplus$ denotes infinite direct sum; the groups T_p are known to be torsion groups. If R is an arbitrary subring of the rationals, then one has a similar sequence except that the sum $\oplus T_p$ is taken over all primes that are not units in R. These localization sequences are natural with respect to the inclusion $\mathbf{Z} \subset R$; specifically, there is a commutative ladder with exact rows as follows:

$$\begin{array}{ccccc} T'' \oplus T' & \xrightarrow{\partial} & L^h_{2q+1}(\mathbf{Z}[G]) & \longrightarrow & L^h_{2q+1}(\mathbf{Q}[G]) \\ \text{proj.}\downarrow & & \downarrow & & \downarrow = \\ T' & \xrightarrow{\partial'} & L^h_{2q+1}(R[G]) & \longrightarrow & L^h_{2q+1}(\mathbf{Q}[G]) \end{array} \tag{9.3}$$

Here T' is the sum of all T_p where p is a prime such that $p^{-1} \notin R$ and T'' is the sum of all T_p where p is a prime such that $p^{-1} \in R$.

We claim that $L^h_{2q+1}(\mathbf{Q}[G]) = 0$ if G has odd order. This result has been understood for some time, and in fact it is implicit in [**PasPet**, Section 4], but for the sake of completeness here is a proof. A result of Connolly states that $L^h_1(\mathbf{Q}[G]) \cong 2L^h_3(\mathbf{Q}[G]) \cong 0$ for any finite group G (see [**Co**, Theorem B]). In fact, by [**Co**, Theorem 3.13] and the splitting

$$\mathbf{Q}[\mathbf{Z}_{2s+1}] \cong \bigoplus_{\substack{0<q \\ q|2s+1}} \mathbf{Q}[\exp(2\pi i/q)]$$

we know $L^h_3(\mathbf{Q}[G]) = 0$ if G is cyclic of odd order. One can now use a Dress induction argument as in one of

(1) the second paragraph following Theorem 1, page 293, of [**Dr**],
(2) Theorem 6.33 on pages 169–170 of [**tD2**], or
(3) Chapter IV, Section 4 of [**ACH**]

to conclude the same result for an arbitrary group G of odd order. The previously cited vanishing theorems now imply that (9.3) reduces to

$$\begin{array}{ccccc} T'' \oplus T' & \longrightarrow & 0 & \longrightarrow & 0 \\ \downarrow & & \downarrow & & \downarrow \\ T' & \longrightarrow & L^h_{2q+1}(R[G]) & \longrightarrow & 0 \end{array}$$

and therefore a simple diagram chase shows that $L^h_{2q+1}(R[G]) = 0$.■

References for Chapter V

[ACH] J. P. Alexander, P. E. Conner, and G. C. Hamrick, "Odd Order Group Actions and Witt Classification of Inner Products," Lecture Notes in Mathematics Vol. 625, Springer, Berlin-Heidelberg-New York, 1977.

[Ba1] A. Bak, *Odd dimension surgery groups of odd order groups vanish*, Topology **14** (1975), 367–374.

[Ba2] ———, *The computation of even dimensional surgery groups of odd torsion groups*, Comm. in Algebra **6** (1978), 1393–1458.

[CS] S. Cappell and J. Shaneson, *The codimension two placement problem and homology equivalent manifolds*, Ann. of Math. **99** (1974), 277–348.

[CM] G. Carlsson and R. J. Milgram, *Some exact sequences in the theory of Hermitian forms*, J. Pure and Appl. Algebra **18** (1980), 233–252.

[Co] F. Connolly, *Linking numbers and surgery*, Topology **12** (1973), 389–409.

[DH] M. Davis and W. C. Hsiang, *Concordance classes of regular $U(n)$ and $Sp(n)$ actions on homotopy spheres*, Ann. of Math. **105** (1977), 325–341.

[DHM] M. Davis, W. C. Hsiang, and J. Morgan, *Concordance of regular $O(n)$-actions on homotopy spheres*, Acta Math. **144** (1980), 153–221.

[tD1] T. tom Dieck, *Orbittypen und äquivariante Homologie II*, Arch. Math. (Basel) **26** (1975), 650–662.

[tD2] ———, "Transformation Groups and Representation Theory," Lecture Notes in Mathematics Vol. 766, Springer, Berlin-Heidelberg-New York, 1979.

[Do1] K. H. Dovermann, "Addition of equivariant surgery obstructions," Ph.D. Thesis, Rutgers University, 1978 *(Available from University Microfilms, Ann Arbor, Mich.: Order Number DEL79-10380.)*—Summarized in Dissertation Abstracts International **39** (1978/1979), 5406..

[Do2] ———, *Addition of equivariant surgery obstructions*, in "Algebraic Topology, Waterloo 1978 (Conference Proceedings)," Lecture Notes in Mathematics Vol. 741, Springer, Berlin-Heidelberg-New York, 1979, pp. 244-271.

[Do3] ———, *Dihedral group actions on homotopy spheres*, in "Current Trends in Algebraic Topology (Conference Proceedings, Univ. of Western Ontario, 1981)," Canad. Math. Soc. Conf. Proc. **2 Pt. 2** (1982), pp. 67–88.

[DP1] K. H. Dovermann and T. Petrie, *G-Surgery II*, Mem. Amer. Math. Soc. **37** (1982), No. 260.

[DP2] ———, *An induction theorem for equivariant surgery (G-Surgery III)*, Amer. J. Math. **105** (1983), 1369–1403.

[DPS] K. H. Dovermann, T. Petrie, and R. Schultz, *Transformation groups and fixed point data*, Proceedings of the A.M.S. Summer Research Conference on Group Actions (Boulder, Colorado, 1983), Contemp. Math. 36 (1985), 161–191.

[DS] K. H. Dovermann and R. Schultz, *Surgery on involutions with middle dimensional fixed point set*, Pac. J. Math. **130** (1988), 275–297.

[Dr] A. Dress, *Induction and structure theorems for orthogonal representations of finite groups*, Ann. of Math. **102** (1975), 291–325.

[MP] M. Masuda and T. Petrie, *Lectures on transformation groups and Smith equivalence*, Proceedings of the A.M.S. Summer Research Conference on Group Actions (Boulder, Colorado, 1983), Contemp. Math. 36 (1985), 193–244.

[Mrg] J. Morgan, *A Product Formula for Surgery Obstructions*, Mem. Amer. Math. Soc. **14** (1978), No. 201.

[OL] R. Oliver, *G–actions on disks and permutation representations I*, J. Algebra **50** (1978), 44–62.

[OP] R. Oliver and T. Petrie, *$G-CW$ surgery and $\widetilde{K}_0(\mathbf{Z}[G])$*, Math. Zeit. **179** (1982), 11–42.

[Pa1] W. Pardon, *Local Surgery and the Exact Sequence of a Localization*, Memoirs Amer. Math. Soc. **12** (1977), No. 196.

[Pa2] ———, *The exact sequence of a localization for Witt groups II: Numerical invariants of odd–dimensional surgery obstructions*, Pac. J. Math. **102** (1982), 123–170.

[PasPet] D. S. Passman and T. Petrie, *Surgery with coefficients in a field*, Ann. of Math. **95** (1972), 385–405.

[Pe1] T. Petrie, *G–maps and the projective class group*, Comment. Math. Helv. **51** (1976), 611–626.

[Pe2] ———, *G-Surgery I-A survey*, in "Algebraic and Geometric Topology (Conference Proceedings, Santa Barbara, 1977)," Lecture Notes in Mathematics Vol. 644, Springer, Berlin-Heidelberg-New York, 1978, pp. 197–223.

[PR] T. Petrie and J. Randall, *Spherical isotropy representations*, I. H. E. S. Publ. Math. **62** (1985), 5–40.

[Ra] A. Ranicki, *The algebraic theory of surgery I: Foundations*, Proc. London Math. Soc. (3) **40** (1980), 87–192.

[St] R. Stong, *Determination of $H^*(BO(k,\ldots,\infty);\mathbf{Z}_2)$ and $H^*(BU(k,\ldots,\infty);\mathbf{Z}_2)$*, Trans. Amer. Math. Soc. **107** (1963), 526–544.

[Sw] R. Swan, *Induced representations and projective modules*, Ann. of Math. **71** (1960), 552–578.

[W0] C. T. C. Wall, *Finiteness conditions for CW complexes*, Ann. of Math. **81** (1965), 56–69.

[W1] ———, "Surgery on Compact Manifolds," London Math. Soc. Monographs Vol. 1, Academic Press, London and New York, 1970.

[W2] ———, *Classification of Hermitian forms: VI—Group rings*, Ann. of Math. **103** (1976), 1–80.

[Ya] M. Yan, *Periodicity in equivariant surgery and applications*, Ph. D. Thesis, University of Chicago, in preparation.

[Yo] T. Yoshida, *Surgery obstructions of twisted products*, J. Math. Okayama Univ. 24 (1982), 73–97.

INDEX TO NUMBERED ITEMS

Some numbered items that do not affect later portions of the book are not listed. Entries marked with an asterisk are amended versions of items appearing earlier.

I.1.1 10
I.1.2 10
I.2.1 14
I.2.2 15
I.2.3 15
I.2.5 16
I.2.6 16
I.2.7 17
I.4.1 20
I.4.2 20
I.4.3 21
I.4.4 22
I.5.1 23
I.5.1* 25
I.5.2 26
I.5.3 26
I.5.4 27
I.5.5 27
I.6.1 29
I.6.2 29
I.6.3 30
....

II.1.0 39
II.1.0* 39
II.1.1 40
II.1.2 41
II.1.3 43
II.1.4 43
II.1.5 43
II.1.6 44
II.2.0 46
II.2.1 47
II.2.2 48
II.2.3 48
II.2.4 48
II.2.5 49
II.2.6a 50
II.2.6b 50
II.3.1.A 51
II.3.1.B 51
II.3.3 53
II.3.4 53
II.3.5 55
II.3.6 55
II.3.7 55
II.3.8 57
II.3.9 59
II.3.10 60
II.3.11 60
II.3.12 61
II.3.13 62
II.3.14 64
II.3.15 64
II.4.1 70
II.4.1 70
II.4.2 70
II.4.3 71
II.4.4 71
II.4.5 72
II.4.6 73
....

III.B1–2 84
III.1.1 84
III.1.2 85
III.2.1 86
III.2.2 87
III.2.3 87
III.2.4 87
III.2.5 88
III.2.5A 88
III.2.6 88
III.2.7 89
III.2.8 90
III.2.9 90
III.2.10 92
III.2.11 92
III.3.1 93
III.3.1.A 94
III.3.2 93
III.3.3 93
III.3.4 94
III.3.5 95
III.3.6 96
III.3.7 96
III.4.1 98
III.4.2 98
III.5.1 102
III.5.2 102
III.5.3 103
III.5.4 105
III.5.5 107
III.5.6 107
III.5.7 108
III.5.8 108
III.5.9 109
III.5.10 109
....

IV.1.1a–e 116
IV.1.2 117
IV.1.3 118
IV.1.4 118
IV.1.5 118
IV.1.6 119
IV.1.7 119
IV.1.8 119
IV.1.9 119
IV.1.10 120
IV.1.11 121
IV.2.1 121
IV.2.2a–c 122
IV.3.1 125
IV.3.1* 126
IV.3.2 126
IV.3.3 127
IV.4.1a–b 129
IV.4.2 129
IV.4.3 129
IV.4.4 129
IV.4.5 130
IV.4.6 130

IV.4.7.........130
IV.4.8.........130
IV.4.9.........131
IV.5.1.........132
IV.5.2.........132
IV.6.1.........133
IV.6.1*........134
IV.6.2.........134
IV.6.2*........134
IV.6.3.........136
IV.6.4.........136
....

V.1.0.A–D.....145
V.1.0.D*......147
V.1.0.E........147
V.2.1..........148
V.2.2..........149
V.2.3..........150
V.2.4..........150
V.2.5..........151
V.2.6..........151
V.3.1..........152
V.3.2i–vi......153
V.3.3..........154
V.3.4..........154
V.3.5..........155
V.3.6..........157
V.3.7..........157
V.3.8..........159
V.4.1..........161
V.4.2..........161
V.4.3..........162
V.4.4..........162
V.4.5..........162
V.4.6..........162
V.4.7..........162
V.4.8..........162
V.4.9..........162
V.5.1..........164
V.5.2..........164
V.5.3..........164
V.5.4..........165
V.5.5..........166
V.5.6..........167
V.6.0..........171
V.6.1..........171
V.6.2..........172
V.6.3..........173
V.6.4..........173
V.6.5..........174
V.7.1..........175
V.7.2..........175
V.7.3..........176
V.7.4..........176
V.7.5..........176
V.7.6..........177
V.7.7..........178
V.7.8..........178
V.7.9..........179
V.7.10.........179
V.7.11.........180
V.8.1..........181
V.8.2..........181
V.8.3..........181
V.8.4..........181
V.8.5..........181
V.8.6..........181
V.8.7-*d*........181
V.8.8..........181
V.8.9-*d*........181
V.8.10.........184
V.8.11.........184
V.9.1..........187
V.9.2..........188
V.9.3..........188

INDEX TO NOTATION

The entries in this index are basically arranged alphabetically, with Roman letters followed by Greek letters and other symbols at the end. Brief explanations of entries are included whenever possible. A small box (□) in the right hand column indicates that no specific page references seemed appropriate.

SYMBOL		PAGE
A	(self-conjugate subgroup of $K_i(\mathbb{Z}[G])$)	129
$\mathrm{Aut}(x,H)$	(set of self-maps in equivariant fundamental groupoid)	135
B	(abbreviation for $B_0(G,\mathcal{F}(\lambda))$)	156
B	(indeterminacy subgroup for stepwise obstructions)	27
$B(I^a)$	(specialization of preceding to theory I^a)	60
$B(G,n)$	(classifying space for $G-\mathbb{R}^n$ bundles)	58
$\mathbb{B}_n$	(equivariant fundamental groupoid of $B(G,n)$)	58
b_0	(stable vector bundle map)	21
$\hat{b}_0$	(completed stable vector bundle map)	187
$B_0(G,\mathcal{F}(\lambda))$	(Oliver-Petrie subgroup of $\widetilde{K}_0(\mathbb{Z}[G])$)	156
$B_0(G,\lambda)$	(abbreviation for preceding object)	156
c	(unstable vector bundle system map)	21
CAT	(any of the standard manifold categories)	46
$\mathrm{Cay}\mathbb{P}^2$	(Cayley projective plane)	94
$C_G(W)$	(orthogonal centralizer of the orthogonal representation W)	13
CLASS[1,0]	(equivalence class of [1,0])	72
CLS	(image of stable class of projective module)	164
$\mathrm{CLS}^\#$	(quotient of preceding)	179
$\mathrm{Cone}(f)$	(mapping cone of f)	99
$\mathbb{CP}^k$	(complex projective k-space)	3
Cyl	(mapping cylinder)	39
$\mathbb{CP}^{2n}\uparrow G$	(product of copies of $\mathbb{CP}^{2n}$ with group action by permuting factors)	3, 92
$\mathrm{data}(Y)$	(reference data for identity on Y)	17
D^k	(k-dimensional disk)	□
D^k_+	(northern hempsphere of D^k)	□
D^k_-	(southern hempsphere of D^k)	□
$D(G)$	(defect subgroup of $\widetilde{K}_0(\mathbb{Z}[G])$)	50, 129
$D(W)$	(unit disk in orthogonal representation W)	12
d_X	(dimension data for X)	16

$\exp(2\pi i/q)$ .... (standard primitive q-th root of unity) 188
E_β .... (fundamental group of orbit space substratum associated to β) 21, 49

$\check{f}$ .... (map of geometric posets associated to f) 17
${}_1\check{f}$ .... (map of geometric-posets-with-fundamental-group-data associated to f) 66
$\tilde{f}$ .... (map of complete geometric posets associated to f) 147
$\check{f}_{\mathcal{H}}$ .... (analogous map of $\mathcal{H}$-restricted geometric posets) 143
$\mathcal{F}$ .... (subset of complete G-poset) 145
$\mathcal{F}^a$ .... (preferred family of G-manifolds for theory I^a) 46
$\mathrm{FGPD}^G(X)$ .... (equivariant fundamental groupoid) 135
$\mathcal{F}_f$ .... (specific subset of complete G-poset) 161
$\mathcal{F}_f|\alpha$ .... (analog for restriction to substratum α) 161
$f \circ g$ .... (composition of functions) .... □
$F_G(W)$ .... (space of equivariant self-maps of the unit sphere in W) 13
F_G .... (stabilized version of the preceding object) 109
$\mathcal{F}^h$ .... (preferred family of G-manifolds for I^h) 144
$\mathcal{F}^{h(R)}$ (preferred family of G-manifolds for $I^{h(R)}$ 144
$\mathcal{F}^{ht,CAT}$ .... (preferred family of G-manifolds for $I^{ht,CAT}$) 47
$F^*_{H,V}$ .... (closure of the orbit space stratum $M^*_{H,V}$) 38
$\mathrm{Fix}\,(H, M)$ .... (fixed set of H in M) 14
$\widehat{f}_\alpha$ .... (canonical extension of a stratified map to the compactification of an open substratum) 44
$f^{\#}_\beta$ .... (induced map on open substratum given by β) 136

$|G|$ .... (cardinal number of G) 1
G_x .... (isotropy subgroup of G at the point x) 14
G^α .... (generic isotropy subgroup for α) 16
$G\{\alpha\}$ .... (G-orbit of α) 26
$\mathcal{G}^p_1(G)$ .... (specific family of subgroups of G) 161

H .... (hyperbolization map) 184
$\widehat{H}_*$ .... (Tate homology) 55
$\widehat{H}^*$ .... (Tate cohomology) 61
$\mathcal{H}^a$ .... (preferred family of subgroups for theory I^a) 46
$\mathcal{H}^h$ .... (preferred family of subgroups for theory I^h) 57, 145
$\mathcal{H}^{h(R)}$ .... (preferred family of subgroups for theory $I^{h(R)}$) 57, 145
$\mathcal{H}^{h|\alpha}$ .... (preferred family of subgroups for theory $I^{h|\alpha}$) 161
$(h, DIFF)$ .... (index for specific equivariant surgery theory) 57
$(ht, DIFF)$ .... (index for specific equivariant surgery theory) 22

I .... (unit interval) 1
I^a .... (some particular equivariant surgery theory) 46

$I^a(G;X)$ .... (equivariant surgery groups for the preceding theory) 46
$I^a(G;X;\Sigma)$ .... (same, Σ-adjusted versions) 46
$I^a(G;X;\Sigma\subset\Sigma')$ ... (same, relatively $(\Sigma\subset\Sigma')$-adjusted versions) 46
$\mathbb{I}_0^{c,CAT}$ .... (zeroth space in Quinn-Ranicki surgery spectrum for Dovermann-Rothenberg theory) 56
$I^{a,\infty}$ .... (periodic stabilization of I^a) 102
$I^{a,\infty}(G;X)$ .... (same, more explicit notation) 102
$I^{c,CAT}$ .... (some choice of Dovermann-Rothenberg equivariant surgery theory) 47, 48
$\mathbf{id}_Y$ .... (identity map for Y) 80
I^h .... (theory of equivariant surgery up to equivariant pseudoequivalence) 49, 148
$I^h(G;\lambda;\gamma,\delta)$ .... (full notation for surgery obstruction groups in I^h) 148
$I^h(G;\lambda;\gamma,\nabla)$ .... (same as before; ∇ and δ determine each other) 148
$I^{h(R)}$ .... (analog of I^h with coefficients in R) 49, 148
$I^{h/Sm}$ .... (analog of I^h with Smith bundle data) 187
$I^{h|\alpha}$ .... (abbreviation for next entry) 161
$I^{h|\alpha}(W_\alpha;\lambda|\alpha;\gamma|\alpha,\nabla|\alpha)$
.... (surgery groups for restriction to α substratum) 160, 161
$I^{h,\Gamma}$ .... (analog of I^h using Cappell-Shaneson homology Γ-surgery) 51, 187
$I^{ht,CAT}$ .... (Dovermann-Rothenberg equivariant surgery theory for CAT manifolds) 9, 22, 47
$I^{ht,DIFF}$ .... (Dovermann-Rothenberg smooth equivariant surgery theory) 9, 22
$I^{ht,DIFF,\infty}$ .... (periodic stabilization of the preceding object) 106
$I^{ht,DIFF}(G;\lambda)$ ..... (Dovermann-Rothenberg equivariant surgery obstruction group) 23
$I^{ht,DIFF}(G;\lambda;\Sigma)$ ... (Σ-adjusted Dovermann-Rothenberg equivariant surgery obstruction group) 23
$I^{ht,DIFF}(G;\lambda;\Sigma\subset\Sigma')$
.... (relatively $(\Sigma\subset\Sigma')$-adjusted analog of the previous group) 23
$I^{ht,PL}$ .... (piecewise linear analog of $I^{ht,DIFF}$ (*q.v.*)) 47
$I^{ht,TOP}$ .... (same, topological analog) 47
$I^{ht(R),CAT}$ .... (Dovermann-Rothenberg equivariant surgery theory with coefficients for CAT manifolds) 48, 133
$I^{s,CAT}$ .... (Dovermann-Rothenberg simple equivariant surgery theory for CAT manifolds) 47, 48
$I^{s,DIFF}$ .... (Dovermann-Rothenberg simple smooth equivariant surgery theory) 48
$I^{s,PL}$ .... (same, piecewise linear analog) 48
$I^{s,TOP}$ .... (same, possible topological analog) 48

$\mathcal{J}_k(G;X)$ .... (a specific set of Dovermann-Rothenberg normal maps) .. 104
J_Σ .... (forgetful map) 25

$\mathcal{K}$ .... (indeterminacy of stepwise obstruction) 32
K^h_Σ .... (abbreviation for next entry) 154
$K^h_\Sigma(G;\lambda;\gamma,\nabla)$ ...... (Σ-adjustable classes in $I^h(G;\lambda;\gamma,\nabla)$ 153
$K_*(f)$ .... (surgery kernel for f) 165
$\mathbf{KP}^n$ .... (quaternionic projective n-space) 94
$\widetilde{K}_0(R[G])$ .... (reduced projective class group) 5

$L^A(\mathbb{Z}[G],w)$ .... (intermediate Wall group for A) 129
$L^B(\mathbb{Z}[G],w)$ .... (same thing) 129, 156
$L_n^{BQ,h}(X)$ .... (Browder-Quinn surgery obstruction groups) 42
$L_n^{BQ,h}(X;\Sigma)$ .... (Σ-adjusted version of preceding
object) 42
$L_n^{BQ,h}(X;\Sigma\subset\Sigma')$ .. (relatively ($\Sigma\subset\Sigma'$)-adjusted version
of preceding object) 42
$L^{BQ,h(R)}$ .... (Browder-Quinn groups for surgery with
R-localized coefficients) 64
$L^{BQ,T(a)}$ .... (Browder-Quinn theory related to I^a) 61
$\mathbb{L}_0^{BQ,c}$ .... (zeroth space in Quinn-Ranicki surgery spectrum for
Browder-Quinn theory) 56
$\mathcal{L}^c$ .... ($\mathcal{L}^{c,CAT}$, where $CAT = DIFF$ or PL) 49
$\mathbb{L}_0^c$ .... (zeroth space in ordinary Quinn-Ranicki
surgery spectrum) 56
$\mathcal{L}^{c,CAT}(G;R_X)$ .... (Lück-Madsen equivariant
surgery obstruction group) 49
$\mathcal{L}^{c,CAT}(G;R_X;\Sigma)$ .. (Σ-adjusted
analog of the previous group) 49
$\mathcal{L}^{c,CAT}(G;R_X;\Sigma\subset\Sigma')$
.... (relatively ($\Sigma\subset\Sigma'$)-adjusted
analog of the previous group) 49
$L^{D(G)}(\mathbb{Z}[G],w)$ ..... (intermediate projective Wall group for $D(G)$) 57
$L^*_{G,X}(\mathbb{Z})$ .... (Yoshida's equivariant symmetric Wall group) 120
$L_n^{-i}(\mathbb{Z}[G],w)$ .... (subprojective Wall surgery obstruction group) 115, 130
$L_n^h(G,w)$ .... (homotopy Wall surgery obstruction group) 21
$L_n^h(\mathbb{Z}[G],w)$ .... (same as $L^h(G,w)$) .... □
$L_n^p(\mathbb{Z}[G],w)$ .... (projective Wall surgery obstruction group) 115
$L_n^s(G,w)$ .... (simple homotopy Wall surgery obstruction group) 48
$L_n^s(\mathbb{Z}[G],w)$ .... (same as $L^s(G,w)$) .... □
$L_n^{s(i)}(\mathbb{Z}[G],w)$ ...... (supersimple Wall surgery obstruction group) 131
${}^RL_*^X$ .... (Ranicki quadratic L-group) 115
${}^RL_X^*$ .... (Ranicki symmetric L-group) 119
$\ell(\alpha)$ .... (integer invariant of group action for closed
substratum α) 152

$\mathcal{L}\Gamma_n^{BQ,c}(G;R;\psi:\mathbb{Z}[\mathrm{FG}\mathrm{PD}^G(X)]\to A)$
.... (analogs of Lück-Madsen groups for Cappell-Shaneson Γ-surgery) 137
$\mathcal{L}\Gamma_n^{BQ,c}(-"-;\Sigma)$.... (adjusted analog of the previous object) 137
$\mathcal{L}\Gamma_n^{BQ,c}(-"-;\Sigma\subset\Sigma')$
.... (same but relatively adjusted) 137
$L^1=L^2$ .... (label for surgery-theoretic statement) 49
$L_n^{-\infty}(\mathbb{Z}[G],w)$ ...... (infinite subprojective Wall surgery obstruction group) .. 131

$\mathcal{M}$ .... (maximal order) 50
M_π .... (maximal order for $\mathbb{Z}[\pi]$) 129
$M_{(H)}$ .... (all points with isotropy subgroup conjuage to H) 14
$M_{(H,V)}$ .... (all points with isotropy subgroup conjuage to H and slice type V) 14
$M_{H,V}$ .... (same as above) 38
$M^*_{H,V}$ .... (image of above in the orbit space) 38
$M\langle\alpha\rangle$ .... (result of cutting and pasting) 11
M_α .... (functionally determined subcomplex) 146
$\widehat{M}_\alpha$ .... (canonical compactification of open stratum) 44

$NH(x)$ .... (subgroup of the normalizer of H, where $H=G_x$) 108
Nonsing M .... (nonsingular set of group action) 11
NonSing X^*_α .... (nonsingular part of orbit space stratum) 136
$N(X_\alpha)$ .... (subset of normalizer of G^α (*q.v.*)) 16
$N\!\uparrow\!\mathbb{Z}_2$ .... (special case of $X\!\uparrow\!G$) 96
$n(\lambda)$ .... (Oliver-Petrie dimension bound for certain complexes) ... 145

$\mathcal{O}_+(G)$ .... (specific, canonical G-poset) 146

P^* .... (dual/conjugate module to projective $\mathbb{Z}[G]$-module P) ... 164
$\mathfrak{P}_G$ .... (family of p-subgroups of G plus the trivial subgroup) ... 145
proj_X .... (projection onto the X factor in a product) .... □
$\mathbf{P}_Y$ .... (product map in equivariant surgery) 84
$\mathbf{P}_Y(\Sigma\subset\Sigma')$ .... (same thing for relatively adjusted groups) 97
$\mathbf{P}(\beta)$ .... (closed substratum associated to β) 103
$\mathbf{P}_\infty$ .... (periodic stabilization) 100

$\mathbb{Q}[\pi]$ .... (rational group ring) 129

REL .... (relative to, leaving the object unchanged) 11
RelSing C .... (relative singular set on closed substratum) 11
res $\mathbf{E}$ .... (restriction to $\mathbf{E}$) 144
$res(\alpha)$ or res_α .... (restriction to α substratum) 27, 162
$R[G]$ .... (group ring of G over the ring R) .... □
$R(G)$ .... (complex representation of G) 81

$\mathsf{R}[G]$ .... (same as above for $R = \mathsf{R}$, also describable as the regular representation of G over the reals) 91
(R_M, R_N, Φ) .... (dynamic Lück-Madsen geometric reference) 66
RP^k .... (real projective k-space) 68
R_X .... (simple Lück-Madsen geometric reference for X) 49

$S^{CAT,s}$ .... (surgery-theoretic CAT simple structure set) 11
$S^*_{H,V}$ .... (lower strata in the closure of the orbit space stratum $M^*_{H,V}$) 38
Sing M .... (singular set of group action) 11
S^k .... (standard k-sphere) .... □
$\mathsf{S}^{\pm k}$ .... (suspension or desuspension of indexing data, geometric references, etc.) 17
S^{k+W} .... (one point compactification of the representation $\mathsf{R}^k \oplus W$) 12
$S^{k+W}\langle\alpha\rangle$ .... (result of cutting and pasting on S^{k+W} via α) 12
$ST(M/G)$ .... (stratified tangent bundle of orbit space) 41
$S(W)$ .... (unit sphere in orthogonal representation W) 12
s_X .... (oriented slice data for X) 17
$|s_X(\cdots)|$ .... (unoriented slice data for X) 17
S^1 .... (unit circle) .... □

$\mathbf{T}$ .... (control/tubular data for a stratification) 72
$TFib$ .... (bundle of tangents along the fibers) 41
Th .... (Thom complex) 118
t_X .... (transport of identity map of X) 58
T_X .... (tubular neighborhood of stratum X) 69
$T_{X,Y}$ .... (intersection of previous object with Y) 69
$T(x; X)$ .... (tangent space of x in X) 68
$T_x(X)$ .... (same as above) 72
T_α .... (set of closed substrata defined by α) 103
T_0 .... (another object of the same type) 103

$\mathcal{UI}^{c,CAT}_k(G; X)$ ..... (unstable analogs of Dovermann-Rothenberg groups) 104
$\mathcal{UI}^{t,sp}_k(D(V), S(V))$. (stably tangential analog of the preceding object) 104

$V^\perp$ .... (orthogonal complement) 19

$Witt(G, X)$ .... (G-equivariant Witt group invariant) 184
$Witt_H(X)$ .... (H-equivariant Witt group invariant) 86
$Witt_+(R)$ .... (symmetric Witt group of ring R) 86
$Wh(\pi)$ .... (Whitehead group of π) 55
$\mathbf{W}_k(R, \Gamma\ etc.)$ ...... (Bak variant of Wall group) 28
$\mathcal{W}(X)$ .... (generalized (equivariant) Whitehead group) 61
w_X .... (orientation data for X) 16, 17

W_α or $W(\alpha)$ .... (quotient group $N(X_\alpha)/G^\alpha$ (*qq.v.*)) 16
$W_*^\chi(G,\mathbf{Z})$ .... (Witt groups for $\mathbf{Z}[G]$ with orientation character χ) 120

$[X]$ .... (fundamental orientation class) 119
$X\uparrow G$ .... (product space with group action by permuting the factors) 3, 81
X^g .... (fixed set of g in X) 17
X^H .... (fixed set of H in X) 17
$X\uparrow S$ .... (analog of $X\uparrow G$ with S a finite G-set) 92
X_Σ^* .... (union of all strata in Σ) 39

$Y\#$ .... (fixed set of a certain subgroup on space Y) 99

${}^w\mathbf{Z}$ .... (integers with coefficients twisted via w) 119
$\mathbf{Z}[\mathrm{FGPD}^G(X)]$ .... (equivariant fundamental groupoid ring) 135
$\mathbf{Z}_{(\ell)}$ .... (integers localized away from ℓ) 152
$\mathbf{Z}_n$ .... (integers mod n) .... □
$\mathbf{Z}/2$ .... (integers mod 2) 61
$\mathbf{Z}_{(\alpha)}$ .... (integers localized away from $\ell(\alpha)$) 152
$\mathcal{Z}(\lambda)$ .... (indeterminacy of I^h stepwise obstruction) 157
$\mathcal{Z}_\Sigma(\lambda)$ .... (analog for Σ-adjusted maps) 171

α .... (closed substratum) 16
$\alpha_{P(B)}$ .... (twisted product maps) 130

β .... (closed substratum) 17

γ .... (extra slice data for surgery up to pseudoequivalence) 144
$\gamma|\alpha$ .... (analog for restriction to substratum α) 161
Γ_n^c .... (abbreviation for $\Gamma_n^c(\varphi : B \to A)$) .... □
$\Gamma_n^c(\varphi : B \to A)$ .... (Cappell-Shaneson homology surgery obstruction groups for the local epimorphism φ) 66
$\Gamma_n^{BQ,c}$ .... (abbreviation for items listed below) .... □
$\Gamma_n^{BQ,c}(G; \Lambda_X; \psi : \mathbf{Z}[\mathrm{FGPD}^G(X)] \to A)$
.... (analogs of Browder-Quinn groups for Cappell-Shaneson Γ-surgery) 135
$\Gamma_n^{BQ,c}(-"-;\Sigma)$ .... (adjusted analog of the previous object) 136
$\Gamma_n^{BQ,c}(-"-;\Sigma\subset\Sigma')$ (same but relatively adjusted) 136
$\Gamma W_\beta(X_\beta)$ .... (certain set of group elements) 28
γ_β .... (inverse to stepwise surgery obstruction) 27, 61
(γ,δ) .... (additional data for addition in I^h) 143, 144

δ .... (extra slice data for surgery up to pseudoequivalence) 144
$\Delta(G,\mathcal{F})$ .... (Oliver-Petrie subset of $\Omega(G,\mathcal{F})$) 147
$\Delta^{h|\alpha}(W_\alpha,\mathcal{F}|\alpha)$ .... (analog for restriction to α substratum) 161
$\Delta(Z_\gamma)$ .... (Gap Hypothesis balance for substratum Z_γ) 96

$\theta^c_{X,\Sigma\subset\Sigma'}$ .... (map from Browder-Quinn theory to Lück-Madsen theory) 64

$\Theta^{G,s}_{k+W}$ .... (equivariant Kervaire-Milnor group) 12

$\Theta_{k+\varepsilon}$ .... (ordinary Kervaire-Milnor group) 13

θ_α .... (obstruction homomorphism) 171

Λ .... (simplification of Λ_X) 89

Λ_X .... (indexing data or geometric reference (*qq.v.*) for X) 84

λ .... (simplification of λ_X or $\lambda(f)$) 89

$\lambda(f)$ .... (functional indexing data) 142

λ_X .... (indexing data for X) 16

$\widehat{\lambda_X}$ .... (indexing data for pseudoequivalence surgery) 57

$\lambda_X|\alpha$ .... (analog of λ_X for restruction to substratum α) 161

$\mu^B(Y), \mu^c(Y)$ .... (Yoshida twisted product with Y) 136, 178

μ^h_Y .... (Yoshida twisted product with Y for L^h) 99

μ^s_Y .... (Yoshida twisted product with Y for L^s) 99

$\mu(\beta)$ .... (degree data for substratum β) 142

$\nu_{H,V}$ .... (tubular neighborhood projection) 39, 40

ν_X .... (stable normal bundle/fiber space) 119

Ξ .... (vector bundle system) 18

$\xi\downarrow X$ .... (vector bundle over X) 58

$(\xi\oplus etc.)|_y$ .... (fiber of vector bundle over y) 144

Π .... (partially ordered set (poset) with G-action) 18

$\mathit{\Pi}$ .... (G. Anderson's indexing data for surgery with coefficients) 116

$\mathit{\Pi}^\chi$ .... (similar to preceding with orientation data twisted by χ) 121

Π-bundle .... (vector bundle system) 18

$\pi(X)$ .... (geometric poset) 16

${}_1\pi(X)$ .... (geometric poset with fundamental group data) 58

$\pi(X;\mathcal{H}^a)$ .... ($\mathcal{H}^a$-restricted geometric poset) 46

$\Pi(X)$ .... (complete G-poset for X) 145

π_X .... (local retraction onto stratum X) 69

$\pi_{X,Y}$ .... (restriction of previous object to Y) 69

$\pi-\pi$ .... (label for fundamental surgery-theoretic statement) 23

$\Pi(\xi)$ .... (vector bundle system associated to vector bundle ξ) 18

ρ .... (Witt invariant) 120

$\rho^c_{X,\Sigma\subset\Sigma'}$ .... (map from Browder-Quinn theory to Dovermann-Rothenberg theory) 64

$\rho^{h(R)}_{X,\Sigma\subset\Sigma'}$ ... (same with coefficients in ring $R \subset \mathbb{Q}$) ... 66
$\rho_{H,V}$ ... (tubular function on orbit space stratum) ... 39, 40
ρ_X ... (tubular function on tubular neighborhood of X) ... 69
ρ_X ... (Pontryagin-Thom collapsing map for X) ... 119
$\rho_{X,Y}$ ... (restriction of previous object to Y) ... 69

Σ ... (subset of geometric poset) ... 24
$\Sigma|\alpha$ ... (analog for restriction to α substratum) ... 162
$\sigma_*(f,b)$ ... (quadratic signature of a normal map) ... 119
$\sigma^*(X)$ ... (symmetric signature of a Poincaré complex) ... 120
σ_β ... (stepwise surgery obstruction) ... 26
$\sigma_\beta(I^a)$ ... (stepwise surgery obstruction for theory I^a) ... 60
$\sigma^\Gamma_*(f,b)$ ... (quadratic signature of a normal map for Cappell-Shaneson Γ-surgery) ... 132
σ_0 ... (final stepwise surgery obstruction) ... 176

$\varphi(X;\Sigma \subset \Sigma')$ ... (transformation of equivariant surgery theories) ... 51, 137

χ ... (orientation character) ... 120
$\chi_G(Y)$ ... (Burnside ring Euler characteristic of Y) ... 96
$\chi(K_*)$ ... (Euler characteristic of graded projective module K_*) ... 164
$\chi(N)$ ... (Euler characteristic of space N) ... 96

$\Psi(\alpha)$ ... (subset of complete G-poset) ... 146
ψ ... (data for equivariant local coefficients) ... 135

$\mathit{\Omega}$ ... (all nonmaximal strata in $\pi(X;\mathfrak{P}_G)$) ... 156
$\Omega(G)$ ... (Burnside ring) ... 146
$\Omega(G;\mathcal{F})$ ... (variant of the Burnside ring) ... 146
$\Omega^\chi_*(G)$ ... (G-bordism group with orientation character χ) ... 120
$\Omega_n(\mathit{\Pi},d)$ ... (bordism groups for degree d normal maps with indexing data $\mathit{\Pi}$) ... 116
$\Omega^t_n(\mathit{\Pi},d)$ ... (analog of the preceding for stably tangential degree d normal maps) ... 127
$\Omega^\Gamma_n(\mathit{\Pi},d)$ ... (analog of $\Omega_n(\mathit{\Pi},d)$ for Cappell-Shaneson Γ-surgery) ... 132

∂M ... (boundary of M) ... □
$\partial_\pm M$ ... ($\pm$ portion of the boundary of M) ... □
∇ ... (extra stable slice data for surgery up to pseudoequivalence) ... 144
$\nabla|\alpha$ ... (analog for restriction to α substratum) ... 161
$[\cdots]$ ... (greatest integer function) ... 31
$[\cdots]$ ... (stable class of projective module) ... 164, 165
$*$ ... (group-theoretic free product) ... 122
$*_G$ ... (basepoint class in $\mathcal{O}_+(G)$) ... 147

$\varnothing$ .... (empty set) .... □

:= .... (equal by definition) .... □

II, $\coprod$ .... (disjoint union) .... □

$\overrightarrow{(uv)}$ .... (shifted secant line parallel to line through u and v) 68

□ .... (completion of step in proof) .... □

■ .... (completion of proof or discussion) .... □

SUBJECT INDEX

Terms beginning with Greek letters are listed as if the letter names were spelled out in the usual way. For example, "σ-compact" would be interpreted as "sigma-compact" and placed between "short exact sequence" and "signature."

A

abstract prestratification (sense of Mather), 70
______, equivariant, 70
Additional Standing Hypothesis for Section V.3, 154
adjustable maps for surgery up to pseudoequivalence, 152
adjusted, ALSO SEE: relatively adjusted
adjusted G-normal cobordisms, 24
______, sense of Browder-Quinn, 43
______, sense of Dovermann-Rothenberg, 24
______, sense of Lück-Madsen, 59
______, surgery up to pseudoequivalence, 145, 147–148
adjusted G-normal maps, 24
______, for a G-surgery theory, 47
______, sense of Browder-Quinn, 43
______, sense of Dovermann-Rothenberg, 24
______, sense of Lück-Madsen, 59
______, sense of surgery up to pseudoequivalence, 145, 147, 152
adjusted equivariant surgery obstruction groups, 2, 3, 23, 24, 42, 46–47, 49, 136, 137
______, Browder-Quinn, 43–44, 56, 65–66, 136
______, ______, exact sequence for, 43
______, ______, Orbit Sequence for, 44
______, Dovermann-Rothenberg, 23, 24, 48
______, ______, exact couple, 28
______, ______, Orbit Sequence, 27
______, Lück-Madsen, 49, 58-59, 137
______, for surgery up to pseudoequivalence, 50
______, ______, Orbit Sequence, 49
adjusted maps, SEE: adjusted G-normal maps
adjusted product maps, 97–98, 100, 175–1807
adjusted stratified maps, sense of Davis, 40
algebraic Poincaré complex, 119
______, quadratic, 119
______, symmetric, 119
algebraic formation (sense of Ranicki), 118, 130, 139
algebraic structures (in surgery), 9
algebraic surgery (theory of Ranicki), 1, 115, 119, 132
algebraic surgery obstruction groups, sense of Ranicki, 115
______, quadratic, 119
______, symmetric, 119

Anderson indexing data (for surgery with coefficients), 116
applications of equivariant surgery (to transformation groups), 9–10
atomic (subset pairs of a geometric poset), 53

B

background references, 1, 6–8
Bak groups (variants of Wall groups), 28
______, and stepwise surgery obstructions, 28, 31–32
Balance, Gap Hypothesis, 96
base space, universal, 10
bordism groups for surgery with coefficients, 116, 121, 132
______, proof of nonemptiness, 121
Browder-Quinn Γ-groups, 135–137
______, $\pi - \pi$ Theorem, 136
Browder-Quinn groups, adjusted, Orbit Sequence for, 43
______, adjusted, exact sequence for, 43
______, for localized surgery, 66–67, 134
______, fourfold periodicity for, 44
______, $L^1 = L^2$ Theorem, 43
______, splitting theorem for odd order groups, 108–109
______, stepwise obstructions, indeterminacy considerations, 160
Browder-Quinn surgery theory, 2, 3, 41–43, 62–65
______, adjusted version, 43
______, Davis-Hsiang-Morgan groups, 66
______, simple equivalence version, 43
______, theories with localized coefficients, 65-66
bundle data , equivariant, 21
______, ______, for Browder-Quinn theory, 42
______, ______, ______, relation to other notions of bundle data, 62--63
______, ______, for surgery up to pseudoequivalence, 147
______, ______, Smith (sense of Petrie-Randall), 50
______, ______, completed, 187
______, ______, sense of Lück-Madsen, 49
______, ______, sense of Dovermann-Rothenberg (usual sense), 21
bundle map, R-stable, 58–59
______, equivariant, completed stable, 187
bundle, principal, 10
Burnside ring, 129, 146, 175, 184
______, $\Omega(G;\mathcal{F})$ variant, 146
______, ______, Oliver-Petrie subset, 147
______, ______, product maps, 181–183
______, operation on projective class group (for finite groups), 176

C

canonical stratification of a smooth G-manifold, 14, 38–40, 71–73
Cappell-Shaneson groups, 51, 66, 115, 131, 159

Cayley projective plane, 94, 151
centralizer (of a representation), 13
chain level Umkehr homomorphism, 128
change of coefficients morphisms, 56–57
closed subset of a geometric poset or G-poset, 20, 43, 151
closed substrata, induced map of (associated to an equivariant map), 17
cobordism, fork/pair-of-pants, 105–106
Codimension ≥ 2 Gap Hypothesis, 15
Codimension ≥ 3 Gap Hypothesis, 23, 49
coefficients, change of, morphisms, 56–57
cohomology, Tate, 61
collar neighborhoods, compatible with stratifications, 40
commutativity relations (for maps of equivariant surgery theories), 52
Comparison Theorem II.3.4 (for maps of equivariant surgery theories), 53, 91
______ , relative version, 54
compatibility relation, for product maps, 102
______(-s) , for Mather prestratifications, 69
complete G-poset, 87, 145
______ , essential elements, 146
complete geometric poset, 87
completed bundle data, 187
completed stable bundle map, 187
Condition, Frontier (for a prestratification), 69
Conditions A and B (for a prestratification), 68
Conditions, Weak Gap (Lück-Madsen), 49
cone, mapping, 99
configuration of invariant submanifolds, 13
conjugation on projective class group, 164
connected open substratum, 39
connected sum, equivariant, 12
connectivity hypotheses, 5, 151
consistency conditions (for a stratified vector bundle), 42
control data (for a stratified object), SEE: tubular data
controlled vector field, 70–71
corners, smooth manifolds with, 39
covering space, oriented double, 42–43
______ , universal, 11
covering transformations, 11, 127
$\mathbf{CP}^2\uparrow G$-preperiodic, 88
CS stratification (sense of Siebenmann), 14
cutting and pasting, equivariant, 11
CW complex, equivariant, 15

D

data, degree, 142
______ , dimension, 16, 142

——, dynamic, 66, 92
——, functional, 142
——, (oriented) normal slice representation, 16, 142
——, orientation, 16, 142
——, reference, 5
——, supplementary for constructing Dovermann sums, 143–144
Davis-Hsiang-Morgan equivariant surgery obstruction groups, 66, 160
defect group $D(G)$, 50, 129, 160
degree > 1, special condition on normal map, 119
disk, unit (in an orthogonal representation), 12
dimension indices, 16
domination, equivariant (by a finite $G - CW$ complex), 15
Dovermann-Rothenberg theories, SEE: equivariant surgery theories, Dovermann–Rothenberg type
Dovermann sum construction, 24, 86, 148, 149, 153, 183
——, abstract characterization, 86, 153
——, and products, 85, 102, 149, 183
——, associated group structure on equivariant surgery groups, 24, 86
——, ——, for surgery up to pseudoequivalence, 143–144
——, ——, ——, nilpotence condition, 148
——, behavior under restriction, 162
——, for ordinary surgery with coefficients, 121, 122–124
——, important geometric properties, 153
——, relation to disjoint union sum, 59, 64, 86, 153
dual projective module, 164
dynamic geometric references and indexing data, 66
——, periodicity theorems for, 92
dynamic vs. static reference data, 66

E

ensemble stratifié (sense of Thom), 70–71
equivalence, ALSO SEE: pseudoequivalence AND simple equivalence
——, for surgery with coefficients, 118
——, in Browder-Quinn theory, 43, 45
——, localized, 65, 133
——, with twisted coefficients (in Λ), 132
equivalences (for a G-surgery theory), 47
equivariant abstract prestratification (sense of Mather), 70
equivariant completed bundle map, 187
equivariant connected sum, 12
equivariant cutting and pasting, 11
equivariant domination (by a finite $G - CW$ complex), 15
equivariant fundamental groupoid (system), 135
–"– ring, 135
equivariant homotopy equivalence, 2, 20
——, localized, 65, 133

equivariant Kervaire-Milnor groups, 13
equivariant local coefficients, 135
______, and equivariant surgery, 135–137
equivariant localization, 2, 56, 133
equivariant map, isogeneric, 17, 87
______(-s) , transverse linear, 40
______, ______, relation to maps of vector bundle systems, 40
______, ______, relation to stratified maps, 40
equivariant normal bundle, 144
equivariant normal maps, stably tangential with special framings, 107–108
equivariant orientation, 17
equivariant $\pi - \pi$ Theorem, 23 ALSO SEE: $\pi - \pi$ Theorem
______, for Dovermann-Rothenberg theory, 23
______, ______, failure outside Gap Hypothesis range, 29–30
equivariant $\pi - \pi$ trivialization, 171
equivariant prestratification, abstract (sense of Mather), 70
equivariant product formulas, open questions, 100–101
equivariant pushouts, analog of R. Brown's result, 20
equivariant S-duality, 127
equivariant simple homotopy equivalence, 48
______, topological analog, 48
equivariant smooth triangulation theorem, 52
equivariant surgery, and equivariant local coefficients, 135–137
______, applications to transformation groups, 9–10
______, equivalences in a given theory, 97
______, infinite periodicity, 90
______, isomorphisms in a given theory, 97
______, localized, 65–66
______, periodically stabilized, 101–109
equivariant surgery obstruction groups, 16 ALSO SEE: equivariant surgery theories
______, algebraic (sense of Yoshida), 120
______, Davis-Hsiang-Morgan version, 160
______, global, 22
______, ______, spectral sequences for, 22
______, necessary indexing data, 16, 142
______, unstable (no Gap Hypothesis), 104–110
______, ______, example of nonstable summand, 108–109
______, ______, relation to Browder-Quinn groups, 105–106
______, ______, relation to Dovermann-Rothenberg groups, 106
______, ______, splitting theorems, 107, 109
______, ______, stabilization maps, 106
______, stepwise, 21, 22
equivariant surgery obstruction, global, 22
______, stepwise, 21, 22
______, relations between, exceptional cases, 27–29
______, relations between local and global, 24–27

equivariant surgery sequence, (infinite) periodically stabilized, 82, 101
equivariant surgery space/spectrum, 56
equivariant surgery theories, 9
———, Browder-Quinn type, 41–43 ALSO SEE: Browder-Quinn ⋯
———, ———, for simple equivalences, 41–43
———, ———, localized, 134
———, ———, $L^1 = L^2$ Theorem, 43
———, ———, relation to Dovermann-Rothenberg theory, 64–65
———, ———, with coefficients, 134–137
———, ———, with twisted coefficients, 135–137
———, Dovermann-Rothenberg type, 9, 23–27
———, ———, for coefficients in a ring, 48, 134
———, ———, for nonsmooth categories, 47–48
———, ———, for simple equivalence, 48
———, ———, ———, stepwise obstruction, 55
———, ———, change of manifold categories, 52
———, formal properties, 45–47
———, for positive dimensional groups, 50
———, for pseudoequivalence, 49–50 ALSO SEE: equivariant surgery up to pseudoequivalence
———, general product formulas, 82
———, Lück-Madsen type, 48–49,58–61
———, ———, $L^1 = L^2$ Theorem, 49
———, ———, for simple equivalences, 61
———, ———, relation to Browder-Quinn theory, 64–65, 137
———, ———, relation to Dovermann-Rothenberg theory, 59–62
———, ———, simple version, 49
———, maps of, 51
———, ———, Comparison Theorem, 53
———, ———, commutativity relations (III.1.A-B), 52
———, periodic stabilizations, 80, 101
———, periodicity theorems, SEE: Periodicity Theorem/periodicity theorems
———, products for free actions, 81
———, restrictions to subgroups, 67
———, Type I, 83
———, Type II, 83
———, Type III, 83
———, typical features, 47
———, underlying idea, 9
———, versions with nonsimply connected strata, 48–49
———, with coefficients, Browder-Quinn type, 65–66, 133–137
———, ———, Dovermann-Rothenberg type, 48, 56–57, 133
———, ———, Lück-Madsen type, 137
equivariant surgery up to pseudoequivalence, 141–148
———, normal cobordisms, Euler characteristic condition (1.0.E), 148
———, normal maps, Euler characteristic condition (1.0.F), 147

equivariant surgery up to pseudoequivalence, normal maps (*continued*)
______ , ______ , conditions (1.0.A–D), 145
equivariant tangent bundle, stable, 108
equivariant transversality, 30
equivariant tubular neighborhood theorem, 18
equivariant Whitehead Theorem, 20
equivariant Whitney prestratification, 69
equivariantly oriented, 12
Euler characteristic, 96
______ , equivariant, 175
______ , relation to signature, 96
Euler characteristic conditions for surgery up to pseudoequivalence, 5, 146
______ , anomalous feature, 148
______ , behavior under restriction, 161–162
______ , behavior under product constructions, 181–183
______ , extension to even order groups, 148–149
Euler class, 97
exact couple(s) in equivariant surgery, 28
exactness conditions, for a G-surgery theory, 46
______ , for Browder-Quinn theory, 43
______ , for Dovermann-Rothenberg theory, 26
______ , for Lück-Madsen theory, 49
exact sequence for the obstruction homomorphism, 173
extension, top hat, 100

F

faces, (smooth) manifolds with, 39, 104
failure of equivariant $\pi - \pi$-Theorem outside the Gap Hypothesis range, 29–30
Farrell-Hsiang correction to Wall group definitions, 42
$\mathcal{F}$-complex, 146
______ , restriction to a subobject/substratum, 161
______(-s) , $\Omega(G;\mathcal{F})$-equivalent, 146
Five Lemma, refined version, 54
fiber transports, ALSO SEE: transports
______ , sense of Lück-Madsen, 43, 58
fibration, spherical SEE: spherical fibration
finiteness assumptions for Chapter I, 14
finitely dominated CW complex, 119
first Stiefel-Whitney class, data for a stratified orbit space, 42
______ , Lück-Madsen equivariant generalization, 58
first author (Dovermann), 1
forgetful maps (of partial Dovermann-Rothenberg surgery spaces), 55–56
forgetful transformations, in Dovermann-Rothenberg theories, 49, 55
______ , in Lück-Madsen theories, 49
______ , to pseudoequivalence, 57
formation, geometric, 165

——, sense of Ranicki, 118, 130, 139, 164
fourfold periodicity, 1
——, in equivariant surgery, 89–90, 133–137, 150–151
——, for Browder-Quinn theory, 44
——, of periodic stabilization, 102
——, of the topological surgery sequence, 92
free action (of a group), 10, 37
——, determination by orbit space and extra data, 10–11, 37, 38
free product (of groups), 121
Frontier Condition (for a prestratification), 69
functional indexing data for surgery up to pseudoequivalence, 142
——, $\mathcal{H}$-restricted, 143
functor, half exact, 28, 51
fundamental groupoid (ordinary), 58
fundamental groupoid ring, equivariant, 135
fundamental groupoid system, equivariant, 135

G

Γ-groups, ALSO SEE: Cappell-Shaneson groups
——, Browder-Quinn, 135–137
——, Lück-Madsen, 137
Gap Conditions, Weak (sense of Lück-Madsen), 49
Gap Hypothesis, 1, 20
——, behavior with respect to products, 94
——, borderline cases, 31–32
——, Codimension ≥ 2, 15
——, Codimension ≥ 3, 23
——, default convention, 20
——, Standard, 20
Gap Hypothesis Balance, 96
G-bundle, principal, 10
$G - CW$ complex, 15
general position, 20
generalized Whitehead group, 61
generalized Whitehead torsion, 64
——, and Browder-Quinn theory, 64
generic isotropy subgroup (of a closed substratum), 15
geometric Umkehr map, 127
geometric formation, 165
geometric poset (of a compact G-space), 15
——, closed subset, 20, 43
——, complete, 87
——(-s), maps of, 17
geometric product constructions, 130
geometric reference (sense of Lück-Madsen), 59, 83
——, dynamic version, 66, 92

geometric sums, 105
G-equivariant normal cobordism, sense of Browder-Quinn, 42–43
______, sense of Dovermann-Rothenberg, 22
______, sense of Lück-Madsen, 59
______, sense of surgery up to pseudoequivalence, 144–148
______, ______, condition (1.0.E), 147
______, ______, conditions (1.0.A–D), 145
G-equivariant normal map, sense of Browder-Quinn, 42–43
______, sense of Dovermann-Rothenberg, 22
______, sense of Lück-Madsen, 58–59
______, sense of surgery up to pseudoequivalence, 144–147
______, ______, condition (1.0.F), 147
______, ______, conditions (1.0.A–D), 145
______, specially framed, 108
______, stably tangential, 108
G-equivariant surgery problem, SEE: *G*-equivariant normal map
G-invariant tubular neighborhoods and transverse linearity, 40
global equivariant surgery obstruction, 22
–"– groups, 22
______, spectral sequences for, 22
G-manifold, locally linear, 46
______, ______, topological CS stratification, 14
______, ______, with piecewise linear action, 46, 52
______, ______, with topological action, 46, 52
______, locally smooth(able), SEE: *G*-manifold, locally linear
______, piecewise linear (PL) locally linear, 46, 52
______, smooth, canonical stratification, 14, 38–40, 71–73
G-normal maps, SEE: *G*-equivariant normal maps
G-poset, 50 ALSO SEE: geometric poset
______, closed subset, 151
______, complete, 87, 145
______, ______, essential elements, 146
G-Poset pair (*note capitalization*), 66
G-Poset (*note capitalization*) vs. *G*-poset, 67
greatest integer function, 31
group action, free, 10
______, semifree, 12
______(-s), locally linear, 10
group structure, defined by Dovermann, SEE: Dovermann sum construction
______, on equivariant surgery obstruction groups, SAME AS ABOVE
______, on unstable surgery obstruction groups, 105
$G-\mathbb{R}^n$-bundles, locally linear, 47, 52
______, ______, stable 47, 52
______, ______, unstable 47, 52
______, piecewise linear locally linear, 52
G-signature, 3, 81

———, and multiplicative induction, 93–97
———, formula for involutions, 97

H

half exact functor, 28, 51
high connectivity, surgery up to (sense of Kreck), 51
$\mathcal{H}$-isogeneric, 143
homology equivalence, local, 56
homomorphisms of vector bundle systems, 18
homotopically stratified space (sense of Quinn), 45
homotopically stratified surgery (sense of Weinberger), 45
———, relation to G-surgery, 45
homotopy equivalence, equivariant, 2
———, ———, localized, 65
homotopy theory (in surgery), 9
$\mathcal{H}$-restricted functional indexing data, 143
$\mathcal{H}$-restricted geometric poset, 46
hyperboliztion map, 184
hypotheses, connectivity, 5, 151
———, Euler characteristic, 5
———, finiteness in Chapter I, 14
———, gap SEE: Gap Hypothesis
———, inductive, for stepwise surgery problems, 20
Hypothesis, Simplifying, for Section III.4, 97
———, ———, for Section V.7, 175
———, Standing, for Section V.3, 151, 154
———, ———, for Section V.8, 181
———, ———, for Sections V.5–6, 163, 171

I

indeterminacy of stepwise obstructions, 27–28, 31–32
indexing data, 16, 83
———, conventions for, 17
———, dimension indices, 16
———, dynamic, 66, 92
———, orientation data, 16
———, (oriented) normal slice representation data, 16
———, preperiodic, 87, 88, 97
———, product construction for, 17
———, sense of Anderson (for surgery with coefficients), 116
———, suspension and desuspension of, 17
indexing data for identity maps, 17
indexing data for surgery up to pseudoequivalence, 142
———, degree data, 142
———, dimension data, 142
———, functional data, 142

indexing data for surgery up to pseudoequivalence (*continued*)
———, orientation data, 142
———, restricted degree data, 142
———, slice data, 142
———, supplementary data for constructing Dovermann sums, 143–144
indices, dimension, 16
induced map of closed substrata (associated to an equivariant map), 17
induction, multiplicative, 81
———, ———, and G-signatures, 93–97
———, ———, and periodicity manifolds, 92–94
inductive hypothesis for a stepwise surgery problem, 20
infinite subprojective Wall groups ($L_*^{-\infty}$), 131
intermediate surgery obstruction groups, 115, 128
involutions, G-signature formula for, 97
isogeneric (equivariant) map, 17, 87
———, and strongly saturated orbit structure, 87
———, $\mathcal{H}$–, 143
———, weakly, 143, 154
isotropy subgroup, generic (of a closed substratum), 15
———, ———, 40
———, ———, relation to maps of vector bundle systems, 40
———, ———, relation to stratified maps, 40
isovariant stratified surgery groups, of Browder-Quinn, 2, 3, 41–43
———, of Weinberger, 4, 45, 92
isovariant surgery structure sequence, of Browder-Quinn, 45
———, of Weinberger, 4, 92

K

kernel, surgery, SEE: surgery kernel
Kervaire-Milnor groups, equivariant, 13
———, nonequivariant/ordinary, 13

L

$L^1 = L^2$ Theorem, 43, 49, 64
Λ-equivalence (*i.e.*, with twisted coefficients), 132
L-group, SEE: surgery obstruction group
limiting subprojective Wall groups ($L_*^{-\infty}$), 131
local coefficients, equivariant, 135
———, ———, and equivariant surgery, 135–137
local homology equivalence, 56
local linearity, 2
———, and strongly saturated orbit structure, 88
———, piecewise linear, 2
———, topological, 2
local retractions for Mather prestratifications, 69
localization, equivariant, 56

——, exact sequence in L-theory, 187–188
localizations of the integers, 152
localized equivariant homotopy equivalence, 65, 133
localized equivariant surgery, 65–66, 133–134
localized pseudoequivalence, 49–50, 144, 161, 186
localized ring, 50, 152
——(-s) , obtained from the integers, 152
locally finite prestratification, 69
locally linear G-manifold, with piecewise linear action, 46, 52
——, with topological action, 46, 52
——, topological CS stratification, 14
locally linear $G - \mathbf{R}^n$-bundles, 47, 52
——, piecewise linear, 52
——, stable 47, 52
——, unstable 47, 52
L-theory, localization sequence, 187–188
Lück-Madsen theory, ALSO SEE: equivariant surgery theories, Lück-Madsen type
——, Γ-groups, 137
——, outside the Gap Hypothesis range, 49, 109–110

M

manifolds, periodicity, 87
mapping cone, 99
maps of equivariant surgery theories SEE: equivariant surgery theories, maps of
maps of geometric posets, 17
Mather prestratification, 69
——, relation to Whitney prestratification, 70
maximal order, 50, 129, 160
May-McClure-Triantafillou equivariant localization, 2, 56, 133
metabolic quadratic form, 93
morphisms of equivariant surgery theories, SEE: equivariant surgery theories, maps of
multiplicative induction, 81
——, and G-signatures, 93–97
——, and periodicity manifolds, 92–94

N

negative Wall groups, 115, 128
nilpotent groups, restriction to, 50
$n(\lambda)$ bound for dimensions of Oliver resolutions, 145, 150
——, associated difficulties involving periodicity theorems, 151
——, for nilpotent groups, 145, 150
nonfinite generation of pseudoequivalence surgery obstruction groups, 174
nonsingular set (of a group action), 11
normal bundle, equivariant, 144
normal cobordism, for surgery with coefficients, 117
——, G-equivariant, SEE: G-equivariant normal cobordism

normal map, degree > 1, special algebraic condition, 119
______ , (equivariantly) specially framed, 108
______ , for a G-surgery theory, 47
______ , G-equivariant, SEE: G-equivariant normal map
______ , of normal spaces (in the sense of Quinn), 118
______ , stably tangential, 125
______ , stably tangential, corresponding bordism theory, 126
______ , stably tangential (equivariant sense/ordinary sense), 108
______(-s) , for surgery with coefficients, 116
______ , ______ , group structure on bordism classes, 116
normal slice representation data, oriented, 16, 66
normal slice types, 14, 38–40, 71–73
normal space (sense of Quinn), 119
______ , triple, 119
______(-s) , normal map of, 119

O

objects, restricted (sense of Wall), 116 ALSO SEE: $L^1 = L^2$ Theorem
obstruction (to stepwise surgery), 21 ALSO SEE: stepwise equivariant surgery obstructions
obstruction homomorphism, 171–172
______ , additivity, 173
______ , associated exact sequence, 173
odd Wall groups, vanishing theorem for odd order groups, 187–188
Oliver-Petrie subset of a generalized Burnside ring, 147
Oliver resolutions, $n(\lambda)$ bound for dimensions, 145, 150
______ , ______ , associated difficulties involving periodicity theorems, 151
______ , ______ , for nilpotent groups, 145, 150
$\Omega(G;\mathcal{F})$-equivalent $\mathcal{F}$-complexes, 146
open substratum, 39
orbifolds, 73, 74
Orbit Sequence, 27, 46, 97
______ , for Browder-Quinn theory, 44
______ , for Dovermann-Rothenberg theory, 27
______ , for Lück-Madsen theory, 49
______(-s) , commutativity properties for transformations of theories, 51, 54
orbit space (of a group action), 10
______ , and additional data, 37, 38
______ , smooth, canonical smooth structure for free actions, 10
______(-s) , ______ , triangulation theorem for, 67
orbit space data, structured/weighted, 38
orbit structure, strongly saturated, 87
______ , ______ , and isogeneric maps, 87
ordinary surgery obstruction group, 3, 21, 48
______ algebraic formulation (sense of Ranicki), 115, 119
orientation, equivariant, 17

orientation data, 16
oriented (G-manifold), equivariantly, 12
oriented normal slice representation data, 16

P

periodic stabilization, 4, 29, 80, 82, 94, 101, 102
———, failures for even order groups, 82, 96, 104
———, fourfold periodicity of, 102
———, statement of underlying idea, 101
periodically stabilized equivariant surgery obstruction group, 102
———, fourfold periodicity, 102
periodically stabilized equivariant surgery exact sequence, 82
periodicity manifolds, 86–87
———, and multiplicative induction, 92–94
———, and the Gap Hypothesis, 94
———, background information, 80–82
———, examples, 92–94
periodicity principle, rough statement, 81
Periodicity Theorem, equivariant, abstract version (III.2.7), 89–90
———, ———, ———, proof, 100
———, ———, ———, variants, 91–92
———, ———, main steps in the proof, 91
———, ———, products with $\mathbf{C}P^2\uparrow G$, 90–91
———, ———, products with $\mathbf{C}P^{2n}$, 90
———, ———, Yan's extension to Weinberger's theory, 92
periodicity theorems, for surgery up to pseudoequivalence, 141, 150, 151
———, ———, proofs, 183–186
———, for surgery with coefficients, 115
———, fourfold, in equivariant surgery, 3, 90, 102, 133, 134, 136, 150
———, ———, in ordinary surgery, 1, 3
———, ———, ———, for topological surgery sequences, 92
———, further developments, 82
periodicity, algebraic, 3
———, finite, 3
———, geometric, 3
———, infinite, 3
Π-bundle, 18
$\pi - \pi$ Question, Unstable, 29
$\pi - \pi$ Theorem, 23
———, failure outside the Gap Hypothesis range, 30–31
———, for Browder-Quinn theory, 43
———, for Browder-Quinn Γ-groups, 136
———, for Cappell-Shaneson Γ-groups, 132
———, for Dovermann-Rothenberg equivariant surgery groups, 23
———, for surgery with coefficients, 117
———, for Wall surgery groups, 23

$\pi - \pi$ trivialization, equivariant, 171
piecewise linear local linearity, 2
piecewise linear locally linear $G - \mathbf{R}^n$ bundles, 52
$\Pi(X)$-structure, 146
______, essential subcomplex, 146
Poincaré complex, algebraic, 119
______, normalized, 127
______, quadratic algebraic, 119
______, symmetric algebraic, 119
Pontryagin-Thom class (of a normal space), 119
poset (*i.e.*, partially ordered set), ALSO SEE: geometric poset AND G-poset
______, geometric (of a compact G-space), 15
Poset (*note capitalization*), 66, 67, 92
Poset vs. poset, 67
position, general, 20
positive-dimensional compact Lie groups, equivariant surgery theories for, 50
preferred family of subgroups (for a G-surgery theory), 46
preperiodic indexing data, 87, 88, 97
prestratification, 69, 68
______, Conditions A and B, 68
______, Frontier Condition, 69
______, Whitney, 69
______, abstract (sense of Mather), 69
______, ______, germ equivalence, 70
______, ______, germ equivalence, 70
______, ______, local retractions, 69
______, ______, relation to Whitney prestratification, 70
______, ______, tubular data, 69
______, ______, tubular functions, 69
______, ______, tubular neighborhoods, 69
prestratification, locally finite, 69
principal G-bundle, 10
Principal Orbit Theorem, 11
product construction, twisted, for s, h, p, and subprojective groups, 131
______(-s) , geometric, 130
product formulas, for projective class surgery invariants, 177–180
______, equivariant, open questions, 100–101
______, in ordinary surgery, 83
______, Ranicki, 199–120, 132
______, twisted, 115, 129
______, Yoshida, 81
product maps, adjusted, 97
______, and stepwise surgery obstructions, 98
______, and the Dovermann sum construction, 85, 102, 149, 183
______, for surgery up to pseudoequivalence, 149, 150, 180–183
______, ______, additivity, 149, 183

______ , ______ , and Yoshida twisted products, 175–180
______ , in Browder-Quinn theories with coefficients, 134, 136
______ , in equivariant surgery, 84, 102, 106
______ , ______ , additivity, 84
______ , ______ , and Yoshida twisted products, 98–100
______ , ______ , compatibility relations, 102, 107
______ , on variants of the Burnside ring, 181–183
products, twisted SEE: twisted product maps
projective class group, 5, 128–130, 156
______ , conjugation, 164
______ , Oliver-Petrie subgroups $B_0(G, \mathcal{F}(\lambda))$, 156
______ , module structure over Burnside ring, 176
______(-s) , Swan finiteness theorem, 175
projective class map (on projective surgery groups), 164
projective class of a surgery kernel, 156, 164–165
______ , product formulas, 177–180
projective module, dual, 164
projective surgery groups, 115, 128
______ , projective class map, 164
______ , hyperbolization map, 184
proof of III.2.7, 100
-"- IV.1.2, 123
-"- IV.1.3, 124
-"- IV.1.8, 127
-"- IV.1.9, 127
-"- V.2.1, 180–183
-"- V.2.2, 183
-"- V.2.3, 183–186
pseudoequivalence, 5 49–50, 57–58, 141–142
______ , equivariant surgery theories for, 49–50, 142–149
______ , localized, 49–50, 144, 161, 186
______ , passage from other notions of equivalence to, 57
______ , surgery obstruction groups, nonfinite generation 174
pullbacks of vector bundle systems, 18

Q

quadratic algebraic Poincaré complex, 119
quadratic form, of a highly connected surgery problem, 117
______ , metabolic, 93
______(-s) , in surgery, 9
quadratic signature (of a normal map), 119, 132
______ , for Cappell-Shaneson surgery, 132
______ , product formula, 120, 132
______ , ______ , for Cappell-Shaneson surgery, 132
quadratic surgery obstruction groups, 119
quaternionic projective space, 94, 151

Quinn surgery spaces/spectra, 56

R

Ranicki sequence, 164, 184 ALSO SEE: Rothenberg sequences
reference data, 5, 66 ALSO SEE: geometric reference AND indexing data
______, dynamic, 66
______, static, 66
regular representation, 91
relationships between connected open substrata, 39
relative nonsingular set (of a stratum), 136
relatively adjusted equivariant surgery obstruction groups, 25
______, exact couple, 28
______, exact sequence, 26
relatively adjusted G-normal maps, 25
representation, (real) regular, 91
______, one-dimensional unitary, standard notation, 143
restricted, ALSO SEE: adjusted
restricted (G-normal) cobordisms, sense of Lück-Madsen, 59
restricted (G-normal) maps, sense of Lück-Madsen, 59
restricted degree data, 143
restricted functional data, 143
restricted objects (sense of Wall), 116 ALSO SEE: $L^1 = L^2$ Theorem
restricted stepwise surgery problem, 20
restricted surgery obstruction groups, 2
restriction maps, for passage to subgroups, 67
______, in surgery up to pseudoequivalence, 160
ring, Burnside, 129, 146, 175, 184
ring, localized, 50
Rothenberg sequences, for change of K-theory, 55, 130
______, ______, sense of Browder-Quinn, 56
______, ______, sense of Dovermann-Rothenberg, 55, 56
______, ______, sense of Lück-Madsen, 61
______, relating L^s and L^h, 55
______, relating L^h and L^p, 130
______, relating L^h and L^{DG}, 130, 164
Rothenberg-Weinberger examples, 29–31
$\mathbb{R}$-stable (equivariant) bundle map, 58–59

S

S-duality, equivariant, 127
secant line, shifted, 68
second author (Schultz), 1
Shaneson splitting, 99, 129
shifted secant line, 68
Σ-adjusted, SEE: adjusted AND relatively adjusted
Σ-good, SEE: adjusted AND relatively adjusted

signature, equivariant/G-, SEE: G-signature
———, ordinary, of a closed manifold, 80
———, ———, of a compact bounded manifold, 29, 30
———, ———, relation to Euler characteristic, 96
———, quadratic (of normal map), 119
———, ———, identification with Wall surgery obstruction, 119
———, symmetric (of a Poincaré complex), 119
simple equivalence, in Browder-Quinn theory, 44, 45
———, ———, and generalized Whitehead torsion, 64
———, in Dovermann-Rothenberg theory, 48
simple homotopy equivalence, equivariant, 48
simple homotopy equivalence, equivariant, topological analog, 48
simple stratified homotopy equivalences, 43
simple transverse linear isovariant homotopy equivalences, 43, 44
———, and generalized Whitehead torsion, 64
Simplifying Hypothesis, for Section III.4, 97
———, for Section V.7, 175
singular set (of a group action), 11
slice extension property, 108
slice types, normal, 14, 66, 71–73
Smith bundle data for surgery up to pseudoequivalence, 187
smooth G-manifold, canonical (smooth) stratification, 14, 38–40, 71–73
———, closed substratum (*pl.* substrata), 14
———, standard concept of stratum (*pl.* strata), 14
smooth orbit spaces, triangulation theorem for, 67
smooth stratification, 2
———, types of objects admitting, 67, 71–73
smooth stratification theorem for orbit spaces, 2, 71–73
smooth submersions and stratification data, 42, 69, 72, 73
smooth triangulation theorem, equivariant (for finite group actions), 52
spacification, 56
specially framed G-normal maps, 108
spectrum (stable homotopy-theoretic sense), associated to a surgery space, 56
sphere, unit (in an orthogonal representation), 12
spherical fibration, 119
———, associated Thom complex, 119
$S\pi$-duality, 127
splitting theorem, for periodic stabilization, 107
———, Shaneson, 129
———(-s) , for Browder-Quinn groups, 108–109
———, for Lück-Madsen groups, 108–110
———, for stepwise obstructions, 108–109
———, for unstable equivariant surgery groups, 107, 109
stabilization, periodic, 4 ALSO SEE: periodic stabilization
stable bundle map, completed, 187
stable equivalence of vector bundle systems/Π-bundles, 18

stable equivariant tangent bundle, 108
stably tangential G-normal maps, 108
stably tangential (nonequivariant) normal maps, 125
_____, corresponding bordism theory, 126
Standard Gap Hypothesis, 20
Standing Hypothesis, for Section V.3, 151, 154
_____, for Section V.8, 181
_____, in Sections V.5–6, 163, 171
static vs. dynamic reference data, 66
stepwise equivariant surgery obstruction group, 21, 22
stepwise equivariant surgery obstructions, 3, 21, 22
_____, behavior under restriction, 162
_____, excision property, 99
_____, for an abstract G-surgery theory, 46–47
_____, for Browder-Quinn groups, 43–44
_____, _____, indeterminacy considerations, 160
_____, for simple Dovermann-Rothenberg theory, 55
_____, for Lück-Madsen theory, 49, 59–60
_____, for surgery up to pseudoequivalence, 155, 163–175
_____, _____, additivity, 158
_____, indeterminacy, 157
_____, _____, examples, 160
_____, and products, 98
_____, and relatively adjusted equivariant surgery groups, 26
_____, _____, exceptional cases, 26
_____, nonstandard (Morimoto's results), 31
_____, _____, changes needed for certain papers/books, 31–32
_____, periodically stabilized, 101–103
_____, _____, difficulties for even order groups, 104
_____, summary, 97
stepwise surgery problem, 20
_____, inductive hypothesis, 20
stepwise surgery problem, restricted, 20
stepwise surgery, 20
Stiefel-Whitney class, SEE: first Stiefel-Whitney class
stratification, ALSO SEE: prestratification
_____, canonical (of a smooth G-manifold), 14, 38–40, 71–73
_____, sense of Thom, 70–71
_____, smooth, 14
_____, topological CS, of a locally linear G-manifold, 14
_____, _____, sense of Siebenmann, 14
stratification of smooth orbit spaces, 39
stratification theorem for smooth orbit spaces, 72
_____, proof, 71–73
stratified homotopy equivalences, 43
_____, simple, 43

stratified maps (sense of Davis), 40
——, adjusted, 40
——, relation to transverse linear maps, 40
stratified sets, surgery theory for, sense of Browder-Quinn, 38
stratified surgery, SEE: isovariant stratified surgery
stratified tangent bundles/vector bundles, 42
stratum (of a G-manifold) with given isotropy subgroup and slice types, 14, 39
stratum (*pl.* strata), of a smooth G-manifold, 38, 39
——, of the orbit space of a smooth G-manifold, 38, 39, 71–73
stratum, relative nonsingular set, 136
strongly saturated orbit structure, 87
——, and isogeneric maps, 87
structure of manifolds, surgery-theoretic means of study, 9
structure set, simple relative (for the category CAT), 11
subgroup, generic isotropy (of a closed substratum), 15
——, isotropy, 14, 15
subprojective Wall groups, 115, 128, 130
——, infinite/limiting case, 131
substrata, ALSO SEE: substratum (*sing.*)
substrata, closed, induced map of (associated to an equivariant map), 17
——, connected open, relationships between, 39
substratum, connected open, 39
sums,
——, Dovermann, SEE: Dovermann sum construction
——, (ordinary) geometric, 105
supersimple surgery obstruction groups, 115, 131
surgery, homotopically stratified, sense of Weinberger, 45
——, ——, ——, relation to equivariant surgery, 45
surgery kernel, 156, 158, 159
——, projective class, 156, 163–175
——, ——, product formulas, 177–180
surgery obstruction groups, adjusted, 2, 3, 23, 24, 43–50, 56, 58–59, 65–66, 136, 137
——, algebraic (Ranicki), 115
——, Cappell-Shaneson, SEE: surgery obstruction groups, Γ-groups
——, equivalence of algebraic and geometric, 115
——, for homology surgery (Cappell-Shaneson), 116, 131 ALSO SEE: surgery obstruction groups, Γ-groups
——, Γ-groups (Cappell-Shaneson), 23, 51, 66, 115, 131, 159
——, geometric, 115
——, intermediate, 115, 128
——, limiting subprojective, 131
——, negative, 115, 128
——, projective, 115, 128
——, restricted, 2
——, sense of Cappell-Shaneson, SEE: surgery obstruction groups, Γ-groups (Cappell-Shaneson)

surgery obstruction groups (*continued*)
______, sense of Wall, 21, 22, 23
______, ______, Farrell-Hsiang correction, 23
______, ______, homotopy, 21
______, ______, simple, 48
______, subprojective, 115, 128, 130
______, supersimple, 115, 131
______, symmetric, 119
______, ______, algebraic equivariant (sense of Yoshida), 120
surgery obstructions, stepwise, 3
______, ______, nonstandard (Morimoto's results), 31
______, ______, ______, changes needed for certain papers/books, 31–32
surgery problem, SEE: normal map
surgery space/spectrum, 56
______, equivariant, 56
surgery structure sequence (of Sullivan-Wall), 45
______, equivariant, sense of Dovermann-Rothenberg, 34
______, ______, ______, infinite extension by periodic stabilization, 82
______, fourfold periodicity, 92
______, isovariant/stratified, sense of Browder-Quinn, 45
______, ______, sense of Weinberger, 4, 92
surgery theories, equivariant, 9
surgery theory, capsule description, 9
______, equivariant, 9
______, ______, adjusted version, 43
______, ______, sense of Browder-Quinn, 41–43
______, ______, ______, for simple equivalences, 41–43
______, ______, sense of Lück-Madsen, 58–61
______, ______, ______, for simple equivalences, 61
______, for transverse linear isovariant maps, 41–43
______, sense of Cappell-Shaneson, 51
______, sense of Kreck, 51
______, sense of Wall, background references, 1, 6–8
______, ______, Farrell-Hsiang patch for, 42
______, ______, $L^1 = L^2$ Theorem, 43
______, ______, surgery up to homotopy equivalence, 21
______, ______, surgery up to simple homotopy equivalence, 48
______, transverse linear isovariant (Browder-Quinn), 2, 3
surgery up to high connectivity (sense of Kreck), 51
surgery up to pseudoequivalence, 49–50, 57–58, 141–149
______, adjustable maps, 152
______, adjustable subgroups, 152
______, ______, quotients, 154
______, ______, ______, and stepwise obstructions, 155
______, adjusted maps, 152
______, bundle data for, 147

———, conditions (1.0.E–F), 147
———, indexing data, 142 ALSO SEE: indexing data
———, localized versions, 186–187
———, normal cobordisms, Euler characteristic condition, 148
———, normal maps, Euler characteristic condition, 147
———, ———, conditions (1.0.A–D), 145
———, ———, conditions (1.0.E–F), 147
———, Periodicity Theorems, 141, 150, 151
———, ———, proofs, 183–186
———, product maps, 149, 150, 180–183
———, ———, additivity, 149, 183
———, projective class invariants, 156, 163–175
———, ———, product formulas, 177–180
———, restriction maps, 160
———, Smith bundle data for, 187
———, stepwise obstructions, 155
———, ———, additivity, 158
———, ———, behavior under restriction, 162
———, ———, indeterminacy, 157
———, ———, indeterminacy, examples, 160
———, with Cappell-Shaneson stepwise obstructions, 187
———, with Smith bundle data, 187
surgery with coefficients, bordism groups for, 116, 121, 132
———, equivariant, sense of Dovermann-Rothenberg, 133
———, ———, sense of Lück-Madsen, 137
———, localized coefficients, 49, 56, 65–66, 133, 134
———, twisted coefficients, 135–137
Swan finiteness theorem for projective class groups, 175
symmetric Witt group, 86
symmetric algebraic Poincaré complex, 119
symmetric signature (of a Poincaré complex), 119
symmetric surgery obstruction groups, 119
———, algebraic equivariant (sense of Yoshida), 120

T

tangent bundle, stable equivariant, 108
———(-s), stratified, 42
Tate cohomology, 61
Thom complex, 119
top hat extension (of a normal cobordism), 100
topological CS stratification (sense of Siebenmann), 14
topological local linearity, 2
topological surgery sequence, fourfold periodicity (of Kirby-Siebenmann), 92
transfer on bordism groups, 120
transformations of equivariant surgery theories SEE: equivariant surgery theories, maps of

transformations, covering, 11, 127
transports (sense of Lück-Madsen), 43, 58
_____, relation to Browder-Quinn reference maps, 43
transversality, equivariant, 30
transverse linear isovariant homotopy equivalences, 43, 44
_____, simple, 43, 44
transverse linear isovariant surgery theory, 2, 3 ALSO SEE: Browder-Quinn
transverse linear maps, 40
_____, relation to maps of vector bundle systems, 40
_____, relation to stratified maps, 40
transverse linearity and G-invariant tubular neighborhoods, 40
triangulation theorem, for smooth orbit spaces, 67
_____, smooth equivariant (for finite group actions), 52
tubular data for Mather prestratifications, 69
_____, local retractions, 69
_____, tubular functions, 69
_____, tubular neighborhoods, 69
tubular functions for orbit spaces, 39–40
tubular neighborhood (of a G-orbit), 38
tubular neighborhood theorem, equivariant, 18
tubular neighborhoods for Mather prestratifications, 69
twisted product construction/map, 81, 98–99, 120, 129
twisted product formulas, 115, 121, 132–133
twisted product maps, 4, 81
_____, and equivariant surgery, 81, 98–100
_____, L^h version, 99
_____, L^p version, 130–131
_____, $L^{negative}$ version, 130–131
_____, versions for intermediate L-groups, 129, 130
_____, Yoshida's formula/theorem, 4, 81, 100, 120–121
twisted products, SEE: twisted product maps
Type I equivariant surgery theories, 83
Type II –"– , 83
Type III –"– , 83

U

Umkehr homomorphism, chain level, 128
Umkehr map, geometric, 127
unitary representation, one-dimensional, standard notation, 143
universal base space, 10
unstable equivariant surgery groups, 105
_____, splitting theorems, 107, 109
unstable Lück-Madsen groups, splitting theorem, 109
Unstable $\pi - \pi$ Question, 29

V

vector bundles, stratified, 42
vector bundle systems (= Π-bundles), 18
———, homomorphisms of, 18
———, pullbacks of, 18
———, stable equivalence of, 18
vector field, controlled, 70–71

W

Wall groups, equivariant symmetric (sense of Yoshida), 120
———, homotopy, 21
———, infinite subprojective, 131
———, negative (= subprojective), 115, 128
———, odd, vanishing theorem for odd order groups, 187–188
———, projective, 115, 128
———, simple, 48
———, subprojective, 115, 128, 130
———, with coefficients, 118–119
Weak Gap Conditions (sense of Lück-Madsen), 49
weakly isogeneric, 143, 154
Whitehead Theorem, equivariant, 20
Whitehead groups, generalized (equivariant), 61
———(-s), ordinary, and Rothenberg sequences, 55, 130
Whitney prestratification, 69
———(-s), equivariant, 69
———, relation to Mather prestratification, 70
Witt group, associated to an equivariant bordism group, 120
———, associated to an orientation homomorphism, 120
———, symmetric, 86
Witt invariant, 86, 184
———, homomorphisms on equivariant bordism groups, 120

Y

Yoshida's theorem (on twisted products), 4, 100, 119–120
Y-preperiodic indexing data, 87